D1071791

# Permanent Magnets in Theory and Practice

# Permanent Magnets in Theory and Practice

Malcolm McCaig, *PhD, FInstP*

Permanent Magnet Consultant.
Previously Assistant Director of
Research of the former Permanent
Magnet Association

A HALSTED PRESS BOOK

JOHN WILEY & SONS
New York — Toronto

Published in the U.S.A. and Canada by Halsted Press,
a Division of John Wiley & Sons, Inc., New York
ISBN 0-470-99269-7





**Library of Congress Cataloging in Publication Data**

McCaig, Malcolm.
Permanent magnets in theory and practice.

"A Halsted Press book."
Bibliography: p.
Includes index.
1. Magnets, Permanent. I. Title.
QC757.9.M3 538'.22 77-23949
ISBN 0-470-99269-7

Typeset by Mid-County Press, Wimbledon, London SW19
Printed in Great Britain by Billing & Sons Ltd, Guildford and London

# DEDICATION

In grateful memory of the Permanent Magnet Association. Formed by six Magnet Manufacturers in 1930 as the Cobalt Magnet Association with the object of standardizing the compositions and prices of cobalt-steel magnets, it soon became involved in developing Al–Ni–Fe and Al–Ni–Co–Fe alloys. In 1935 it changed its name to the Permanent Magnet Association: It carried out a policy of sharing technical information, and maintained a patents pool. Eventually it embarked on co-operative research, and built its own research laboratory.

The PMA reached its peak with 14 members in the 1950s. Publishing over 100 scientific papers, bulletins and patents, The Permanent Magnet Association gained an international reputation. The staff of the Central Research Laboratory answered innumerable enquiries from members and others. During the 1960s the membership began to decline as a result of firms being amalgamated or ceasing to make magnets. New methods of making magnets did not always harmonize with the rest of their business. Finally the remaining seven members agreed to disband on the 31st March 1975.

I was offered an appointment in the Association's Central Research Laboratory on the morning of 12th April 1946, a few hours before the official opening ceremony. Apart from a few months delay, before I could leave my previous appointment, I worked in the Laboratory for the whole time that it was open.

If the Permanent Magnet Association had not existed, this book could not have been written. One of my objects in writing is to ensure that as much as possible of what seems valuable in the accumulated knowledge of the former Permanent Magnet Association may not be allowed to perish.

M.M.

## PREFACE

I commenced writing this book at the end of twenty-eight years work in a Research Laboratory concerned with most aspects of permanent magnets. Twenty-eight years ago it would have been quite unthinkable to use the pronoun 'I' in a scientific book. Times have changed, and recently the editor of 'Nature' replaced longer phrases by 'I' in a letter from me. In this book 'I' is used to warn the reader that an observation described or an opinion expressed is my own, and that some of my peers may disagree.

Some of the material in this book is based on notes and reports accumulated over the years, which I rescued from destruction. This work cannot always be attributed to any particular individual, and 'we' followed by the past tense should be understood as short for 'My colleagues and myself in the Central Research Laboratory of the former Permanent Magnet Association'. 'We' followed by the present tense is used in the normal way in mathematical or other deductions.

My ideal is that even the writer of a scientific book should aim to use the great heritage of different forms and expressions provided by the English language not just for literary variety, but to convey as accurately as possible the precise shade of meaning. I realize only too well that I must often have failed to achieve this object.

The book is written in SI units, but whenever possible the CGS equivalent is given or the method of calculating it is indicated. To the best of my knowledge all the main symbols and names for units are at present internationally accepted, although sometimes a less preferred alternative is used. So long as I can remember, the name symbol or definition of at least one important magnetic quantity has been altered by international agreement on the average every three or four years. The only exception was the period from 1939 to 1945, when the war prevented international committees from meeting. It is thus almost impossible for a reader ever to handle a book that conforms to all the current international agreements. I have no reason to believe that the situation will be any different in the future.

I thank all those who have helped by giving me permission to reproduce diagrams. Acknowledgements are made as they arise. I also thank Dr A. G. Clegg of the Magnet Centre, Sunderland Polytechnic and my son Andrew for reading the type-script and making helpful suggestions, and Dr H. Lemaire, of Aimants Ugimag SA, for lending me his personal list of magnet manufacturers and their products.

Malcolm McCaig,
121 Ringinglow Rd.,
Sheffield S11 7PS.

# CONTENTS

# Chapter 1

# Historical introduction and fundamental concepts

This introductory chapter is intended to put the development of permanent magnetism into historical perspective, and to explain some of the fundamental concepts that are necessary for the understanding of the remainder of the book.

Permanent magnetism has a long history, the ability of the lodestone to attract iron having been noted in ancient Greece. The compass is believed to have been invented in China, and to have been introduced into Europe by 1200 AD or earlier. A number of accounts of magnetism had already been written when Gilbert (1600) wrote his famous book 'De Magnete' or 'On the Magnet'. The original Latin title was longer. Andrade has described this book as 'The first systematic treatise of experimental physics in the modern sense'. This description is well deserved: Gilbert was a careful and objective observer, who was quick to detect and denounce false beliefs. It is rather depressing to note that some statements about the miraculous powers of magnets, rejected by Gilbert as superstitions, are still met from time to time today.

Andrade (1958) wrote an excellent account of the early history of the permanent magnet. This account was based on a lecture given in Sheffield, and rather remarkably published in Endeavour and also incorporated as the first chapter of two of the most important books on permanent magnetism that have appeared in the English language: Hadfield (1962) and Parker and Studders (1962).

Before the discovery of the magnetic effect of an electric current by Oersted in 1820, and the invention of the electromagnet by Sturgeon in 1825, the production of permanent magnets must have been a tedious process. The only materials available were the naturally occurring lodestone, a form of magnetite $Fe_3O_4$, and various forms of iron carbon alloys, that would now be called carbon steel or possibly cast iron.

These were materials capable of being hardened for making tools, cutlery, or weapons. Magnetizing was carried out by stroking with another magnet, or sometimes by cooling in the Earth's magnetic field. The latter process is now dignified by the name thermoremanence. The success achieved by such methods was remarkable; Gowin Knight (1713–72) for example made a magnet capable of lifting 28 times its own weight.

Many of these early magnets were built up from wires or strips; material of small cross-section was more efficiently hardened by quenching, and more easily magnetized by stroking with another magnet.

Special alloy steels were developed towards the end of last century. These steels contained tungsten and rather later chromium. Their introduction proceeded slowly, and knowledge did not pass immediately from country to country. Germany appears to have been the pacemaker until 1914.

The cobalt steels were discovered in Japan in 1917, and represented a big step forward. Alloys based on Fe, Ni and Al, which still represent about 50% by value of magnet production, were also discovered in Japan in 1931, although many of the subsequent improvements in these alloys were made elsewhere, particularly in Britain and Holland. Ferrites are a cheaper material and account for more than half the weight of permanent magnets produced today. They were developed in Holland and Germany in the early 1950s. In recent years the major research effort has been directed to the rare-earth-cobalt alloys. There has been an intense study of these materials throughout the world, but the most significant results have come from the USA. Magnets of these alloys are expensive, but have the best properties so far achieved.

The properties and methods of production of these materials are described in Chapter 4, and their stability under various circumstances in Chapter 5. Theoretical understanding of ferromagnetism and magnetic hysteresis, which are discussed in Chapters 2 and 3 respectively, advanced alongside the improvement in materials, which also made possible new methods of design and new applications, the subjects of Chapters 6 and 7.

To some extent the reputation of permanent magnets suffers from this long history. Obsolete materials persist tenaciously, and continue to be made after better and possibly even cheaper magnets are available. As late as 1940 an enormous laminated assembly similar to those described prior to 1820 was in use for demonstration experiments at Manchester University. It was not regarded as a museum piece, but as the strongest magnet available. As recently as 1970 at our laboratory in Sheffield, we were approached by a local school to remagnetize a motley collection of magnets, most of which could have been manufactured last century, although they may well have been made much later.

Some schools and colleges are equipped with modern magnets, but it is clear that many engineers and scientists have a false idea that magnets are very unreliable, because they saw only obsolete materials in their younger days.

The long history of magnetism has had other subtle consequences. Presumably because permanent magnets predated electric currents, the electromagnetic system of units involved the concept of the magnetic pole. I had decided to write this book from the point of view that free magnetic poles are entirely ficticious. Then only a week before I commenced writing, I heard in a radio news bulletin an announcement that the monopole predicted by Dirac had been discovered. If this claim is substantiated, such monopoles may become technologically important. At present they appear to be rare objects for which one fishes in the heavens with balloons. So far as the subject of this book is concerned, they can be ignored, although a magnetic pole is a useful mathematical fiction that helps to solve some problems.

The properties of permanent magnet materials cannot be discussed without first explaining the units in which they are measured. The next section explains the system of units which is used in this book.

## 1.1 MAGNETIC UNITS

Most students today are taught SI units, but most papers on permanent magnetism are still published in CGS units. By the time this book is in print the change to SI units is likely to have progressed somewhat further. Unfortunately the correct SI units, required to describe some aspects of permanent magnetism, are still not agreed. The policy in this book is to use SI units. When a choice has to be made between rival SI units, the choice is made on the basis of convenience. At the same time some substantial help is given to readers more familiar with CGS units.

As far as possible the alternative possibilities and the background to the arguments are not discussed in this chapter, the object of which is to enable a busy reader to understand the rest of the book. The arguments and difficulties are relegated to Appendix 1. The reader who disagrees with any of the procedures outlined in this chapter should refer to this appendix.

## 1.2 SI UNITS

The SI unit of magnetizing force $\boldsymbol{H}$ can be derived in the following manner. Suppose we have an infinitely long solenoid uniformly wound with $N$ turns per metre. Let the current flowing in this solenoid be $I$

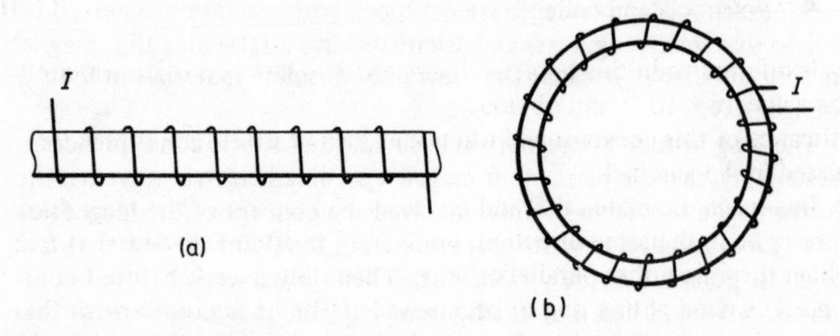


*Fig.1.1. (a) Infinite and (b) toroidal solenoid with N turns per metre*

ampere. Then the magnetizing force $\boldsymbol{H}$ in this solenoid is $NI$ ampere-turns per metre. It is generally recommended that the word turns should be omitted and that the units of $\boldsymbol{H}$ should be quoted as $\text{Am}^{-1}$. The infinite solenoid can also be replaced by a toroidally wound solenoid as shown in Fig.1.1.

Magnetic flux $\phi$ is defined by considering the flux linking a single turn coil. If this flux changes an emf $E$ is induced in the coil such that

$$E = -\mathrm{d}\phi/\mathrm{d}t \text{ V} \tag{1.1}$$

The significance of the minus sign in Equation 1.1 is that if the emf causes a current to flow, it is in a direction that produces a flux that tends to neutralize the change in $\phi$. The unit of flux is called the weber (Wb).

If a flux $\phi$ cuts a small area $A$ metre$^2$ normal to the direction of the flux, there is said to be a flux density:

$$\boldsymbol{B} = \phi/A \tag{1.2}$$

The unit of $\boldsymbol{B}$ is clearly $\text{Wbm}^{-2}$, but the special name tesla (abbreviation T) has now been approved for this unit. Most formulae for the effects of a magnetic field are now expressed in terms of the flux density $\boldsymbol{B}$, rather than the magnetizing force $\boldsymbol{H}$; e.g. the force on a straight wire of length $L$ metre in a flux density $\boldsymbol{B}$ at right angles to the wire is given by:

$$\boldsymbol{F} = \boldsymbol{B}IL \text{ newton} \tag{1.3}$$

If Equation 1.1 is used as a definition then Equation 1.3 must be regarded as an experimental result.

$\boldsymbol{B}$ and $\boldsymbol{H}$ are both vector quantities having direction as well as magnitude. In free space $\boldsymbol{B}$ and $\boldsymbol{H}$ are not independent, but are related by the equation:

$$\boldsymbol{B} = \mu_0 \boldsymbol{H} \tag{1.4}$$

$\mu_0$ is referred to in this book as the magnetic constant. It has precisely the value $4\pi \times 10^{-7}$ and obviously has dimensions $\text{TmA}^{-1}$. The significance of this constant and other methods of describing it are discussed in Appendix 1.

In a magnetic material $\boldsymbol{B}$ and $\boldsymbol{H}$ can vary independently. They do not have to be in the same direction, but theory is often limited to cases in which they are either parallel or antiparallel. Suppose the toroid in Fig.1.1 is filled with a ring of permanent magnet material, having a uniform cross-section. Suppose the material is initially unmagnetized, and a magnetizing force is applied and gradually increased. If $\boldsymbol{B}$ is plotted against $\boldsymbol{H}$ a curve such as OPQ in Fig.1.2 is obtained. If $\boldsymbol{H}$ is now reduced to zero $\boldsymbol{B}$ still has a positive value $\boldsymbol{B}_r$, and it is necessary to apply a negative magnetizing force $H_c$ to reduce $\boldsymbol{B}$ to zero. If $\boldsymbol{H}$ is reversed several times between the same positive and negative limits, a symmetrical loop $QB_rH_cQ'B_r'H_c'Q$ results. This is known as a hysteresis loop. Hysteresis implies a lagging of the changes in $\boldsymbol{B}$ behind those of $\boldsymbol{H}$. Hysteresis occurs in some other physical phenomena such as the application and removal of tension to certain materials.

Up to a point the larger the maximum value of $\boldsymbol{H}$ the larger the area of the hysteresis loop, but when $\boldsymbol{H}$ reaches a value sufficient to saturate the magnetic material, the area of the loop ceases to grow. $B_r$ and $H_c$ reach constant values known as the remanence (residual magnetization

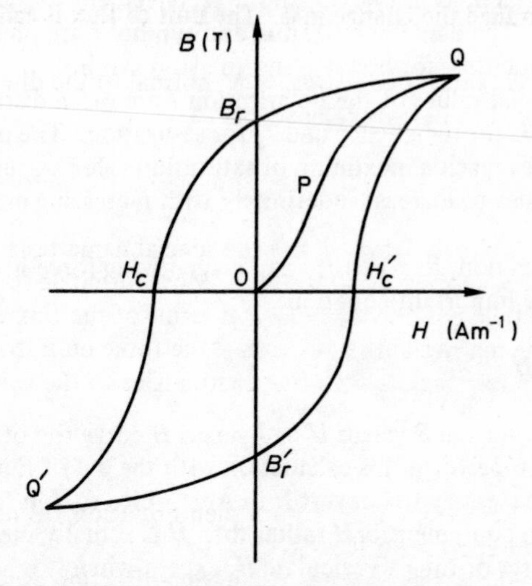


*Fig.1.2. Initial magnetization curve and hysteresis loop*

in the USA) and coercivity respectively. The second quadrant of the hysteresis loop with $\boldsymbol{B}$ positive and $\boldsymbol{H}$ negative is particularly important for describing the quality of a permanent magnet material. There is obviously one point between $B_r$ and $H_c$ for which the product of $\boldsymbol{B}$ and $\boldsymbol{H}$ has a maximum value. Actually the product is negative, but it would normally be pedantic to insist on this point. The importance of this product is explained in Section 6.5. $(BH)_{max}$ has the dimensions of joules per metre$^3$ ($Jm^{-3}$), but there are objections to regarding it as numerically equal to the energy per unit volume. The energy of a permanent magnet is a complicated problem on which opinions differ. The subject is discussed in Section 6.10.

The flux density $\boldsymbol{B}$ is not a measure of the state of a permanent magnet material, since it includes contributions from external magnetizing forces. There are differing views about the units in which the intrinsic properties should be measured, and two units have received official recognition. The two quantities are the magnetization $\boldsymbol{M}$ measured in $Am^{-1}$ and the magnetic polarization $\boldsymbol{J}$ measured in tesla; the latter quantity may also be called the intrinsic flux density $\boldsymbol{B}_i$. These quantities are not independent for

$$\boldsymbol{J} = \boldsymbol{B}_i = \mu_0 \boldsymbol{M} \tag{1.5}$$

Some people attribute considerable theoretical significance to the choice of $\boldsymbol{J}$ or $\boldsymbol{M}$, and this argument is examined in Appendix 1. In this book it is frequently desirable to plot the intrinsic curve and the $\boldsymbol{B}$ curve on the same graph, as in Fig.1.3, and it is convenient to plot both quantities in tesla. The name polarization and symbol $\boldsymbol{J}$ are preferred to intrinsic flux density for brevity, and to avoid symbols with double suffixes. Special values of the polarization $\boldsymbol{J}$ are often distinguished by suffixes; e.g. $J_r$ for remanence and $J_s$ for saturation. The polarization of a material does reach a maximum or saturation value, $J_s$, unlike $\boldsymbol{B}$, which continues to increase indefinitely with increasing magnetizing force.

The polarization, flux density and magnetizing force in a material are related by the important equation:

$$\boldsymbol{B} = \boldsymbol{J} + \mu_0 \boldsymbol{H} \tag{1.6}$$

Thus given either the $\boldsymbol{B}$ versus $\boldsymbol{H}$ or $\boldsymbol{J}$ versus $\boldsymbol{H}$ curve the other can be calculated. To perform this calculation with the aid of Equation 1.6, it is obviously necessary to convert $\boldsymbol{H}$ in $Am^{-1}$ into $\mu_0\boldsymbol{H}$ in tesla. In much modern equipment $\mu_0\boldsymbol{H}$ rather than $\boldsymbol{H}$ is actually measured, and it seems a waste of time to calculate $\boldsymbol{H}$, except when it is actually required, as is sometimes the case.

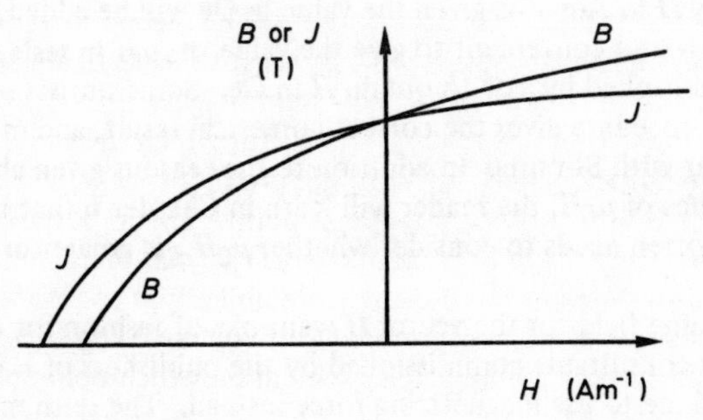


*Fig.1.3. B and J plotted against H*

In some modern materials $\boldsymbol{J}$ remains positive, when the demagnetizing force greatly exceeds $H_c$, the value at which $\boldsymbol{B}$ changes sign. The demagnetizing force at which $\boldsymbol{J}$ becomes zero is called the intrinsic coercivity. In this book it will be denoted by $H_{ci}$, the second suffix *i* meaning intrinsic. In many writings including some by the author, ${}_JH_c$ has been used, but $H_{ci}$ appears neater.

## 1.3 OLDER UNITS

When the author was a student $\boldsymbol{B}$ was normally called the magnetic induction and $\boldsymbol{H}$ the magnetic field. Both were measured in gauss (G), although the terms CGS units and lines/cm$^2$ were often met. Some engineers used lines/in$^2$. In the 1930s the name oersted (Oe) was adopted for the unit of magnetic field $\boldsymbol{H}$. In CGS units the $(BH)_{max}$ value of modern materials is expressed in mega-gauss-oersted (MGOe).

This is the system with which many readers will be most familiar, and clearly some concession must be made to them. The change from gauss to tesla need cause little difficulty. The reader need only remember that one tesla = 10 000 gauss, or write $x \times 10^4$ gauss for $x$ tesla. To relate $(BH)_{max}$ values in Jm$^{-3}$ to one's background knowledge of values in MGOe is more difficult, and whenever reasonably convenient values will be given in both systems.

The CGS equivalent of Equation 1.6 is

$$\boldsymbol{B} = \boldsymbol{H} + 4\pi \boldsymbol{M} \tag{1.7}$$

However, permanent magnet technologists often used the value of $4\pi\boldsymbol{M}$ rather than $\boldsymbol{M}$ itself, and this is obtained by multiplying $\boldsymbol{J}$ in tesla by $10^4$. (In CGS units $\boldsymbol{I}$ and $\boldsymbol{J}$ were also often used for the intensity of magnetization.)

If only $\boldsymbol{H}$ in $\mathrm{Am}^{-1}$ is given the value in Oe will be added, but if it has been found convenient to give the value of $\mu_0\boldsymbol{H}$ in tesla, this value can be multiplied by $10^4$ to obtain $\boldsymbol{H}$ in Oe. Some purists may object, but this procedure gives the correct numerical result, and may help those unfamiliar with SI units. In addition to the reasons given above for using values of $\mu_0\boldsymbol{H}$, the reader will learn in Chapter 6 that the magnet designer often needs to consider whether $\mu_0 H_{ci}$ is greater or less than $B_r$.

The name field for the vector $\boldsymbol{H}$ went out of fashion for a time. The academic consultants commissioned by the publishers of Hadfield (1962) requested me to use magnetizing force instead. The term magnetic field strength now seems to be in fashion again, and is recommended by Rayner and Drake (1970). I accepted the term magnetizing force reluctantly, and am equally reluctant to change again. Perhaps I have emulated permanent magnets and acquired the property of hysteresis. In this book flux density denotes $\boldsymbol{B}$ in tesla, magnetizing force denotes $\boldsymbol{H}$ in $\mathrm{Am}^{-1}$, and the word field is used when either $\boldsymbol{B}$ or $\boldsymbol{H}$ or even both may be implied.

Conversion factors between SI and CGS units are given at the end of Appendix 1.

## 1.4 PERMEABILITY

Engineers and scientists familiar with soft magnetic materials may be surprised that the term permeability has not been mentioned before. The idea of a constant quantity, permeability relating $\boldsymbol{B}$ and $\boldsymbol{H}$ has only a very limited usefulness in dealing with materials exhibiting a large amount of hysteresis. If it is used it must nearly always be qualified by some word such as initial, specifying a particular condition of magnetization. The most common use of permeability in connection with permanent magnets is in connection with recoil lines, as explained with the

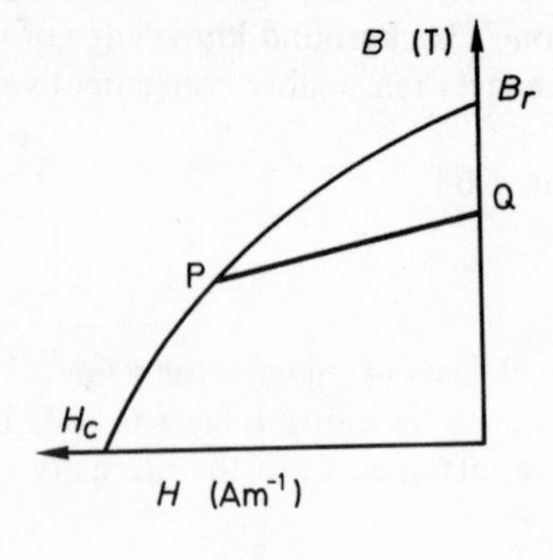


*Fig.1.4. Recoil line and recoil permeability*

aid of Fig.1.4. A magnet is magnetized and a demagnetizing force is applied to take it to point P on the demagnetization curve. If the demagnetizing force is removed, the flux density does not return to $B_r$ but recoils up PQ, which is approximately a straight line. Its slope is $\mu_r\mu_0$, where $\mu_r$ is the relative recoil permeability.

Susceptibility causes some difficulty, because volume, mass and molar susceptibilities are used in CGS units and there are some differences of opinion about the equivalent SI units. Susceptibility is rarely needed in this book, and is explained, when it arises.

# Chapter 2

# Magnetic properties of matter

In everyday language matter is either magnetic or non-magnetic. Magnetic materials are visibly attracted by a magnet of average strength, and non-magnetic materials are not. Actually all materials react in some slight degree to a magnetic field, although a powerful magnet and sensitive equipment is often required to demonstrate this reaction. Such experiments enable apparently non-magnetic materials to be divided into several different classes: diamagnetic, paramagnetic and anti-ferromagnetic materials. Some commercially made permanent magnet materials such as the steels and Al–Ni–Co alloys are described as ferromagnetic, while others such as barium ferrite are described as ferrimagnetic. The words ferromagnetic and ferrimagnetic describe different types of ordering that make a material appear in the popular sense magnetic. There are also some more complicated forms of magnetic order.

In a book of this nature, it is often unnecessary to specify to which sub-class a particular magnetic or non-magnetic material belongs, and often the words non-magnetic and magnetic are used with their everyday meanings.

When studying the behaviour of existing materials and trying to develop better ones, a scientist must seek to understand the processes which he observes. In this chapter I am endeavouring to give a simple account of modern ideas on why different materials respond differently to magnetic fields. One good reason why the account is simple, is that I am not capable of writing a more complicated dissertation. If you are a theoretical physicist, who can write down the mathematical formulae to explain any aspect of solid state physics in terms of quantum mechanics, you would be well-advised to skip this chapter. If, however, you are an engineer or experimental scientist, this chapter may possibly help you to understand some of the terms used in modern magnetism. It may also convince you that theory has at least occasionally gone beyond merely attempting to explain the properties of existing materials, and is helping to direct the search for new and better alloys and compounds. More detailed accounts of the subject are given by Bates (1961) Anderson (1968) Martin (1967) and Hume-Rothery (1962). Some of these books have run to numerous editions; the date given refers to the edition, which I have used, and I cannot be quite certain that the treatment is the same in all other editions.

## 2.1 DIAMAGNETISM

If a magnet approaches a block of an electrical conductor such as copper, eddy currents are induced in the conductor. The direction of the eddy currents is such as to oppose the establishment of magnetic flux in the conductor, and to repel and so oppose the motion of the approaching magnet. These eddy currents are produced by movement of the free electrons in the metal, but ordinary electrical conduction in metals involves resistance, and the eddy currents soon cease as the energy of the electrons is dissipated to raise the temperature of the conductor as a whole. If there were no resistance, the eddy currents would continue, and the magnet and metallic block would continue to repel each other. This phenomenon does actually occur at low temperatures in superconductors, which are becoming of great technological importance. Superconducting magnets can produce large magnetic fields and exert large magnetic forces, but at the time of writing no material has been discovered, which remains superconducting above 23 K. Superconducting magnets have been used to produce the large fields necessary for some operations with rare-earth-cobalt magnets.

Diamagnetism resembles superconductivity, but is an extremely small effect. When a magnetic flux enters a material, it influences the motions not only of the free electrons, but also of the orbital electrons within the atoms. One can imagine the field altering the radius of an orbit or the velocity of an electron, but the most important effect is the Larmor precession. The meaning of precession can be explained by considering the motions of a child's spinning top. Although most tops are constructed so that their equilibrium position is lying sideways, a top is a gyroscope and can remain spinning with its axis vertical. If the top becomes tilted so that it is spinning about an axis inclined to the vertical by an angle $\theta$, gravity exerts a couple tending to increase $\theta$ to $90°$, but this does not happen so long as the top continues spinning. Instead the axis about which the top spins remains inclined at an angle $\theta$ to the vertical, but this axis itself rotates around the vertical direction, although much more slowly than the top spins. This phenomenon is called precession. Of course friction gradually brings the top to rest, and as it does so the angle of tilt $\theta$ increases.

In the atoms of an unmagnetized material, the planes of the electron orbits are oriented at random. The introduction of a magnetic field causes these orbits to precess around the field direction. Here there is no friction and the orbits precess so long as the external field is maintained. The precession of all the electron orbits produces a very small field in opposition to the applied field. The material behaves as if it had a negative susceptibility or relative permeability less than unity. Bismuth is one of the most strongly diamagnetic substances. In CGS units its volume susceptibility is $-1.35 \times 10^{-5}$, and according to the definition accepted by Rayner and Drake (1970), the volume susceptibility in SI units is $M/H$ and is obtained by multiplying the CGS value by $4\pi$.

The mathematical theory of diamagnetism can be developed from the now classical model of the Bohr atom, or from the more sophisticated methods of quantum mechanics, but the practical conclusions differ only in detail.

All materials possess electrons and must therefore give the diamagnetic reaction to a magnetic field. Many materials also possess atomic magnetic moments, and these produce much greater reactions such as paramagnetism, which completely mask the diamagnetism.

## 2.2 PARAMAGNETISM

Paramagnetism depends on whether a material possesses atomic magnetic moments, and on how these moments are oriented by a magnetic field. An atom may possess a magnetic moment arising from its nucleus, the motion of electrons in orbits around the nucleus, and the spinning of electrons. Nuclear magnetic moments are the subject of intensive research in the study of magnetic resonance, but their contribution to magnetism is small compared with that of electrons, and can be neglected in the context of this chapter.

The magnetic moment of an atom is usually the vector sum of the moments produced by all its electrons orbits and spins. The term vector sum is emphasized, because the orbital and spin moments need not all be parallel or antiparallel. In the inert gases, He, Ne, Ar etc. the vector sum of all the electron moments is zero; these gases possess no magnetic moment and are diamagnetic. Any other element can be regarded as having the electronic structure of the inert gas immediately below it in the periodic table of elements, together with a number of additional electrons. Thus Fe with atomic number 26 has 18 electrons arranged in a central core as in Ar with atomic number 18, plus 8 electrons in outer shells. The inner 18 electrons do not contribute to the magnetic moment of Fe, which is determined by the 8 outer electrons.

Applying a small magnetic field causes the atomic moments to precess around it as in diamagnetism. The magnetic moments do, however, extend their influence to other atoms and interfere with one another, and as a result two opposing effects arise. At very low temperatures the predominant effect is damping, which acts rather like friction on a child's top. This damping brings the precession to an end, the moments become aligned, and the material strongly magnetized. This strong magnetism at low temperatures is used to produce temperatures below 4 K. At normal temperatures thermal agitation as the atoms collide with one another predominates over damping. Thermal agitation tends to keep the atomic moments disordered, and prevents them aligning with the field. The interaction of the field and thermal agitation allow the substance to be slightly magnetized by normal fields available in the laboratory. The susceptibilities of paramagnetic solids can be 10 to 100 times greater than those of diamagnetic materials. Paramagnetism remains a fairly small effect, although it is much easier to demonstrate than diamagnetism.

An isolated atom or molecule behaves as if it had one magnetic moment, if the orbital and spin moments interact by what is known as Russell–Saunders coupling. Russell–Saunders coupling can be assumed to operate in gases at normal fields, although very high fields may align spin and orbital moments separately. It does not apply in solids. In the following paragraphs up to equation 2.5, CGS units are used, because there are special difficulties and disagreements about the correct expressions in SI units.

Langevin developed a theory of paramagnetism based on classical Maxwell–Boltzmann statistics, according to which the energies of the molecules are distributed as an exponential function of the energy divided by the absolute temperature. Each atom is supposed to have a dipole moment $\boldsymbol{j}$, and the only interaction considered between the atoms is the thermal agitation. The potential energy of an atom with its moment aligned at an angle $\theta$ to an applied field is $\boldsymbol{jH}(1 - \cos\theta)$. The mean moment $\bar{j}$ in the direction of $\boldsymbol{H}$ is found to be given by:

$$\bar{j}/j = \coth a - 1/a \tag{2.1}$$

where $a$ is written for $\boldsymbol{jH}/kT$. Here $k$ is Boltzmann's constant and $T$ is the temperature in $K$. The function $(\coth a - 1/a)$ is known as Langevin's function, and for small values of $a$ such as arise practically simplifies to

$$\frac{\bar{j}}{j} = \frac{jH}{3kT} \tag{2.2}$$

Measurements of paramagnetism are often given in terms of the molar susceptibility $\chi_M = N\bar{j}/H$ where $N$ is Avogadro's number, the number of molecules per mole. From Equation 2.2

$$\chi_M = \frac{N\bar{j}}{H} = \frac{Nj^2}{3kT} = \frac{N^2j^2}{3RT} \tag{2.3}$$

where $R$ is the gas constant and is equal to $Nk$.

Equation 2.3 indicates that

$$\chi_M T = \text{constant} = C \tag{2.4}$$

Equation 2.4 is called the Curie or Curie–Langevin law. Its derivation should be valid only for a gas such as oxygen, but it also holds for some solids in which the magnetic ions are well isolated. More generally the Curie–Weiss law.

$$\chi_M = C/T - \theta \tag{2.5}$$

where $\theta$ is a constant often holds. This modification can arise from a more precise quantum treatment of the orbital and spinning electrons in a molecule. In the quantum theory the angle between the magnetic moment and applied field cannot have any value, as assumed in the Langevin theory, but only certain discrete values determined by quantum conditions. The interactions between the various atomic or molecular moments in a solid also modify the theory.

The Bohr magneton is a unit often used in describing the magnetic properties of materials from the standpoint of atomic theory. The magnetic moment of a small plane current loop of area $\Delta S$, carrying current $I$ is $I\Delta S$. Different writers at different times have regarded this statement as a definition, an experimental fact, or as a theorem deduced from other laws. Here we illustrate with a simple special case that the statement is in harmony with the law that the force on a conductor of length $L$, carrying current $I$ is $BIL$, when $B$ is the component of the flux density at right angles to $L$. Consider the rectangular current loop PQRS shown in Fig.2.1. Let the direction of the flux density $B$ be parallel to the PQ and RS. If the current $I$ flows in the direction shown by the arrows, there is a force $F$ on QR out of the plane of the paper, and a similar force on SP into the plane of the paper. In SI units with $I$ in ampere, $L$ in metre and $B$ in tesla, $F = \text{QR}\,.\,IB$. The moments about the centre line YOY′ are ½PQ . $I$ . QR . $B$ and ½PQ . $I$ . PS . $B$. The total couple is PQ . QR . $IB = \Delta SIB$ newton metre. By definition of the magnetic moment $j$ in this special case the couple is also given by $jH$ or more generally by $\boldsymbol{J} \times \boldsymbol{H}$. Hence $j = IS$, where $S$ is the area PQ . QR. The

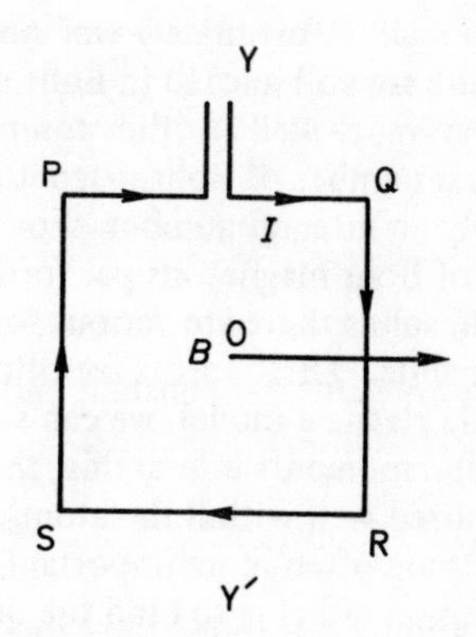


*Fig.2.1. Rectangular current loop in place of uniform flux density B*

mathematically minded reader can easily generalize the theorem for different shapes and fields with different orientations.

The fundamental postulate of the Bohr model of the atom is that the angular moment of an electron in its orbit is $nh/2\pi$, where $n$ is an integer and $h$ is Planck's constant ($6.625 \times 10^{-34}$ Js). If an electron of mass $m_e$ kg moves with angular velocity $\omega$ in an orbit of radius $a_n$ metre:

$$nh/2\pi = m_e a_n^2 \omega \tag{2.6}$$

The magnetic moment of the orbiting electrons is given by the area of the orbit multiplied by the effective current. That is the dipole moment is given by:

$$\begin{aligned} j &= \pi a_n^2 e\omega/2\pi \\ &= a_n^2 e\omega/2 \end{aligned} \tag{2.7}$$

where $e$ is the electronic charge in coulomb. Eliminating $a_n^2$ from Equations 2.6 and 2.7, we find that when $n = 1$

$$j = \frac{he}{4\pi m_e} \tag{2.8}$$

This magnetic moment for a circular orbit of quantum number 1 is the unit known as the Bohr magneton; its value in SI units is $1.165 \times 10^{-29}$ Wbm ($0.927 \times 10^{-20}$ CGS units).

The idea of the Bohr magneton was that it would indicate how many orbiting electrons were contributing to the atomic magnetic moment.

Although unfortunately this simple object has not been achieved, many results are still quoted in Bohr magnetons. From the vector model of the atom, as well as other complications to be discussed further on, the actual number of Bohr magnetons measured for a particular material is rarely an integral number. For materials in a molecular state the number of Bohr magnetons per formula unit may be quoted.

In solids there are various forces interacting between the atomic moments. These forces are often related to certain crystal axes. In a crude classical model, we can see that there are more possibilities of orbital moments interacting, than there are of spin moments, which are localized well within the atoms. Electrostatic forces between the electrons are often more important than the magnetic interactions. One common effect is to bind the orbital moments to the crystal axes, so that the orbital moments are only slightly if at all rotated by a magnetic field. In such a material the magnetic field can rotate the spin moments, although interaction between the spin and orbital moments may provide some opposition to such rotation. This situation is obviously very different from the Russell–Saunders coupling in gases. The magnetic properties are determined primarily by the electron spin moments, and the orbital moments are said to be quenched, as they make little or no contribution to the observed magnetization.

The conclusion, that the magnetic properties of common ferromagnetic materials was accounted for mainly by electron spins, was first deduced from measurements of the gyromagnetic effect. A bar of magnetic material is suspended vertically in a solenoid, and is suddenly magnetized. The sudden alignment of the magnetic moments of the electrons causes an alignment of their angular momenta, and imparts angular momentum to the bar as a whole. As a consequence the bar undergoes a small transient rotation about its axis of suspension, and this can be observed with a mirror, lamp and scale as with a ballistic galvanometer. This deflection was measured and found to be only half as great as had been calculated on the expectation that the magnetization process would result from orienting electron orbits, but could be explained if the process was actually the orientation of electron spins. These results were obtained with iron and nickel. It is now known that in cobalt there is a very small but measurable contribution to the magnetization from electron orbits, and in some rare-earth compounds the orbital contribution is considerable.

## 2.3 FERROMAGNETISM

To explain the strong magnetic properties that are particularly obvious in Fe, and therefore classified as ferromagnetism, Weiss postulated a 'molecular field'. He assumed that there is a field proportional to the magnetization, and that this field tends to keep the magnetic moments of neighbouring atoms parallel. This hypothesis made by Weiss can account for the principal facts of ferromagnetism. It predicts that a material is spontaneously magnetized at temperatures below the Curie point, at which temperature thermal agitation overcomes the molecular field. It also predicts reasonably well the variation of the spontaneous magnetization between absolute zero and the Curie point, as well as the temperature variation of the paramagnetic susceptibility at temperatures above the Curie point.

Weiss was, however, unable to explain the origin of this molecular field. The magnetic dipole–dipole interactions between the magnetic moments of the atoms or molecules is of the order a thousand times too small. Not until 1927 did Heisenberg put forward a satisfactory explanation in terms of quantum mechanics.

Equations based on wave mechanics can be written down to predict the potential energy of atoms whose electron orbits overlap. These equations predict the ordinary electrostatic potentials arising from Coulomb forces, and also certain additional energy terms, which actually depend on the orientation of the electron spins. These terms are described as exchange forces. Rather crudely it may be imagined that electrons in neighbouring atoms may exchange places, and when an electron moves from one atom to another, it carried with it a message about the spin orientation or direction of the magnetic moment in the first atom.

Exchange forces may cause the magnetic moments of neighbouring atoms to be parallel or antiparallel. The sign of the exchange forces determines whether a material is ferromagnetic or not, their magnitude determines the Curie temperature at which ferromagnetism breaks down, while the number of electron spins per atom that can be aligned determines the number of Bohr magnetons and the saturation polarization $J_s$.

The exchange forces fall off much more rapidly with distance than ordinary Coulomb forces, and rarely extend beyond the nearest neighbour. Ferromagnetism is therefore very sensitive to interatomic distance. That is why body centred cubic Fe ($\alpha$Fe) and many alloys of iron in which the body centred cubic structure is preserved are ferromagnetic, while face centred cubic Fe ($\gamma$Fe or austenite) and iron alloys with the face centred structure are not magnetic. The distance between the Fe atoms in the fcc structure is different from that with the bcc structure, and is not correct for the positive alignment of spins that produces ferromagnetism.

Pure Mn is not magnetic, but if Mn is alloyed with suitable proportions of certain elements such as Al or Bi, the interatomic distance between the Mn atoms is modified in such a way that the alloy becomes ferromagnetic.

Attempts to predict ferromagnetism quantitatively are not completely successful, and the subject is somewhat confusing to read. Hume Rothery (1962) gives one of the clearest accounts. He shows how some authors confuse older and more modern versions of the quantum theory, or fail to distinguish between the electronic states of free atoms and solids.

In the presence of a magnetic field 4 quantum numbers are required to specify the possible electron states; $n$ distinguishes different electron shells, $l$ distinguishes different orbits within the shells, $m_l$ is a measure of the magnetic moments of the orbits resolved parallel to the field, and $m_s$ the magnetic moment of the spins resolved parallel to the field. In the modern form of the quantum theory, although we still talk about orbits and shells, these concepts should not be treated too literally. It is not possible to specify the position of an electron, and instead one writes down mathematical functions that indicate the probability of where the electron may be. The quantum numbers do, however, arise from this theory.

Once $n$ is fixed, there are various limitations on the values which $l$, $m_l$ and $m_s$ may assume. Thus $l$ can have any of the values $0, 1, \ldots n-1$, while $m_l$ can have any integral value from $+l$ to $-l$ including zero. The possibility of $l$ being zero illustrates the danger of treating the models too literally, since it suggests an orbit that vibrates along a line passing through the nucleus. The spin quantum number $m_s$ can only have the values $+½$ or $-½$. Pauli's principle requires that no two electrons in an atom can have all four quantum numbers the same, and thus limits the number of electrons with the same value of $n$, or in other words the number of electrons in the same shell. In order to avoid repeating statements such as $n = 2, l = 1$ a kind of shorthand has been adopted. The small letters $s, p, d,$ and $f$ are used to denote electron states with the quantum number $l = 0, 1, 2,$ and 3 respectively, whilst a number in large type preceding these letters indicates the value of $n$. Thus $3p$ represents a state for which $n = 3$, and $l = 1$.

In the first shell $n = 1, l = 0, m_l = 0$, and $m_s = +½$ or $-½$. Thus in the first shell there can be only two electrons. In all atoms except hydrogen, the first shell is full, and there are two $ls$ electrons. This number of electrons is given by an index number after the letter, so that all atoms except $H$ can be said to have $1s^2$ electrons, while $H$ itself has of course $1s^l$. The shell with $n = 2$ can have 8 electrons in all, $2s^2 + 2p^6$. There are 6 electrons with $n = 2$ and $l = 1$, because $m_l$ can have any of the three values $-1$, 0 and $+1$, and for each of these values of $m_l$, $m_s$ can be $+½$ or $-½$. The inert gas Ne with atomic number 10 is the element in which the shell with $n = 2$ is complete with all 8 places full. The effective magnetic moment of a complete shell is zero. The inert gas atoms have only complete shells, and are therefore diamagnetic, and in calculating the moment of other atoms, we can ignore the contribution of any electrons that form the complete shells of the next lower inert gas in the periodic table.

While the spin moments of electrons in complete shells neutralize, those in unfilled shells behave in the opposite manner. As far as the other quantum conditions permit, the spins of electrons in an incomplete shell are arranged so as to maximize the magnetic moment of the atom. One way of expressing this fact is to say that the exchange interactions between electrons in the same atom are positive.

Table 2.1 shows the number of electrons in the various electronic states for elements in two sections of the periodic table that include the magnetically important transition elements, and rare-earth elements. This table really applies to isolated atoms, but it is useful as a rough guide to the behaviour of the solids, in which there are, however, additional complications. The table starts with the inert gas Ar. The next two elements K and Ca have 1 and 2 electrons respectively in the $4s$ state, although the $3d$ shell is still empty. The $3d$ shell fills up gradually in the elements of atomic numbers 21 onwards, until it has its complete complement of 10 electrons in Cu, atomic number 29. In the elements Fe, Ni and Co the shell is rather more than half full, and consequently these elements have a large number of electrons free to align and give them a large magnetic moment. Further on in the periodic table it is the $4f$ shell that is partly full and makes some of the rare-earth elements magnetic. Some of the transuranic elements may also be ferromagnetic.

In order that an element or alloy should be ferromagnetic it is necessary that it should have partly filled shells that allow its spins to align and give it a large moment. In addition the exchange energy must be positive in order to align the moments of different atoms. One theory predicts that the exchange interactions should be positive if the ratio (atomic distance/mean radius of electron orbits) is a little greater than 3. The interatomic distance is related to the radius of the outer electron shells, and for these shells the ratio is likely to be nearer 2 than 3. The interaction between electrons in outer shells is unlikely therefore to be positive, and elements whose atomic moments arise from electrons in the outer shells are not likely to be ferromagnetic. In Fe, Ni and Co, the electrons responsible for the magnetic moment can be seen from Table 2.1 to be in an inner shell, which has a smaller radius. Thus the ratio reaches the value of 3 necessary to produce positive exchange interaction and ferromagnetism. A similar argument can be applied to the ferromagnetic rare-earth elements.

The detailed modern version of the quantum theory does not predict

**Table 2.1** QUANTUM NUMBERS AND NUMBER OF ELECTRONS IN GIVEN STATES FOR TRANSITION AND RARE-EARTH ELEMENTS

| *Atomic No. and symbol* | | *State* | 1*s* | 2*s* | 2*p* | 3*s* | 3*p* | 3*d* | 4*s* | 4*p* | 4*d* | 4*f* | 5*s* | 5*p* | 5*d* | 6*s* |
|---|---|---|---|---|---|---|---|---|---|---|---|---|---|---|---|---|
| | | $n$ | 1 | 2 | 2 | 3 | 3 | 3 | 4 | 4 | 4 | 4 | 5 | 5 | 5 | 6 |
| | | $l$ | 0 | 0 | 1 | 0 | 1 | 2 | 0 | 1 | 2 | 3 | 0 | 1 | 2 | 0 |
| 18 | Ar | | 2 | 2 | 6 | 2 | 6 | | | | | | | | | |
| 19 | K | | 2 | 2 | 6 | 2 | 6 | – | 1 | | | | | | | |
| 20 | Ca | | 2 | 2 | 6 | 2 | 6 | – | 2 | | | | | | | |
| 21 | Sc | | 2 | 2 | 6 | 2 | 6 | 1 | 2 | | | | | | | |
| 22 | Ti | | 2 | 2 | 6 | 2 | 6 | 2 | 2 | | | | | | | |
| 23 | V | | 2 | 2 | 6 | 2 | 6 | 3 | 2 | | | | | | | |
| 24 | Cr | | 2 | 2 | 6 | 2 | 6 | 5 | 1 | | | | | | | |
| 25 | Mn | | 2 | 2 | 6 | 2 | 6 | 5 | 1 | | | | | | | |
| 26 | Fe | | 2 | 2 | 6 | 2 | 6 | 6 | 2 | | | | | | | |
| 27 | Co | | 2 | 2 | 6 | 2 | 6 | 7 | 2 | | | | | | | |
| 28 | Ni | | 2 | 2 | 6 | 2 | 6 | 8 | 2 | | | | | | | |
| 29 | Cu | | 2 | 2 | 6 | 2 | 6 | 10 | 1 | | | | | | | |
| 54 | Xe | | 2 | 2 | 6 | 2 | 6 | 10 | 2 | 6 | 10 | – | 2 | 6 | | |
| 55 | Cs | | 2 | 2 | 6 | 2 | 6 | 10 | 2 | 6 | 10 | – | 2 | 6 | | 1 |
| 56 | Ba | | 2 | 2 | 6 | 2 | 6 | 10 | 2 | 6 | 10 | – | 2 | 6 | | 2 |
| 57 | La | | 2 | 2 | 6 | 2 | 6 | 10 | 2 | 6 | 10 | – | 2 | 6 | 1 | 2 |

*Table 2.1. cont'd*

| | | | | | | | | | | | | | | | |
|---|---|---|---|---|---|---|---|---|---|---|---|---|---|---|---|
| 58 | Ce | 2 | 2 | 6 | 2 | 6 | 10 | 2 | 6 | 10 | 1 | 2 | 6 | 1 | 2 |
| 59 | Pr | 2 | 2 | 6 | 2 | 6 | 10 | 2 | 6 | 10 | 2 | 2 | 6 | 1 | 2 |
| 60 | Nd | 2 | 2 | 6 | 2 | 6 | 10 | 2 | 6 | 10 | 3 | 2 | 6 | 1 | 2 |
| 61 | Pm | 2 | 2 | 6 | 2 | 5 | 10 | 2 | 6 | 10 | 4 | 2 | 6 | 1 | 2 |
| 62 | Sm | 2 | 2 | 6 | 2 | 6 | 10 | 2 | 6 | 10 | 5/6 | 2 | 6 | 0/1 | 2 |
| 63 | Eu | 2 | 2 | 6 | 2 | 6 | 10 | 2 | 6 | 10 | 7 | 2 | 6 | | 2 |
| 64 | Fs | 2 | 2 | 6 | 2 | 6 | 10 | 2 | 6 | 10 | 7 | 2 | 6 | 1 | 2 |
| 65 | Tb | 2 | 2 | 6 | 2 | 6 | 10 | 2 | 6 | 10 | 8 | 2 | 6 | 1 | 2 |
| 66 | Dy | 2 | 2 | 6 | 2 | 6 | 10 | 2 | 6 | 10 | 9 | 2 | 6 | 1 | 2 |
| 67 | Ho | 2 | 2 | 6 | 2 | 6 | 10 | 2 | 6 | 10 | 10 | 2 | 6 | 1 | 2 |
| 68 | Er | 2 | 2 | 6 | 2 | 6 | 10 | 2 | 6 | 10 | 11 | 2 | 6 | 1 | 2 |
| 69 | Tm | 2 | 2 | 6 | 2 | 6 | 10 | 2 | 6 | 10 | 12/13 | 2 | 6 | 0/1 | 2 |
| 70 | Yb | 2 | 2 | 6 | 2 | 6 | 10 | 2 | 6 | 10 | 14 | 2 | 6 | | 2 |
| 71 | Lu | 2 | 2 | 6 | 2 | 6 | 10 | 2 | 6 | 10 | 14 | 2 | 6 | 1 | 2 |
| 72 | Hf | 2 | 2 | 6 | 2 | 6 | 10 | 2 | 6 | 10 | 14 | 2 | 6 | 2 | 2 |

that the atomic moments should be integral multiples of the Bohr magneton. If there are $s$ parallel spins, the effective magnetic moment is predicted to be:

$$j_{\text{eff}} = [2s(s/2 + 1)]^{1/2} \tag{2.9}$$

Thus for $s = 1, 2, 3, 4, 5$, the effective number of Bohr magnetons is 1.73, 2.8, 3.9, 4.9 and 5.9 respectively.

Although some magnetic materials appear to be reasonably well described by a theory which associates magnetic moments with individual atoms, many appear to be better explained by a collective electron theory. In such a theory the electrons are pictured as shared by the atoms of a solid as a whole. The shells of the atomic theory that contained specified small numbers of electrons are replaced by bands for the whole solid that contain an enormous number of possible energy levels. Instead of thinking of a single transition element atom in which electrons begin to occupy the 4$s$ shell before the 3$d$ shell is full, we think of two energy bands that overlap. Electrons begin to occupy the upper band before the lower one is full. So long as this lower band is not full, the quantum conditions allow the spins to align and produce a magnetic moment. In iron this partly full band does correspond to the 3$d$ shell of the Fe atom, but it is not necessary for the average number of electrons per atom in this band to have an integral value.

Although the orbital moments are often quenched, they may still have an important role in permanent magnetism. It has been explained that saying that the orbital moments are quenched means that they are aligned along certain favoured crystal axes. There are certain interactions between the spin and orbital moments. These interactions cause the spin moments also to be aligned along the crystal axes, although not so strongly as the orbital moments, so that the spins can usually be aligned and the material magnetized by a sufficiently large field. As soon as the field is removed the magnetization reverts to the nearest easy crystal direction. This phenomenon is called magnetocrystalline anisotropy. The details of the quantum theory of magnetocrystalline anisotropy are very difficult and not very useful as yet. Purely empirical formulae for the variation of magnetocrystalline anisotropy energy direction are relatively easy to write down and manipulate. A large value of magnetocrystalline anisotropy is often conducive to the occurrence of permanent magnetism.

In rare-earth metals and their alloys, orbital moments appear to make a greater contribution to the magnetic behaviour than in the transition metals. These rare-earth compounds often have very large magnetocrystalline energies.

## 2.4 ANTIFERROMAGNETISM

It has been mentioned that exchange interactions are more often negative than positive. Néel suggested that there should be some substances with an antiferromagnetic order i.e. with the moments in alternate atoms ordered in an antiparallel manner. The magnetic behaviour of antiferromagnetic materials differs only slightly from that of paramagnetic materials. There is, however, a temperature analogous to the Curie temperature of ferromagnetic materials at which the antiferromagnetism is destroyed by thermal agitation. This temperature is now known as the Néel point. Above the Néel point the material behaves like an ordinary paramagnetic material, but at the Néel point the susceptibility is a maximum, and below the Néel point the susceptibility falls as the temperature is reduced. This anomalous behaviour permits antiferromagnetism to be recognized by purely magnetic measurements, but the most conclusive evidence for antiferromagnetism comes from neutron diffraction experiments (see Martin 1967). The Néel point is also marked by anomalies in some physical properties such as thermal expansion and specific heat.

There are now a few materials known in which a transition from antiferromagnetic to ferromagnetic occurs as the temperature is raised. This unusual phenomenon, which makes the material become magnetic with increasing temperature, occurs for example in certain alloys of Cr–Mn–Sb. The transition point can be varied by choice of composition between $-200°C$ and $+150°C$, and one can imagine useful applications, involving this material and permanent magnets.

Remarkable displaced hysteresis loops have been obtained in an epitaxially joined crystal of cobalt and antiferromagnetic cobalt oxide (see Section 3.5). Unfortunately the Néel point of cobalt oxide is just below room temperature. Hopes of finding a useful permanent magnet based on this phenomenon have not as yet been fulfilled.

## 2.5 FERRIMAGNETISM

Materials which are described as ferrimagnetic have considerable commercial importance as permanent magnets, and also in other branches of magnet technology. In a ferrimagnetic material atomic moments are ordered regularly in an antiparallel sense, but the sum of the moments pointing in one direction exceeds that of those pointing in the opposite direction. Figure 2.2 illustrates in a rather crude schematic manner the difference between ferromagnetic, antiferromagnetic and ferrimagnetic ordering. The first two terms have already been explained. Ferrimagnetic ordering requires that the atoms magnetized in one direction have

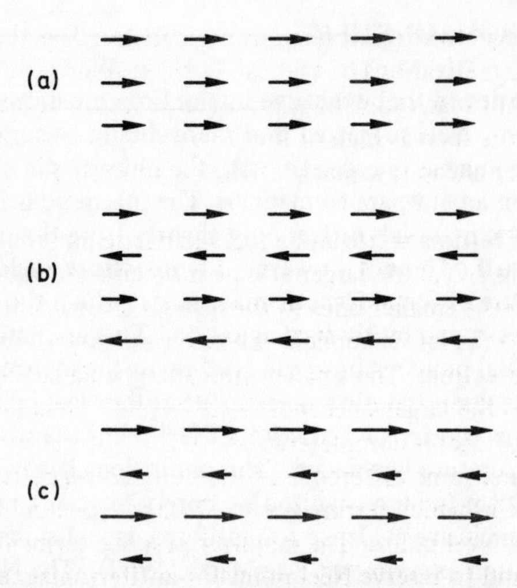


*Fig.2.2. (a) Ferromagnetic ordering with all spins parallel. (b) Antiferromagnetic ordering with spin moments on sites of alternate sub-lattices equal but opposite in direction. (c) Ferrimagnetic ordering with spin moments on sites of alternate sub-lattices antiparallel, but unequal in magnitude, giving a net magnetization in one direction*

a greater effective magnetic moment than those magnetized in the opposite direction. They do not need to be atoms of a different element. In some common ferrimagnetic substances, iron atoms can be found in two different crystallographic sites, so that their nearest neighbours differ in number or type. This difference suffices to make the effective moments on the atoms in the two different sites differ in magnitude. The numbers of Fe atoms in the different kinds of sites need not be the same, but it is necessary that the sites are arranged and repeated in a regular manner throughout the crystal, if ferrimagnetism is to occur. The Fe atoms on the two different sites can be regarded as existing in two different sub-lattices, and in some ferrimagnetic compounds the iron atoms have in effect different valencies. Thus we can suppose that ferrous Fe has 2 electrons in the 4*s* shell and 6 in the 3*d*, while ferric Fe has $4s^3$ and $3d^5$ electrons.

Natural lodestone is actually ferrous ferric ferrite. Barium strontium and lead ferrites make useful permanent magnets. Other ferrites such as those of Mn, Ni and Zn are used for radio frequency transformer cores, ferrites such as those with a mixture of Mg and Mn can be heat treated to have a square loop, and are used for computer memories. The structures of some of these materials are rather complicated, but all of them are ferrimagnetic.

Since the magnetization of ferrimagnetic materials is the difference between the magnetizations of two sub-lattices, their saturation polarization $J_s$ does not reach such high values as in ferromagnetic Fe and Co. Other factors such as high or low coercivity, electrical insulation, or requiring only cheap raw materials are responsible for their commercial viability.

Outwardly a ferrimagnetic and a ferromagnetic material behave similarly. In Fig.2.2(c) the larger atomic moments are shown pointing to the right and the smaller ones to the left. If a sufficient magnetic field pointing to the left is applied, it acts on the larger moments and reverses their direction. The smaller moments, being antiferromagnetically coupled to the larger ones must also reverse. Thus the situation depicted in Fig.2.2(c) is completely reversed.

There has been some difference of opinion and usage as to whether the temperature at which ferrimagnetism disappears should be called a Curie point or a Néel point. The majority practice seems to be to call it a Curie point, and to reserve Néel point for antiferromagnetic materials.

In some materials such as lithium chromium ferrites $J_s$ falls to zero at a temperature at which the magnetization of the two sub-lattices is equal and opposite. If the temperature is increased further the magnetization may arise again, as the magnetic order in one sub-lattice disappears. Such a material is said to have a compensation point.

## 2.6 OTHER FORMS OF MAGNETIC ORDER

Parallel and antiparallel arrangements are not the only forms of magnetic order that can exist. Materials have been found in which the spins order in the form of a helix or have an oscillating canted structure. These are interesting phenomena, but at the present time I am not aware of any important commercial permanent magnet material that has a structure of this nature.

## 2.7 DOMAINS

The theory as it has so far been presented fails to explain how a piece of iron can ever be demagnetized. Weiss explained the possibility of demagnetization by the hypothesis of domains. A domain is a small volume of a substance that is spontaneously magnetized in one direction. In bulk a piece of magnetic material contains many domains magnetized in different directions. The material is demagnetized if these directions are completely random for the material as a whole, so that its net magnetization is zero. Locally the domain patterns cannot be completely random, but they can add up to a net zero magnetization. The study of the behaviour of domains is fundamental to the understanding of permanent magnetism, and is dealt with at length in Chapter 3.

## 2.8 CONCLUSIONS

Much of the theory is very complex and suffers from the disadvantage that each material seems to require special assumptions. Nevertheless some of the pieces in this complicated picture are beginning to fall into place. The theory of localized moments, and the collective electron theory can now be seen as complementary rather than mutually exclusive. In many metals and alloys the electrons do appear to belong to the material as a whole, but in a ferrimagnetic oxide it may be more sensible to consider the electrons to be associated with individual atoms.

In spite of their deficiencies the theories have advanced sufficiently to be useful to the material scientist for the purpose of suggesting how alloys may be modified to improve their magnetic properties, and particularly how to increase their saturation polarization $J_s$. The greatest possibilities of advances appear to involve the rare-earth elements, rather than in alloys containing only the traditional magnetic elements from the transition series.

# Chapter 3

# Fundamental causes of permanent magnetism

This chapter is concerned with the problem of why permanent magnets are permanent. When discussing the theories, which have been proposed to account for the occurrence of permanent magnetism, there is a frequent need to name particular materials as examples. This need may suggest that this chapter should have followed Chapter 4, which describes permanent magnet materials and how they are produced. On balance an even greater difficulty would then have arisen. Readers unfamiliar with ideas about domain processes, magnetic anisotropy, and single domain particles, would find a description of methods of producing magnets as indigestible as a cookery book. Space does not permit a full description of every material at the point where it is mentioned in this chapter, and consequently some forward references to sections in Chapter 4 are unavoidable.

A brief historical outline of the various theories that have been put forward to account for coercivity and hysteresis is given here. Sections then follow in which each possible process is considered in more detail. The way in which these processes may act separately or together in permanent magnets can then be discussed. This treatment may seem a little tortuous, but it is necessary because magnetic hysteresis involves unfamiliar ideas rarely mentioned in the main stream of physics and electrical engineering courses. These ideas include concepts such as shape, crystal and stress anisotropy, magnetic domains and domain walls, together with structural considerations, such as fine particles, precipitates, inclusions and other material defects. The explanation of permanent magnetism has unfolded, with these ideas advancing in parallel. We have to consider how the structure of a material and the various physical phenomena such as anisotropy combine to determine the behaviour of domains and domain walls.

One of the earliest theories of permanent magnetism that still merits consideration today is Ewing's molecular theory. Ewing's model of an array of compass needles appeared to show some hysteresis. Although the behaviour of this model can be discounted as an explanation of permanent magnetism, and its retention for many years in elementary textbooks to the exclusion of any mention of magnetic domains was unfortunate, the reasons why Ewing's theory can be rejected are less obvious than is popularly supposed, and these reasons are therefore examined in Section 3.1.

By 1940 it was customary to explain coercivity in terms of hindrances to domain boundary movements, but it is necessary first to consider various forms of anisotropy. If there is no anisotropy to prevent the magnetic polarization vector rotating into the direction of the applied field, there can be no coercivity. The existence of magnetic anisotropy also influences the structure and behaviour of the domain walls that separate domains with magnetization in different directions. An important book that dealt with these problems was Becker and Döring (1939). At the time when this book was published, internal strains combined with the process of magnetostriction were regarded as the most probable hindrances to the motion of domain walls, although an alternative theory proposed by Kersten attributed coercivity to the pinning of domain walls by non-magnetic inclusions. Neither theory in its original form can account for the coercivities of more modern materials, such as Al–Ni–Co alloys and materials with even higher coercivities. The older theories may have been sufficient to account for the properties of the older steel permanent magnets, but these materials have not been very intensively studied since newer techniques have been available.

Two papers by Stoner and Wohlfarth (1947, 1948) introduced the idea of single domain particles into English scientific literature. Parallel work by Néel in France during the 1939/45 war lead to commercial attempts to make fine particle magnets. The concept of single domain particles dominated permanent-magnet theories for over a decade. New materials were invented and forecasts of the possibilities were very optimistic. Although Stoner and Wohlfarth considered fine particles with strain, crystal or shape anisotropy, their most optimistic predictions were for elongated single domain particles with shape anisotropy.

Many research workers tried to make magnets with these elongated particles, but the results they obtained fell far below the predictions. A new branch of magnetic theory, micromagnetics, was created to account for this difference between theory and practice. During the early 1960s this new theory suggested that attention should be diverted from elongated particles to those with large crystal anisotropy. The early 1960s was, however, a period when permanent magnet research was rather in decline, and little progress was reported in either new theories or materials. Following the disappointments with elongated single domain particles, it was sometimes difficult at a large international conference on magnetism with 20 or 30 sessions, to find sufficient papers on permanent magnetism to occupy one whole session.

The situation was completely changed during the second half of the last decade, when a new family of permanent magnets, based on rare-earth-cobalt compounds was discovered. These compounds possess a particularly high magnetocrystalline anisotropy, and so at least to some extent fulfil the theoretical expectations. At first it was thought that magnets made from these materials should consist of particles fine enough to be single domains, but recently it has emerged that the best magnets are actually made with particles larger than the critical size for single domain behaviour. Theory has thus gone through a full circle and once again pinning and nucleation of domain walls are subjects of study, although the theories and methods of investigation have made great advances since Becker and Döring's book was written.

Judging by the number of papers published, one can estimate that the amount of permanent magnet research now being conducted is five or ten times greater than in the early 1960s, and a reasonable proportion of these papers are concerned with attempts to understand the fundamental processes responsible for magnetic hysteresis.

## 3.1 EWINGS MOLECULAR THEORY AND MODEL

Ewing supposed that that there is a small magnet in each molecule of a substance, and constructed models consisting of plane arrays of compass needles to represent these molecular magnets. An array of this nature does exhibit qualitatively some of the major aspects of magnetic hysteresis. The model depends on the magnetic interactions between the magnets.

The Ewing theory met with the objection discussed in Section 2.2, that most substances that have atomic or molecular moments are only paramagnetic. Calculations show that the magnetic forces between the interacting dipoles can never be sufficient to keep them aligned in competition with the disordering effects of thermal agitation. When therefore the Weiss molecular field, later explained in terms of the quantum mechanical exchange forces, had been accepted as necessary to account for ferromagnetism, the Ewing theory was generally written off, although many years elapsed before Weiss domains replaced Ewing's molecular theory in much elementary teaching.

The reasons for rejecting Ewing's theory of magnetic hysteresis are actually rather more complicated, and it is useful to examine them more closely. Bates (1961) page 14 points out that an atom or molecule occupying a spherical cavity in a magnetized medium experiences a field due to the magnetization of the medium. If the polarization is $J$ tesla the magnitude of this field is $J/3$ if it is expressed in tesla or $J/(3\mu_0)$ if it is expressed as a magnetizing force in $Am^{-1}$. These values exceed the coercivities of many commercial permanent magnet materials, and even the predictions of many theories that have been favourably received. In fact all theories of coercivity predict values less than one hundredth of the Weiss molecular field, and of the same order of magnitude as the dipole interactions predicted by the Ewing theory.

The real flaw in the Ewing model is that it does not allow for soft magnetic materials to exist. Three questions require answers: why does the magnetic interaction not contribute to coercivity, why does the Ewing model show that magnetic interactions do create hysteresis, and why does it not demonstrate the self-demagnetizing effect described in Section 6.4?

The answer to the first question appears to be that the magnetic interactions calculated for a spherical cavity do not apply to cavities of other shapes. The field is greater in a plane cavity transverse to the field and zero in a needle shaped cavity parallel to the field. The consideration of such cavities was formerly used in the definition of $\boldsymbol{B}$ and $\boldsymbol{H}$, but the method is now perhaps rather mistakenly out of fashion. An external field acts on all the atomic magnets, and can induce rotation in a group of any shape that will offer the least resistance. Thus magnetic interactions will offer no resistance to the reversal of magnetization in a long needle shaped volume parallel to the field. In other words magnetic interactions do not prevent the nucleation of domains of reversed magnetization, provided they have this elongated shape.

Some years ago it occurred to me that the Ewing model represents a monatomic thin film that should have zero self-demagnetizing field. In an attempt to demonstrate the self-demagnetizing field, I had a three dimensional model constructed as shown in Fig.3.1. In this model all the magnets were fixed except the centre one. My first attempt failed; the central magnet remained parallel to the others. A calculation suggested that the positive magnetic interactions were enhanced by the proximity of the poles of neighbouring magnets, the magnets being too long compared with their distance apart. This situation almost certainly exists in the old Ewing models, and accounts for them appearing to demonstrate hysteresis.

I had the model rebuilt with shorter high coercivity magnets, and the model now worked as I had hoped. When the central magnet was unclamped, it reversed so that it was antiparallel to the remainder. This model demonstrates the existence of a self-demagnetizing field, although rather imperfectly. The actual field experienced by the central magnet depends on the lattice arrangement of the magnets as well as

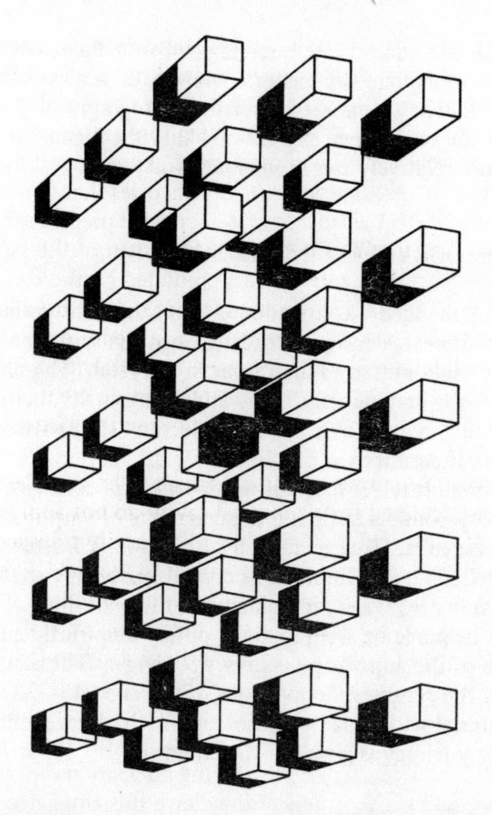

*Fig.3.1. Three dimensional atomic model, consisting of regular array of bar magnets. All bars are parallel except central one, which is free to rotate, and turns in the opposite direction to the remainder, showing that there is a self-demagnetizing field*

the overall shape of the model. With hindsight it would have been instructive to have made a model in which any group of magnets could be unclamped.

Thus the Ewing model can demonstrate a self-demagnetizing field if it is constructed in three dimensions. Its apparent demonstration of coercivity is a spurious effect produced by using magnets that are too long. The magnetic interactions between molecular magnets are quite large and have a considerable influence on domain structure, although they are not responsible for coercivity.

Paine and Luborsky (1960) used a form of the Ewing model to investigate the influence of magnetic interactions between fine particles of powder magnets. They concluded that imperfections in the structure such as 'Cross ties' can initiate buckling and so reduce coercivity. They used this idea to explain the higher coercivity of Alnico VIII compared with Alnico V. They compared electron micrographs of the precipitate structures in the two alloys, and showed that the higher coercivity one had a structure relatively free from such cross ties.

## 3.2 MAGNETOCRYSTALLINE ANISOTROPY

In Section 2.3 the idea was introduced that the quantum-mechanical interactions between electrons, tends to align the spontaneous magnetization along certain preferred directions in a crystal. The phenomenon is called magnetocrystalline anisotropy and it enters the theory of permanent magnetism in various ways. Anderson (1968) attempts to give what he himself describes as a rather crude calculation of crystal anisotropy. Although it refers to quantum-mechanical concepts most of the detailed theory consists in fitting the constants of empirical formulae to experimental results. This account by Anderson is, however, one of the most intelligible explanations of crystal anisotropy that I have met.

Considerable progress in understanding the behaviour of permanent magnets can be made by writing down purely empirical equations for the variation of the anisotropy energy with angle. These equations are chosen to fit the symmetry conditions of the crystal.

For a material with uniaxial anisotropy such as hexagonal cobalt, this anisotropy energy is given by the equation

$$E = K + K_1 \sin^2\theta + K_2 \sin^4\theta \tag{3.1}$$

In this equation $\theta$ is the angle between the direction of magnetization and the single uniaxial easy direction. In addition to cobalt a number of important permanent magnet materials such as barium ferrite and $SmCo_5$ have uniaxial anisotropy. It is unwise to assume that $K_2$ can be neglected unless experiments on the particular material have shown that it is in fact small. There are some materials in which $K_1$ is negative. Such materials have an easy plane of magnetization, and they do not make good permanent magnets.

In body centred cubic iron the preferred direction lies along the cube edges or in the terminology of Miller indices [100] directions. In face centred cubic nickel the cube diagonals or [111] directions are preferred In the absence of a magnetic field, the spontaneous magnetization lies along one of the easy directions, although in an unmagnetized multidomain piece of material, the actual crystal axis varies from domain to domain.

In uniaxial materials domains can be directed in any one crystal in only two directions, and these domains are separated by what are called 180° walls. In body centred cubic Fe the magnetization in neighbouring domains differs by 180° or 90°, and so we have 180° and 90° walls. In face centred cubic nickel there are four easy axes and 8 possible directions of magnetization, and the angles between these directions and the possible types of wall are obviously more complicated.

The variation of crystal energy with direction for a cubic material is usually given in terms of direction cosines. Let OX, OY, OZ be the cube edges of a crystal, and let the magnetization be in the direction OP. Let $\alpha_1 = \cos \text{POX}$, $\alpha_2 = \cos \text{POY}$ and $\alpha_3 = \cos \text{POZ}$. The energy per unit volume of the material if it is magnetized in the direction OP is

$$E = K + K_4(\alpha_1^2\alpha_2^2 + \alpha_2^2\alpha_3^2 + \alpha_3^2\alpha_1^2) + K_6(\alpha_1^2\alpha_2^2\alpha_3^2) \tag{3.2}$$

$K$ is a constant that is logically included but rarely used. In many books $K_4$ and $K_6$ in Equation 3.2 are replaced by $K_1$ and $K_2$. The present nomenclature has been adopted to enable Equations 3.1 and 3.2 to be combined to deal for example with tetragonal materials.

In this book anisotropy constants are given in $\text{Jm}^{-3}$. The CGS unit is erg $\text{cm}^{-3}$, and values given in these units must be divided by 10 to obtain the SI value.

For many materials the term in $K_6$ is small and can be neglected. For nickel and similar materials in which the cube diagonals [111] directions are the easy directions, Equation 3.2 still holds but $K_4$ is negative.

Equations 3.1 and 3.2 are completely symmetrical for a 180° change in the direction of magnetization. There are a number of other magnetic effects, which are unchanged if the direction of magnetization is reversed. English translations of Russian texts frequently refer to such effects as 'even processes'. This description seems logical, but has been little used by writers who think in English. The influence of a magnetic field itself and also exchange anisotropy discussed in Section 3.5 are examples of processes that are not 'even processes'.

A tetragonal crystal may be regarded as a cubic crystal distorted in one lattice direction. If OX is the direction of uniaxial anisotropy, Equations 3.1 and 3.2 may be combined to give:

$$E = K + K_1(1 - \alpha_1^2) + K_2(1 - \alpha_1^2)^2 + K_4(\alpha_1^2\alpha_2^2 + \alpha_2^2\alpha_3^2 + \alpha_3^2\alpha_1^1) \tag{3.3}$$

The term containing $K_6$ has been neglected. An equation such as 3.3 may arise as a result of other processes than tetragonal distortion. For example a cubic alloy such as an anisotropic Al–Ni–Co may acquire a very slight uniaxial anistropy as a result of a very imperfect or even accidental field treatment.

Preferably measurements of crystal anisotropy constants are made on single crystals. One method is to measure the energy of magnetization $\int H dJ$ along different crystal axes. This measurement is accomplished by first determining the $J$,$H$ curve from the demagnetized state to saturation and then the area between this curve and the $J$ axis. Examples are given in Fig.3.2.

The magnetization curve may be measured in a yoke, in which case cuboid samples with edges coinciding with the required crystal axes can be used. Often, however, only small crystals are available, and it is necessary to use an open circuit method. A vibrating sample magnetometer such as that originally described by Foner (1959) is very suitable. On open circuit a correction for the self-demagnetizing factor of the sample, explained in Section 6.4, must be made. Such corrections can only be made accurately for ellipsoids although approximate formulae have been proposed for some other shapes.

For a cubic material the energies required to magnetize to saturation in various crystallographic directions can be written down from Equaation 3.2. In the [100] direction

$$\alpha_1 = 1, \alpha_2 = \alpha_3 = 0$$

therefore $E = K$. In the [110] face diagonal direction

$$\alpha_1 = \alpha_2 = 1/\sqrt{2}, \quad \alpha_3 = 0$$

therefore $E = K_4/4 + K$. In the [111] direction

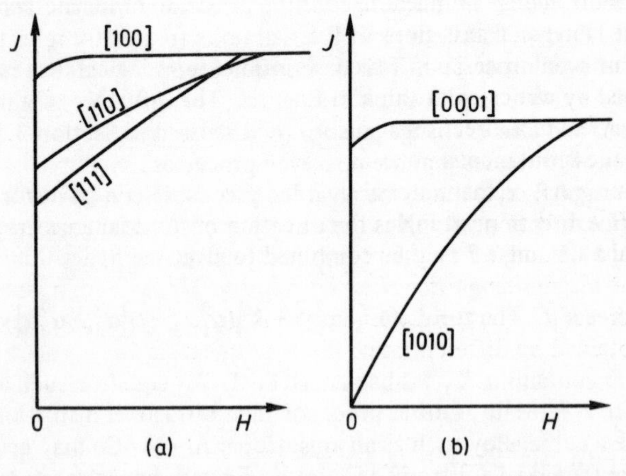


*Fig.3.2. Approximate magnetization curves in various crystal directions for (a) Fe and (b) Co*

$$\alpha_1 = \alpha_2 = \alpha_3 = 1/\sqrt{3}$$

therefore

$$E_{111} = K_4/3 + K_6/27 + K$$

and hence

$$K_4 = 4(E_{110} - E_{100}) \tag{3.4}$$

and

$$K_6 = 27(E_{111} - E_{100}) - 36(E_{110} - E_{100}) \tag{3.5}$$

These calculations eliminate errors, which arise from other causes such as strains, provided they are isotropic and contribute equally to the energy of magnetization in all directions. In this connection it is arguable that the energies should be calculated from the curves between $J_s$ and $J_r$ rather than the initial magnetization curves, since various domain processes not connected with crystal anisotropy contribute to the energy calculated from the latter. Probably the distinction is not very important, provided the same method is used in calculating all the energies used in the same equation.

A number of other methods of deducing $K_4$ from the magnetization curves of cubic materials are included by Becker and Döring (1939) in a general account of the subject.

For a uniaxial material it is clear from Equation 3.1 that the difference in the energy of magnetization perpendicular, $E_\perp$, and parallel $E_\parallel$, to the hexagonal axis gives $K_1 + K_2$.

For a uniaxial material in which the first anisotropy constant predominates, we can calculate an anisotropy field $H_a$ capable of saturating a single crystal in a direction at right angles to the preferred direction. Suppose the spontaneous polarization is $J_s$ tesla, and suppose that in a small volume $\Delta V$ of the material the polarization is held in equilibrium by field $\boldsymbol{H}$ $\mathrm{Am^{-1}}$ at right angles to the preferred direction, so that $J_s$ is inclined at an angle $\theta$ to the preferred direction and hence at $90° - \theta$ to $\boldsymbol{H}$. The magnetizing force $H$ exerts a torque $HJ_s\Delta V \,.\, \cos\theta$ Nm tending to increase $\theta$. The torque tending to return $J_s$ to the preferred direction is obtained by differentiating the expression for the crystal energy: i.e.

$$\frac{\mathrm{d}E}{\mathrm{d}\theta} = \frac{\mathrm{d}}{\mathrm{d}\theta}(K_1 \sin^2\theta)\Delta V = 2K_1 \sin\theta \cos\theta \Delta V$$

Equating these torques

$$HJ_s \Delta V \cos\theta = 2K_1 \sin\theta \cos\theta \Delta V$$

or

$$H = \frac{2K_1 \sin\theta}{J_s} \tag{3.6}$$

The value of $H$ that aligns $J_s$ with the field is obtained by putting $\sin\theta = 1$. Hence the anisotropy field $H_a$ is given by

$$H_a = 2K_1/J_s \tag{3.7}$$

Thus measuring $H_a$ enables $K_1$ to be calculated.

Sometimes it may be more convenient to cite $\mu_0 H_a$ in tesla. The above derivation holds in CGS units with $J$ the intensity of magnetization and $H$ in Oe and torques in dyne-cm.

In cubic crystals the anisotropy field depends on the sign of anisotropy constants, and the direction of the magnetizing field, which cannot simply be specified as perpendicular to the easy direction of magnetization.

In some of the new rare-earth-cobalt materials described in Section 4.5, the value of $H_a$ is so great that it cannot easily be attained in a normal laboratory. The very large anisotropy fields of some $RCo_5$ compounds (R = rare-earth metal) were estimated by extrapolation from experiments with much lower fields. On the basis of the very large values estimated, good permanent magnet properties were predicted for this class of materials. These predicted properties have been realized in some but not all of the $RCo_5$ compounds, as a result of research stimulated by the crystal anisotropy experiments.

Another method of measuring crystal anisotropy constants makes use of a torque magnetometer, which enables the torque required to hold a crystal with its axes inclined at various known angles to an applied field, to be measured. In older instruments this torque was measured by the twist in a calibrated wire or strip, which held the crystal at the required angle. Ideally the sample should be cut in the shape of an oblate ellipsoid, but a thin disc is usually satisfactory, provided a field well in excess of $H_a$ can be applied. The disc is rotated about an axis perpendicular to its plane and also to the applied field. It is most important that the sample should be circular and mounted symmetrically about its centre, as otherwise spurious torques are introduced. It is difficult to interpret the results if the applied field does not saturate the sample, so the torque magnetometer is used mainly for investigating materials that do not have very large anisotropy fields.

In materials with cubic anisotropy, the torque curves expected depend on the crystal plane of the sample. For a sample of cubic crystal cut so that its surface is a [100] plane, let $\alpha_1 = \cos\theta$, then $\alpha_2 = \sin\theta$ and $\alpha_3 = 0$. Inserting these values into Equation 3.2, we obtain

$$E = K + K_4 \cos^2\theta \sin^2\theta = K + K_4(1 - \cos 4\theta/8) \tag{3.8}$$

The torque $T$ is obtained by differentiating this equation:

$$T = \frac{K_1 \sin 4\theta}{2} \tag{3.9}$$

A somewhat more complicated analysis by Brailsford (1960) shows that in a [110] plane containing two face diagonals and a cube diagonal the expression for the torque is

$$T = -\left(\frac{K_4}{4} + \frac{K_6}{64}\right)\sin 2\theta - \left(\frac{3K_4}{8} + \frac{K_6}{16}\right)\sin 4\theta + \frac{3K_6}{64}\sin 6\theta \tag{3.10}$$

In this expression $\theta$ is the angle in a [110] plane measured from a face diagonal. The sign of the torque in all these expressions is a matter of convention.

The torque expressions for uniaxial anisotropy are simpler and are obtained by differentiating Equation 3.1.

$$T = K_1 \sin 2\theta + K_2(\sin 4\theta)/2 \tag{3.11}$$

Figure 3.3 shows the form of the torque curve obtained from Equation 3.11 for $K_1 = 2K_2$.

If large single crystals are not available, crystal anisotropy measurements may sometimes be made on assemblies of small crystals aligned and set in a binder such as wax.

Some Al–Ni–Co alloys described in Section 4.2.5 are difficult to obtain as single crystals, but are made with a columnar structure. All the grains have a [100] cube edge parallel to a given direction, but in the plane perpendicular to this direction the crystal axes are distributed at random. Torque curves were measured on these columnar structures by K. Hoselitz and myself (Hoselitz and McCaig, 1951, 1952). The nature of the torque curves varied greatly with heat treatment, and on whether a magnetic field had been applied during this heat treatment, as is customary in the production of these alloys, and of course on the way in which the sample had been cut and the direction of the applied

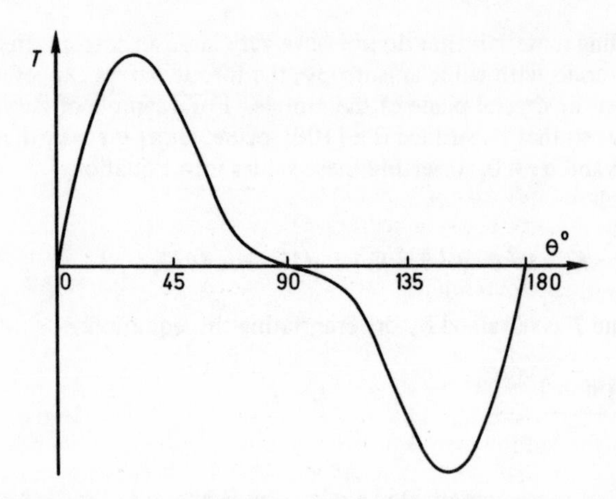


*Fig.3.3. Torque curves calculated from Equation 3.1 for $K_1 = 2K_2$*

field relative to the columnar axis. Originally I supposed that these torque curves represented measurements of crystal anisotropy. It is now clear that the shape anisotropy of a precipitate formed during the heat treatment plays a predominant role. This precipitate is oriented partly by the crystal planes, and partly by the field, if any, applied during the heat treatment. Torque measurements are, however, a legitimate method of investigating the anisotropy caused by these oriented precipitates. A method of separating shape anisotropy and crystal anisotropy effects has been described by Nesbitt and Williams (1955), who found that in Alnico V treated without a field a pseudo-cubic anisotropy is produced. This anisotropy is caused by uniaxial shape anisotropy directed along the different cube edges. Unlike true crystal anisotropy it falls off at high fields. Uniaxial anisotropy produced by directing the shape anisotropy along the nearest cube edge to the field applied during heat treatment does not appear to fall off in this way.

The true crystal anisotropy of Al–Ni–Co alloys is probably positive cubic, similar to iron, but smaller in magnitude. Table 3.1 contains the values of the magneto-crystalline anisotropy of Fe, Ni and Co and a few important permanent magnet materials at room temperature.

A remarkable feature of magnetocrystalline anisotropy is its extreme temperature sensitivity. Anderson (1968) gives a short account of theories of this temperature dependence with references. Zener for example has predicted that the anisotropy constants of Fe should decrease with the tenth power of the absolute temperature, and this is roughly what is found experimentally. In other materials the changes appear almost capricious. MnBi has a large uniaxial anisotropy at room temperature, which falls rapidly until it changes sign as the temperature is reduced. In the technologically important barium and strontium ferrites the anisotropy constants fall with increasing temperature much less rapidly than $J_s$. Consequently the anisotropy fields increase at high temperatures and fall off at low temperatures. The latter effect has disagreeable consequences for the behaviour of ferrite magnets at low temperatures, as explained in Section 5.5.6. In some $RCo_5$ compounds $K_1$ decreases with temperature, and in others it increases or passes through a maximum.

## 3.3 STRESS ANISOTROPY

If the magnetization curves or hysteresis loops of nickel or Permalloy wires are measured, while they are subjected to tension, remarkable changes take place in the shape of the curves, as shown in Figs.3.4 and 3.5. In Ni the permeability is greatly reduced by tension and the remanance and coercivity almost disappear. The direction in which tension is applied becomes a hard or non-preferred direction of magnetization. In Permalloy on the contrary, tension makes the hysteresis loop more square, and both remanance and coercivity increase. Apparently tension makes the direction in which it is applied into an easy or preferred direction of magnetization.

These differences are explained by considering the phenomenon of magnetostriction. When Ni is magnetized it contracts in the direction of magnetization and expands in the transverse direction. The reverse happens when Permalloy is magnetized. In Ni the lattice separation of the atoms is slightly reduced and in Permalloy slightly increased in the direction of magnetization. One can deduce that tension should decrease magnetization in the first case and increase it in the second, by introducing the thermodynamic relation:

$$\left(\frac{\partial J}{\partial Z}\right)_H = \left(\frac{\partial L}{\partial H}\right)_Z \qquad (3.12)$$

where $Z$ = stress and $L$ = length.

Alternatively one may feel that this follows from a plausible general principle, that stress favours a change in magnetization that produces a magnetostriction in the same sense as the stress.

It is understandable from results such as those shown in Figs.3.4, and 3.5, that stresses were invoked to account for the properties of permanent magnets. When I first became concerned with permanent magnets in 1946, internal stresses were considered one of the most probable causes of coercivity, and the anisotropy produced in certain Al–Ni–Co alloys by cooling in a magnetic field was thought to be

**Table 3.1** MAGNETOCRYSTALLINE ANISOTROPY CONSTANTS AT ROOM TEMPERATURE IN $Jm^{-3}$

| *Material* | *Structure* | $K_1$ | $K_2$ | $K_4$ | $K_6$ |
|---|---|---|---|---|---|
| Fe | bcc | | | $4.8 \times 10^4$ | $\pm 5 \times 10^3$ |
| Ni | fcc | | | $-4.5 \times 10^3$ | |
| Co | cph | $4.1 \times 10^5$ | $2.0 \times 10^5$ | | |
| $BaFe_6O_{19}$ | hex | $3.0 \times 10^5$ | | | |
| MnBi | hex | 8 to $9 \times 10^5$ | | | |
| $SmCo_5$ | hex | $10.5 \times 10^6$ | | | |

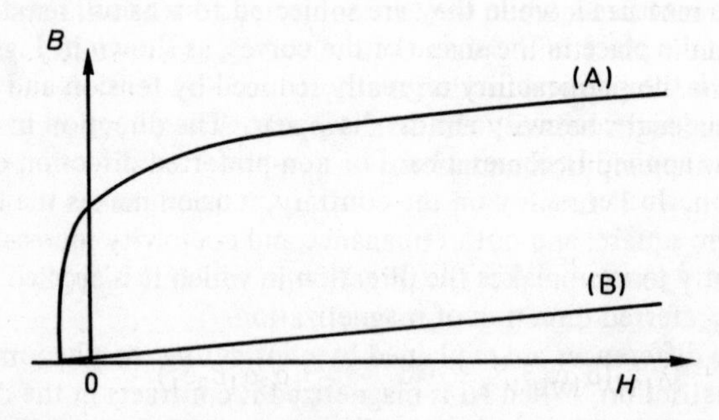


*Fig.3.4. Influence of tension on magnetization curve of Ni. A: No tension. B: 33 kg mm$^{-2}$*

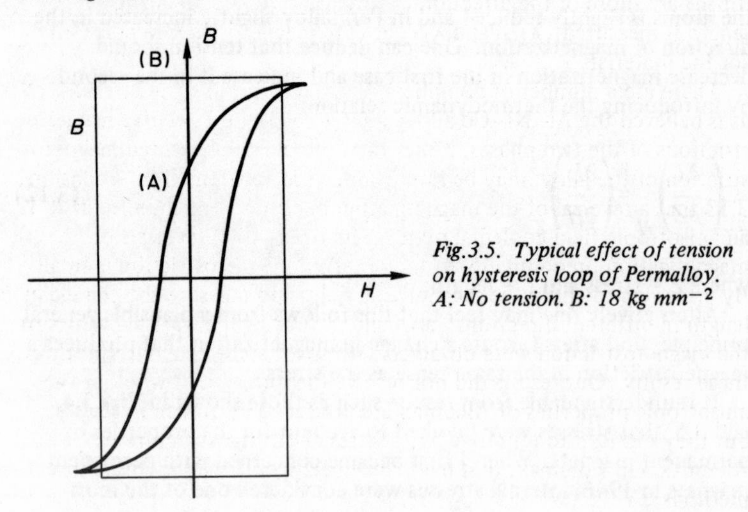


*Fig.3.5. Typical effect of tension on hysteresis loop of Permalloy. A: No tension. B: 18 kg mm$^{-2}$*

produced by 'freezing in stresses' produced by magnetostriction as the alloy cooled through its Curie point.

One of my early tasks was to measure the magnetostriction of a number of permanent magnet alloys, including Alcomax and later the columnar variety of this alloy. The experiments were carried out on samples that had been cooled with and without a field.

The method is illustrated in Fig.3.6 and resembled a laboratory method for measuring thermal expansion. A bridge of non-magnetic material was placed on the magnet. One end of the bridge had a sharp edge; the other rested on a spindle carrying a mirror. Changes in length of the magnet were followed by observing with a travelling microscope a spot of light reflected by the mirror.

Today strain gauge methods would probably be preferred, but at that time it was difficult to make them sufficiently sensitive. The improvement lies more in the availability of reliable solid state amplifiers, than in the strain gauges themselves.

In general magnetostriction is anisotropic and for cubic crystals the change in length between the demagnetized and the magnetized state is given by the empirical equation:

$$\frac{dl}{l} = \frac{3}{2}\lambda_{100}(\alpha_1^2\beta_1^2 + \alpha_2^2\beta_2^2 + \alpha_3^2\beta_3^2 - 1/3)$$

$$+ 3\lambda_{111}(\alpha_1\alpha_2\beta_1\beta_2 + \alpha_2\alpha_3\beta_2\beta_3 + \alpha_3\alpha_1\beta_3\beta_1) \quad (3.13)$$

$\alpha_1\alpha_2\alpha_3$ are the direction cosines of the direction of the applied field, and $\beta_1\beta_2\beta_3$ are those of the direction in which the fractional change in length is measured. $\lambda_{100}$ and $\lambda_{111}$ are the magnetostriction constants for the material.

There are a number of complications in interpreting these results. If as is believed the Al–Ni–Co alloys are two phase, the relative magnetostrictions of the two phases, rather than the overall measured magnetostriction of the alloy, may be important. It is apparent from Equation 3.13 that a reversal of the magnetization does not contribute to $dl/l$. If an anisotropic field cooled alloy is magnetized, 180° reversals of magnetization predominate, so the observed magnetostriction is small. By arranging to apply the magnetizing field, and measure the change in length in different directions, large and at first sight surprising values of the magnetostriction were obtained. We spent some time interpreting these results. Our results did not resolve the question of whether the anisotropic properties of Alcomax are produced by frozen in stresses, but they did lead to tentative hypotheses about the structure of Alcomax and Columax, that have since been confirmed by more direct methods.

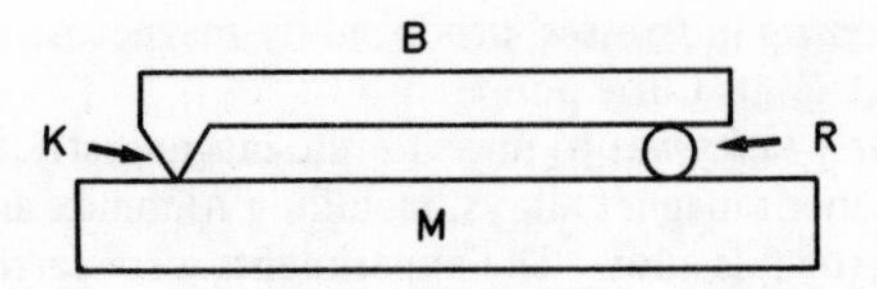


*Fig.3.6. Magnetostriction apparatus. M: Magnet. B: Bridge with knife-edge K. R: Roller carrying mirror*

Measurements of magnetostriction at high temperatures are rather difficult to perform, but eventually I felt able to publish results, showing that the magnetostriction of Alcomax falls to very small values at the temperature at which the field treatment is effective (McCaig, 1952).

By this time the publications of Stoner and Wohlfarth (1947, 1948) on single domain particles had transferred interest from stress to shape anisotropy, and the shape of oriented precipitate particles is now accepted as responsible for the anisotropy produced by cooling in a magnetic field. The possible influence of internal stresses on the coercivity is discussed later in Section 3.8.2 in connection with domain wall processes, but external forces do not produce a change of more than a few percent in the remanence and coercivity of most permanent magnet materials, and in consequence stresses are now rather discounted in the theory of permanent magnetism.

It is difficult to prove or disprove suggestions that stress may play some subsidiary role in determining the properties of some Al–Ni–Co alloys. It is obvious that stresses may arise between the precipitate and matrix or two phases of a spinodal decomposition. In some Russian work remarkable changes were produced by stresses on the properties of Vicalloy; this subject is discussed in Section 4.7.

It is usual as in the preceeding pages to treat stress anisotropy as a separate effect from crystal anisotropy, but it is arguable that it arises from a distortion of the crystal lattice and is therefore a modification of crystal anisotropy.

## 3.4 SHAPE ANISOTROPY

Any magnetic body with an elongated shape possesses anisotropy in the sense that if free to rotate it aligns with its greatest length parallel to an external field. Even more significantly a larger magnetic field is required to magnetize it to saturation parallel to a short rather than a long dimension. For simplicity it is convenient to consider a prolate ellipsoid of revolution. Such an ellipsoid has a self demagnetizing factor $N_a$ parallel to its major axis and $N_b$ parallel to any minor axis. That is to say there is a self-demagnetizing force $N_a J_s / \mu_0$ when the ellipsoid's saturation polarization is directed along its major axis, and $N_b J_s/\mu_0$ when it is directed along a minor axis. It follows that the magnetizing force necessary to produce saturation along a major axis by

$$H_a = (N_b - N_a)J_s/\mu_0 \tag{3.14}$$

Further the excess energy when the ellipsoid is magnetized along a minor axis is $(N_b - N_a)J_s^2/2\mu_0$ per unit volume.

$H_a$ can be regarded as a shape anisotropy field similar to $2K_1/J_s$ for materials with uniaxial crystal anisotropy or $3\lambda Z/2$ for materials with stress anisotropy (Z = stress).

Although shape anisotropy exists in a body of any size, just as magnetocrystalline anisotropy exists in a single crystal of any size, a symmetrically shaped body such as a sphere shows no shape anisotropy unless it consists of fine shaped particles, but a single crystal sphere can have crystal anisotropy. The paper by Stoner and Wohlfarth (1948) drew attention to the fact that a material consisting of fine particles can have anisotropy arising from the shape of these particles. If the particles are small enough to be single domain particles, they can have large coercivities as a result of their size and shape or any other form of anisotropy.

Shape anisotropy is a macroscopic property arising in a particle with millions of atoms. Stress and crystal anisotropy can under certain circumstances hinder the movement of domain walls. Shape anisotropy can lead to a high coercivity only when there are no domain walls. Nevertheless shape anisotropy of relatively large particles can make a material anisotropic, even when its coercivity is still low.

## 3.5 EXCHANGE ANISOTROPY

Exchange anisotropy was first proposed to explain some peculiar displaced hysteresis loops obtained by Meiklejohn and Bean (1956, 1957) with cobalt particles coated with cobaltous oxide. Suppose as depicted in Fig.3.7 an antiferromagnetic phase A is coherently joined to a ferromagnetic material or phase F. It is necessary that there should be some correspondence between the crystal lattices of the two materials. Suppose also that the Néel point, the temperature at which antiferromagnetism disappears in A, is lower than the ferromagnetic Curie point of F. Let F be magnetized at a temperature above the Néel point of A. At this temperature A is paramagnetic and its spins are disordered. Now let the temperature be reduced, while the magnetization of F is maintained by an external field. When the Néel point of A is reached, antiferromagnetic order will spontaneously arise in A. At this point the

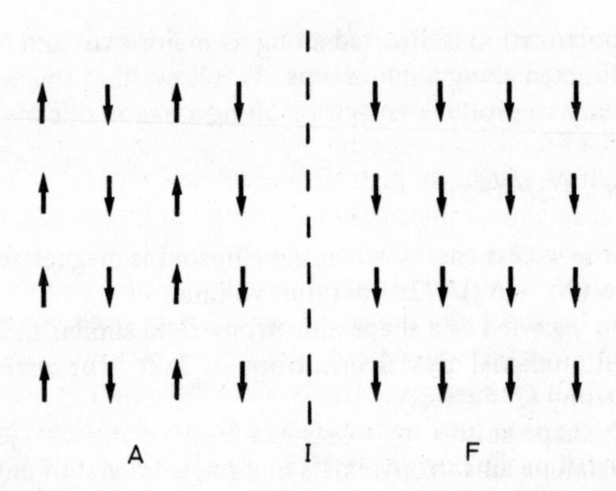


*Fig.3.7. Exchange anistropy. A: Antiferromagnetic material. F: Ferromagnetic material. I: Interface*

spins on the two sides of the interface between A and B may be coupled by exchange interaction. This interaction itself could be antiferromagnetic or ferromagnetic. The consequence in either case is that the antiferromagnetism in A is coupled to the ferromagnetism in F. Although a magnetic field cannot act directly to any significant extent on an antiferromagnetic material, it can influence the manner in which antiferromagnetism is set up in A, through the ferromagnetic material F and the coupling between A and F. Thereafter A influences B, so that its direction of magnetization is strongly bound to its original direction. The further the temperature is reduced below the Néel point of A, the stronger this bond becomes. The resulting displaced hysteresis loop found by Meiklejohn and Bean for the cobalt-cobaltous oxide powder is illustrated in Fig.3.8.

Naturally the discovery caused considerable interest, and raised hopes of a new method of producing improved permanent magnets. The Néel point of cobaltous oxide is close to room temperature, and significantly displaced loops are obtained in this system only at considerably lower temperatures. So far no useful permanent magnets have been found with these displaced loops. Roth and Luborsky (1964) produced displaced loops at room temperature in barium ferrite to which potassium had been added, but the resulting properties were not as good as can be obtained on undoped barium ferrite.

There is no theoretical reason, which would have allowed this failure to obtain useful displaced loops at room temperature to be predicted. There are certainly antiferromagnetic materials with Néel points above room temperature. An intensive search for permanent magnets with these displaced loops was made. Pairs of materials including an antiferromagnetic compound like cobaltous oxide, and also alloys in which it was hoped to precipitate an antiferromagnetic phase were investigated. So far the efforts have been unsuccessful. Of course a considerable number of factors need to be right simultaneously. These factors include besides sufficiently high Néel and Curie points, crystal lattices which match or almost match, and the right kind of interaction between the phases.

Lack of success in this line of research has now rather dampened enthusiasm. Furthermore the advantages that appeared to be promised by displaced loops have now been outstripped by other methods in rare-earth-cobalt compounds. The possibility cannot be excluded that exchange interaction may be responsible for producing some high coercivities without producing displaced loops. It has certainly been considered as a possible explanation of pinning mechanisms in some rare-earth systems.

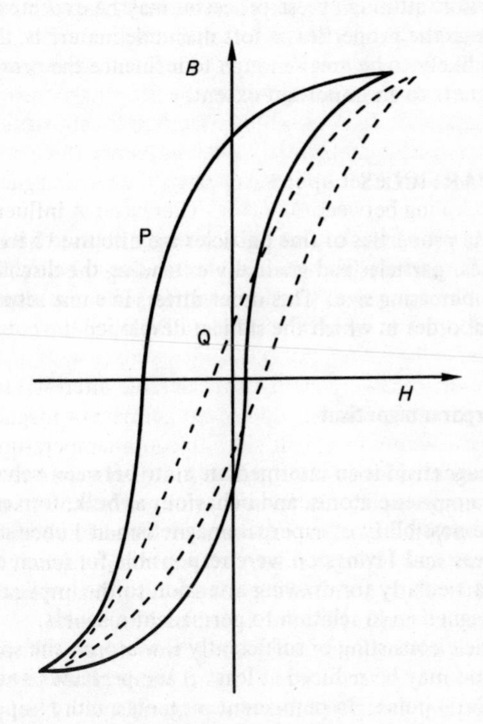


*Fig.3.8. Typical displaced hysteresis loop. Cooled through Néel point of antiferromagnetic phase. P with and Q without field*

## 3.6 OTHER POSSIBLE FORMS OF ANISOTROPY

A number of other possible sources of magnetic anisotropy have been discussed, particularly by Néel.

One of the more interesting suggestions is that if a ferromagnetic alloy is subjected to a magnetic field at a temperature below its Curie point but high enough for diffusion processes to occur freely, atoms may diffuse into some kind of ordered arrangements, in such a manner as to minimize the energy. In this way a preferred direction of magnetization can be formed. This process is presumably regarded as additional to any normal methods by which anisotropy is influenced by strains, differing magnetizations and lattice constants of ordered and disordered phases.

Another possible source of anisotropy is associated with the differing environment of an atom near the surface of a body, which may include a grain or a fine particle, and one in the interior of this body. The latter has neighbours all round it, the former has not.

On the whole although these processes may be expected to contribute significantly to the properties of soft magnetic materials, they are regarded as unlikely to be great enough to influence the properties of permanent magnets to an important extent.

## 3.7 FINE PARTICLES

The magnetic properties of fine particles are discussed here, starting with the finest particles and gradually extending the discussion to particles of increasing size. This order differs in some respects from the historical order in which the subject developed.

### 3.7.1 Superparamagnetism

Superparamagnetism is an intermediate state between behaviour as isolated paramagnetic atoms, and behaviour as bulk, ferromagnetic matter. The possibility of superparamagnetism had been considered by Néel, but Bean and Livingston were responsible for much experimental work and particularly for drawing attention to the implications of superparamagnetism in relation to permanent magnets.

In a particle consisting of sufficiently few atoms, the spontaneous magnetization may be reduced at least at temperatures not very far below the Curie point. In permanent magnets a more important effect is that thermal agitation may change the direction of magnetization spontaneously, even in the absence of a magnetic field, or under the influence of a field smaller than would otherwise be necessary. The effect can be compared to the Brownian movement.

As will be shown in Section 3.7.3 a particle small enough for these thermal effects to be important is too small to contain domain boundaries. If the particle has no anisotropy the magnetization can change freely from one direction to another. Usually the particle is anisotropic because of shape, strain or crystal anisotropy. In the simplest case with only two easy directions of magnetization separated by 180°, the coercivity $H_c$, that is to say the magnetizing force necessary to reverse the magnetization can be used as a measure of the energy required to effect such a reversal. The relaxation time $\tau$ is defined so that the probability of the magnetization vector reversing in time $t$ due to thermal agitation is $1 - e^{-t/\tau}$. This time has been shown to be related to the volume of the particle $V$, and the temperature $T$ K by the approximate formulae:

$$1/\tau = 10^9 \, e^{-VH_cJ_s/2kT} \tag{3.15}$$

In this equation $J_s$ is the saturation polarization and $k$ is Boltzmann's constant. Bean and Livingston (1959) take $\tau \geqslant 10^2$ s as a rough criterion for stable behaviour, unaffected by thermal fluctuations. The number $10^9$ is not very precisely described as a frequency factor. They calculate that for spherical particles in which crystal anisotropy is the controlling factor, the smallest radius for stable behaviour is 0.004 $\mu$m for hcp cobalt and 0.0125 $\mu$m for bcc iron. Elongated particles with even smaller radii are stable. They write 'for an elongated iron particle with the optimum shape anisotropy the critical volume is equal to that of a sphere of radius 0.003 $\mu$m'.

A compact of particles, of a normally ferromagnetic substance, but with particles so small that they are subject to thermal fluctuations is analogous to a paramagnetic substance. The difference is that the individual units, the particles in the former case, are much larger than the atoms in the latter case. A magnetic material of this type in which thermal fluctuations have a large influence is said to be superparamagnetic. Superparamagnetism is inimical to good permanent magnet properties, and may also lead to large magnetic viscosity, which means that changes in magnetization continue for some time after a change in the magnetic field. There are, however, other mechanisms that can produce after effects and magnetic viscosity.

If a demagnetizing force $H$, less than the coercivity $H_c$ is applied it is plausible to rewrite Equation 3.15

$$1/\tau = 10^9 \, e^{-V(H_c-H)J_s/2kT} \tag{3.16}$$

This equation predicts that the magnetic viscosity after-effects, caused by thermal agitation, should increase as the coercivity point is approached. Such an increase is actually observed. Bean and Livingston have used magnetic measurements on superparamagnetic materials to determine particle sizes, referring to the method as magnetic granulometry.

There is little doubt that superparamagnetism is observed in fine iron and iron cobalt particles produced by electrolysis or chemical reduction, as described in Section 4.6. I am, however, sceptical of claims that superparamagnetism occurs in milled powder of barium and strontium ferrite. The average size of such powders is usually about 1 $\mu$m. To be superparamagnetic particles of these ferrites would certainly need to be smaller than 0.01 $\mu$m in size. If for example 10% by volume of a mixture of 1 $\mu$m and 0.01 $\mu$m particles were of the smaller size, there would be $10^5$ small particles for every single large one, and the surface area of the small particles would be 10 times greater than that of the large ones. Almost any method of measuring particle size, except perhaps magnetic granulometry itself, would suggest that this mixture consisted almost entirely of fine particles with a few rogue large ones. The only way in which a powder that appears to have a particle size of about 1 $\mu$m can be influenced by thermal fluctuations is if the superparamagnetic particles in some way nucleate reversals in the average sized particles. Most workers seem now to have rejected the superparamagnetic explanation of the behaviour of overmilled ferrite powders.

Superparamagnetism is probably only a serious problem in connection with the iron and iron cobalt powder magnets already mentioned and occasionally in certain Al–Ni–Co alloys, but even in these materials it arises as the result of some unusual method of production of heat treatment. It rarely occurs as a result of any probable accident in normal production.

### 3.7.2 Thickness of domain boundaries

In this section we consider the energy and thickness of the boundaries or walls, separating domains with different directions of magnetization. These considerations are relevant both for the existence of single domain particles that are so small they they do not have room for a domain boundary, as well as for theories of how coercivity can arise from hindrances to the movement of domain boundaries, when they do exist. Bloch, as long ago as 1932, was probably the first to suggest that walls of finite thickness separate domains, and for this reason the boundary regions between domains are named Bloch walls.

Every writer appears to present the theory in a slightly different form. Some of the differences arise from using different but related quantities such as lattice constant, or number of atoms per unit volume, and others arise from considering special cases, such as cubic or uniaxial crystals, or a predominance of stress anisotropy. The procedure followed below is similar to that used by Hoselitz (1952), as this combines a fair degree of generality with simplicity.

The exchange energy is a function of the square of the angle between the spins in neighbouring atoms. Consequently the exchange energy associated with the boundary between two domains is a minimum if the boundary wall is as thick as possible, and the angle between neighbouring spins is as small as possible. On the other hand while the spins within the domains lie in easy directions of magnetization, as determined by crystal or some other form of anisotropy, within the domain walls the spins lie in unfavourable directions from the point of view of this anisotropy. Thus anisotropy tends to make the walls as thin as possible. The actual thickness of the domain walls is a compromise that minimizes the sum of the exchange and anisotropy energy. The gradual change in the spin directions in a Bloch wall is illustrated schematically in Fig.3.9. Boundaries may separate domains in which the directions of magnetization differ by 90° or 180° in body centred cubic materials such as Fe, but only by 180° in uniaxial materials. Other angles occur in materials with negative cubic anisotropy such as Ni. A 180° boundary is unlikely to be perpendicular to the direction of magnetization of the domains that it separates. A situation such as that shown in Fig.3.10, although such figures are common in early accounts of domains, never exists. It is like bringing two permanent magnets into a repulsion situation, and is energetically unfavourable.

Figure 3.9 shows the gradual transition of spin directions in a 180° wall. In Hoselitz's treatment the exchange energy of a wall is taken as inversely proportional to $\delta$, the wall thickness. The value of $\delta$ is calculated as follows: the energy per unit area of wall arising from anisotropy is written

$$\gamma_F = F\delta = (\tfrac{3}{2}\lambda Z + bK)\delta \tag{3.17}$$

$\lambda$ is the magnetostriction constant, $K$ is the crystal anisotropy constant, $Z$ is the stress, $b$ is a constant of the order unity, but depends on the kind of crystal anisotropy. The energy $\gamma_a$ contributed by exchange forces is

$$\gamma_a = Aa^2/\delta \tag{3.18}$$

$A$ is the exchange interaction energy per unit volume, and $a$ is the inter-

*Fig.3.9. Gradual change in direction of spins in domain wall*

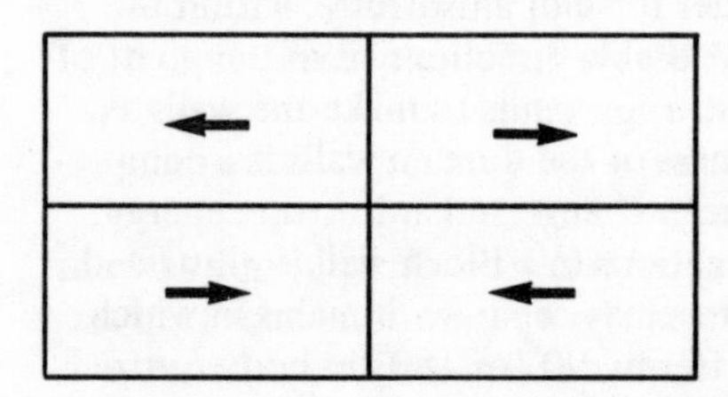

*Fig.3.10. Arrangement of domains that is energetically unfavourable*

atomic distance. Some authors use the exchange integral $I$, where $I = a^3A$, and it is not always clear, which quantity is being used. The total energy per unit area of wall is given by:

$$\gamma = \gamma_F + \gamma_a = F\delta + Aa^2/\delta \tag{3.19}$$

To obtain the minimum value of $\gamma$ put $d\gamma/d\delta = 0$.

$$F - Aa^2/\delta^2 = 0$$

or

$$\delta = a(A/F)^{1/2} \tag{3.20}$$

The values of $\delta$ and $\gamma$ vary considerably from material to material. For nickel and Permalloy the crystal anisotropy constants are small, and stress is the most important component of $F$. For many other materials crystal anisotropy predominates. With $A$ of the order of magnitude $10^9$ and $K = 10^4$ Jm$^{-3}$ as for Fe and Fe alloys, $\delta$ is of the order $100a$, or 0.03 $\mu$m if $a = 0.0003$ $\mu$m. In turn this gives a typical wall energy lying between $10^{-2}$ and $10^{-3}$ Jm$^{-2}$.

In some modern rare-earth alloys crystal anisotropy values are as much as 100 times greater than in Fe. Consequently wall thicknesses are much smaller, and wall energies are somewhat greater. Certain rare-earth alloys have a low Curie point and a smaller value of $A$ (Section 4.5.4). In these materials at liquid helium temperatures domain walls may have effectively zero thickness, and the transition from one domain to another may occur within one lattice parameter.

### 3.7.3 Elongated single domain particles

If the size of a magnetic particle is less than the thickness of a domain wall, it appears that there is no room for a wall within the particle. Such a particle is described as a single domain particle, because it is expected that its magnetization will always remain uniform. In the simple theory the only kind of change in magnetization contemplated is by 'coherent rotation', the simultaneous flipping over of the spins from one preferred direction to another.

A unified expression can be obtained for the coercivity, arising from coherent rotation in single domain particles with uniaxial anisotropy. Let $F_a$ be the difference in anisotropy energy between the difficult (perpendicular) and the preferred direction. Let $\alpha$ and $\theta$ be the angles between the field direction and the preferred axis and direction of magnetization respectively. Let $F_t$ be the total anisotropy and magnetostatic energy. Then

$$F_t = F_a \sin^2(\theta - \alpha) - HJ_s \cos\theta \tag{3.21}$$

$H$ is the applied magnetizing force and $J_s$ the saturation polarization. The polar energy distribution may have one or two minima, according to the value of $H$. The critical value of $H$ at which the polarization must rotate into the field direction, is that at which the second minimum disappears.

The condition for a minimum in $F_t$ is

$$\frac{1}{F_a}\frac{\mathrm{d}F_t}{\mathrm{d}\theta} = \sin 2(\theta - \alpha) + \frac{HJ_s}{F_a}\sin\theta = 0 \tag{3.22}$$

The second minimum disappears at a value of $H$, which also satisfies

$$\frac{1}{F_a}\frac{\mathrm{d}^2F_t}{\mathrm{d}\theta^2} = 2\cos 2(\theta - \alpha) + \frac{HJ_s}{F_a}\cos\theta = 0 \tag{3.23}$$

If the field is applied along the preferred axis, the coercivity $H_c$ is given by:

$$H_c J_s = -2F_a \qquad (3.24)$$

For a random orientation of particles Stoner and Wohlfarth (1948) replace Equation 3.24 by

$$H_a J_s = 0.96 F_a \qquad (3.25)$$

In these equations $F_a$ is replaced by $K_1$ for uniaxial crystal anisotropy, $3\lambda Z/2$ for stress and $J_s^2(N_b - N_a)/2\mu_0$ for shape anisotropy.

A more precise theory involves calculating whether the energy of a particle is lower if it is magnetized as a single domain or two domains separated by a Bloch wall, and assuming that the particle will adopt the state of lower energy. This method involves comparing the energy required to create a Bloch wall with the consequent reduction in magnetostatic energy; i.e. the reduction in the energy of the self-demagnetizing field that follows the division of the particle magnetically into domains. This calculation was carried out by Stoner and Wohlfarth in the papers already cited, but they acknowledge independent development of these ideas by Néel, and also an important paper by Kittel (1946). As a result of their own and other calculations they conclude that a domain wall should not exist in a spherical particle of Fe with a diameter less than 0.015 $\mu$m. In an elongated particle the self-demagnetizing field is less, and a cylindrical rod with length to diameter ratio of 10 should remain a single domain if its diameter does not exceed about 0.075 $\mu$m. If particles of the above sizes do indeed remain single domains, very high coercivities should result. These coercivities should be given by the anisotropy field calculated for shape stress and crystal anisotropy. With elongated iron particles coercivities exceeding $10^6$ $Am^{-1}$ ($\mu_0 H_c = 1.25$ tesla) are predicted by Equation 3.14, and even a departure of 5 to 10% from the spherical shape is sufficient to account for the properties of many common permanent magnet materials.

### 3.7.4 Micromagnetics

In the period 1955 to 1960 a process for producing elongated particles of iron and iron cobalt by electrodeposition into a mercury cathode was developed by the General Electric Co of America (Section 4.8). Although this process is still used for making permanent magnets of certain shapes difficult to produce by other methods, it never succeeded in producing materials of such high coercivities as predicted in Section 3.7.3. Naturally there have been many attempts to explain these discrepancies, and these studies have been dignified by the name micromagnetics.

The first explanation put forward by Bean and Jacobs (1956) was based on electronmicrographs, which showed that the elongated particles were chains of spheres rather than smooth cylinders. They suggested a fanning mechanism, illustrated in Fig.3.11, in which the polarization rotates in opposite directions in alternate spheres. The fanning mechanism is a simple example of an incoherent rotational process, in contrast to the Stoner and Wohlfarth model, in which all the spins throughout the whole particle remain parallel, and is coherent.

The fanning mechanism may well be the correct explanation of the low coercivities, obtained in the particles examined by Bean and Jacobs. The explanation did not destroy faith in the possibility of producing high coercivities, if only a process for producing more perfect particles could be discovered.

Theoretical investigations of the possible modes of rotation of the elementary spins have revealed that even in perfectly smooth ellipsoids or cylinders, coherent rotation of the spins in unison may not be the easiest way in which the magnetization can change. Frei, Shtrikman and Treves (1957) made an important contribution to this study, in one of a series of papers on micromagnetics from the Weizmann Institute of Science, Israel. Figure 3.12 illustrates three possible modes of magnetization change in an infinite cylinder. Coherent rotation in unison is quite easy to understand from Fig.3.12(a), and buckling should also be clear from inspection of Fig.3.12(c). It is in fact rather similar to fanning, but can occur in a uniform cylinder or ellipsoid, and does not require a chain of spheres. Curling is illustrated in Fig.3.12(b), but requires a little imagination to visualize a three-dimensional process. Perhaps it can best be described by saying that each spin rotates about the radius of the cylinder, drawn through its position. The important feature of curling is that it involves no increase in the radial component of magnetization, no production of 'poles' on the cylindrical surfaces, and therefore no increase in magnetostatic energy. On the other hand curling involves more work against exchange forces and as much work

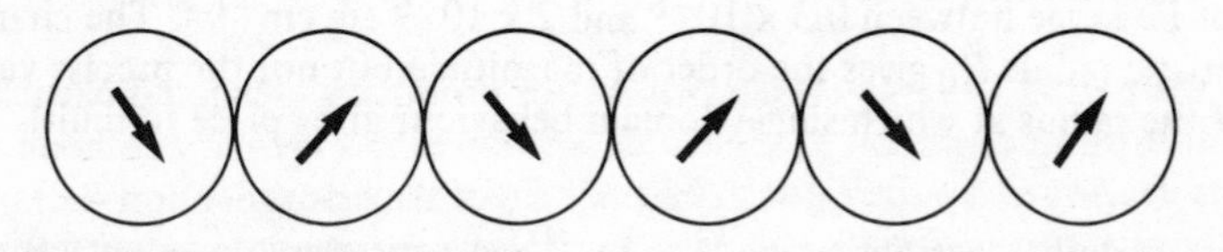

*Fig.3.11. Fanning in chain of spheres*

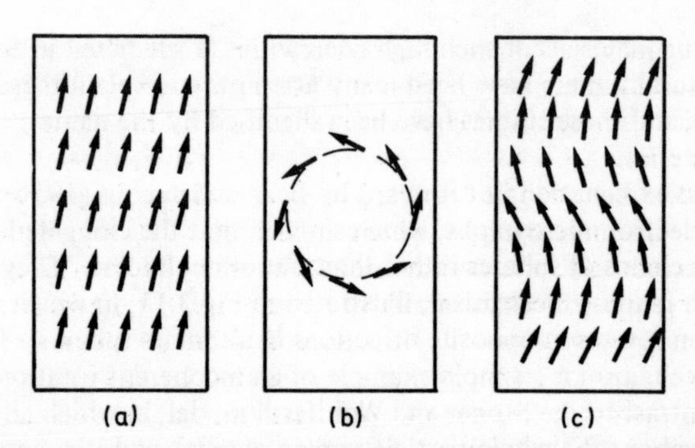


*Fig.3.12. Modes of magnetization change for infinite cylinder (as suggested by Frei, Shtrikman and Treves, 1957). (a) Coherent rotation, (b) curling, (c) buckling*

against crystal anisotropy as does coherent rotation. Buckling involves some increase in magnetostatic, exchange and crystal energy. Frei and his co-workers have calculated the nucleation field required to initiate curling and buckling in an infinite cylinder. Unless there is high crystal anisotropy, once curling or buckling is initiated the magnetization can reverse, so the nucleation field ought to be closely related to the coercivity. Figure 3.13 compares the results of these calculations with the coercivity for coherent rotation in unison. In this figure the reduced nucleation field $h_n$ is plotted against the reduced radius, $R/R_0$. These reduced quantities are frequently used in the literature on this subject. The reduced nucleation field is expressed as a fraction of the maximum value for a long needle shaped ellipsoid, calculated on the basis of shape anisotropy, that is

$$h_n = 2\mu_0 H_n / J_s \tag{3.26}$$

where $H_n$ is the actual nucleation field.

The characteristic radius $R_0$ is defined in CGS units by

$$R_0 = C^{1/2} J_s \tag{3.27}$$

$C$ is described as the 'exchange energy constant' and values suggested for Fe range between $0.3 \times 10^{-6}$ and $2 \times 10^{-6}$ erg $cm^{-1}$.* The characteristic radius $R_0$ gives the order of magnitude but not the precise value of the radius at which single domain behaviour gives place to multi-

* Obviously converting erg $cm^{-1}$ to $Jm^{-1}$ and expressing $J_s$ in tesla will not give the same result.

domain behaviour. The difficulty in inserting numerical values into Equation 3.27 lies in the uncertainty in the value of $C$. Taking $C = 10^{-6}$ erg cm$^{-1}$, somewhere in the middle of the range quoted above, gives $R_0 = 0.006$ μm for Fe. This radius is about an order of magnitude less than Stoner and Wohlfarth predicted for elongated particles, and is quite near the size at which superparamagnetism is expected.

The theory for an infinite cylinder is, however, quite illuminating. For particles with the reduced radius $R/R_0 < 1$, buckling and coherent rotation predict the same coercivity. There is a very small region in which buckling predicts the lowest coercivity and is therefore the expected process. Very soon curling, which is magnetically unfavourable for small radii becomes the easiest mode, and the coercivity calculated on the basis of curling then falls steadily with increase of radius. According to Figure 3.13 a fourfold increase of radius is accompanied by a twofold decrease of coercivity. In this range of sizes, the behaviour of the particles is intermediate between that predicted by single domain and boundary wall theory. Much of the time such particles may be single domains, but when a sufficient field is applied to promote curling, the magnetization becomes incoherent while it is being reversed. Processes such as curling and buckling may indeed be compared with the nucleation of a domain wall. The Stoner and Wohlfarth prediction that an elongated particle is less likely to contain a domain wall than a spherical one of the same diameter, applies only in the absence of an

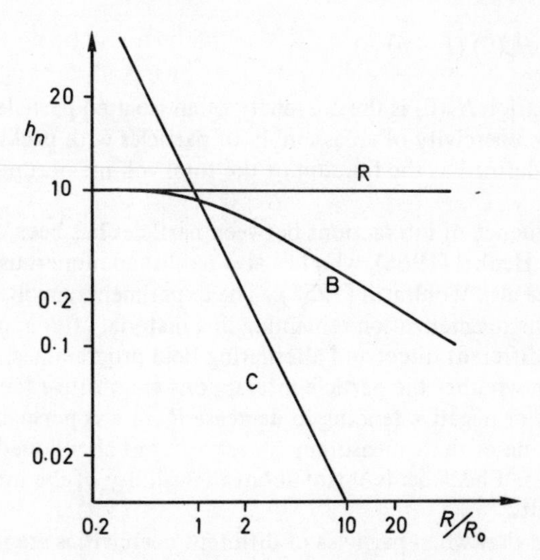


*Fig.3.13. Nucleation field for infinite cylinder plotted against reduced radius. R: Rotation in unison. B: Buckling. C: Curling. (Freil et al, 1957)*

external field. In practice it has not so far been possible to produce material with coercivities based on shape anisotropy reaching more than about 200 $kAm^{-1}$ ($\mu_0 H = 0.24\ T$), and even in particularly perfect particles such as whiskers the values are only a little higher. These values agree reasonably with the predictions of micromagnetics, except for particles possessing very large crystal anisotropy, which are discussed later in Section 3.7.6.

### 3.7.5 Interacting particles

In a real material the particles are so close together that the behaviour cannot be predicted by considering processes in an isolated particle. Although it is conceivable that some arrangements such as a chain of spheres may have a greater coercivity than an individual particle, it is more usual for the coercivity to decrease as the packing increases. In fine particle (ESD) magnets made by electrolysis into mercury it is necessary to choose between a high packing with high $B_r$ and low $H_c$ or low packing with low $B_r$ and high $H_c$. Numerous formulae have been proposed to represent the influence of packing on the coercivity of assemblies of particles with shape anisotropy. These theories are reviewed by Wohlfarth (1959), but it is doubtful whether any are in general more reliable than the simplest formula, proposed originally by Néel, although this formula is admittedly not always reliable.

$$H_c(p) = H_c(0)\,(1 - p) \tag{3.28}$$

In this equation $H_c(0)$ is the coercivity of an isolated particle, and $H_c(p)$ is the coercivity of an assembly of particles with packing $p$. The packing is defined as the fraction of the total volume occupied by particles.

The influence of interactions between particles has been extensively studied by Henkel (1966), who has also published numerous other papers. See also Wohlfarth (1958). The experiments involve measurements of the magnetization remaining in a material after applying a variety of different direct and alternating field programmes. The aim is to establish whether the particle interactions are positive tending to increase $B_r$ or negative tending to decrease $B_r$. I was persuaded to attempt some of these measurements myself, but abandoned the enterprise because I became doubtful about the validity of the interpretation of the results.

It is said that when particles of different coercivities are mixed, the coercivity of the mixture is likely to be nearer that of the softer component.

In some circumstances a material may consist of elongated particles of saturation polarization $J_1$ surrounded by a matrix of smaller saturation polarization $J_2$. It seems possible that the demagnetizing force will act on the stronger polarization $J_1$, while the shape anisotropy depends on the square of the difference of the two polarizations. Thus Equation 3.25 becomes:

$$H_c J_1 = 0.96(J_1 - J_2)^2(N_b - N_a)/\mu_0 \tag{3.29}$$

This equation predicts a rapid decrease in the coercivity with increase in the saturation polarization $J_2$ of the matrix, and this hypothesis helps to explain the behaviour of certain Al–Ni–Co alloys as explained in Section 4.2.6.

### 3.7.6 Particles with crystal anisotropy

Crystal anisotropy has several advantages over shape anisotropy for producing magnets with high coercivity. In the first place crystal anisotropy does tend to prevent curling and buckling in a way that shape anisotropy does not. Curling creates no free poles and no magnetostatic energy. Went, et al (1952) have given the following formula for the critical size of single domain particles with crystal anisotropy:

$$R_c = \frac{18\pi\mu_0}{J_s}\sqrt{\left(\frac{KkT_c}{J_0^2\alpha}\right)} \tag{3.30}$$

In this formula $K$ and $J_s$ are the appropriate crystalline anisotropy constant and saturation polarization at the temperature at which $R_c$ is the critical radius, $k$ is Boltzmann's constant, $T_c$ is the Curie temperature, $J_0$ is the saturation polarization at 0 K, and $a$ is the mean distance between neighbouring atoms, belonging to the appropriate sub-lattice. This formula gives values of 0.25 μm for Co and 1.3 μm for barium ferrite.

Other writers quote much lower values for the size of single domain particles, particularly in Co. They do not, however, explain what is wrong with Equation 3.30. The discrepancy may arise in Went and co-workers' substitution for the exchange energy in terms of the Curie temperature.

Although materials with large crystal anisotropies often have large coercivities, they often seem to have domain boundaries, and coercivities that are less than the anisotropy field. It is rarely possible to obtain small undamaged particles. Fine particles of cobalt produced by electrolysis or chemical reduction, are usually fcc and do not have the large anisotropy of hexagonal cobalt. Powders of rare-earth-cobalt and barium ferrite are normally produced by some form of mechanical attrition, such as ball milling. If this process is continued until the particles are sufficiently fine that they should be of single domain size, the crystal lattice is usually damaged and the coercivity consequently reduced.

In the rare-earth-cobalt compounds the coercivities reach values as high as is desirable without excessive milling. In theory single domain particles ought to have coercivities an order of magnitude higher, but to obtain such high coercivities would be a disadvantage rather than an advantage, since it would create difficulties in magnetization. Fundamental work on the mechanisms controlling coercivity in such materials is described in Section 3.8.5. in connection with the pinning and nucleation of domain walls.

One advantage theoretically predicted and practically achieved in materials with large magnetocrystalline anisotropy coefficients is that high particle compaction and high densities can be obtained with relatively little loss of coercivity. This fact can be understood, if it is remembered that if elongated particles are compressed to 100% density, the shape anisotropy is completely destroyed. If this degree of compaction is applied to particles with large crystal anisotropy, the crystal anisotropy remains. A high coercivity is retained provided sufficient pores or other obstacles remain to prevent the free movement of domain walls. Good ferrite and rare-earth-cobalt magnets can be obtained by pressing and then sintering until 95% of the theoretical density is reached.

## 3.8 DOMAINS AND DOMAIN WALLS

The concept of domains was originally proposed by Weiss to explain how a material with spontaneous magnetization can exist in a demagnetized state. In Section 3.7.2 it was shown that domains are usually separated by walls of finite thickness through which the orientation of the electron spins changes gradually. Several experimental methods are now available for revealing the existence of domain walls and studying their motion. These methods are described in Section 3.8.1. The attempts to explain coercivity by considering factors which hinder the movement of domain walls, or inhibit their formation are then dealt with in subsequent sections.

### 3.8.1 Experimental study of domain walls

The oldest method of making domain walls visible is the powder pattern method originally devised by Bitter. A good description of this method is given by Bates (1961). Bates himself used the method extensively.

A colloidal solution of a magnetic powder in alcohol or other low viscosity liquid is prepared. Powders of iron, ferric oxide or $Fe_3O_4$ have been used. A drop of the solution is placed on the surface of the specimen to be investigated. The surface must be carefully polished without producing stresses likely to distort the patterns. Bates recommends that the final polishing should be carried out electrolytically. The specimen may be placed in the gap of a small electromagnet to enable different fields to be applied, and the powder patterns may be observed through a microscope, or photographed if a permanent record is required. Bates sometimes made cine films of wall movements. Usually the magnetic powder shows up domain walls, because in Bloch walls that normally occur in bulk matter, the spins within the walls are oriented to some extent normal to the surface, forming poles which attract the powder. In very thin films a different type of wall, known as a Néel wall, in which the spin directions remain within the plane of the film, sometimes occurs. Sometimes the domains themselves have a normal component of magnetization, and in such cases it may be possible to show up the domains rather than the walls.

Powder patterns reveal only the domain structure near the surface, and this may differ from that inside the material. Closure domains are often formed on the surface, to reduce magnetostatic energy caused by flux emerging from the surface. Closure domains reduce the magnetostatic energy and are normally formed except in materials with very high anisotropy, in which they would increase rather than decrease the total energy. Figure 3.14 shows how such closure domains are formed, and prevent the occurrence of surface poles.

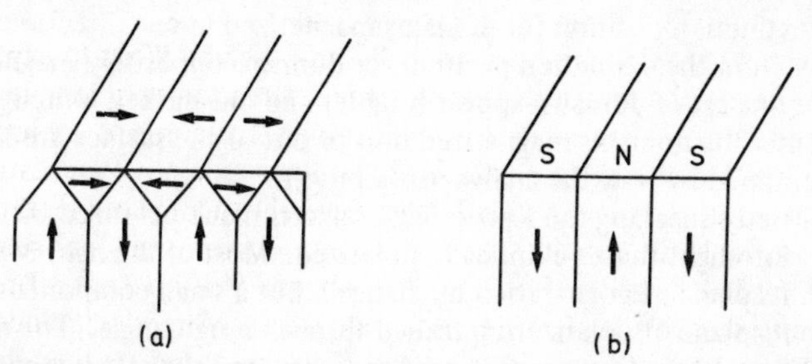


*Fig.3.14. (a) Closure domains eliminate surface poles. (b) No closure domains in high coercivity material*

There are now a number of more sophisticated methods of investigating domain structures. If the actual movement of domain walls is not to be followed, powder patterns can be removed on a replica, and examined in more detail at leisure, by means of an electron microscope at higher magnification than is possible with the optical microscope.

The magnetic fields associated with domains and domain walls can deflect moving electrons, and various techniques have been devised that enable domains to be studied by means of an electron microscope. A reflection electron microscope can be used to observe domain structures near the surface. Transmission electron microscopy can be used to reveal domains within the body of a material, but is limited to thin films. Various techniques have been devised, and these may either contrast domains with different directions of magnetization, or contrast domain walls relative to places of uniform magnetization. In one method the electron microscope is deliberately adjusted to be slightly out of focus, and the deflection of electrons in the film may in some places improve and in others worsen the focus.

The Faraday rotation of the plane of polarization of light, as it passes through a magnetized body, also provides a method of investigating domains in thin films.

The Kerr effect is a small change in the plane of polarization of light that occurs when it is reflected from the surface of a magnetized material. The effect is simplest and largest at normal incidence, but with this arrangement only domains magnetized normal to the surface, or at least with a normal component of magnetization are revealed. In practice this limitation means that the method can only be used with materials, having a large crystal anisotropy. This is a characteristic of the new rare-earth-cobalt materials, which have been so extensively studied in the last few years. Consequently the Kerr effect figures in many recent studies of domains in permanent magnet materials.

The requisite apparatus is a very good metallurgical microscope with a polarizer and analyzer. The analyzer is set close to the position of extinction. The Kerr effect rotates the plane of polarization nearer to the extinction position for domains magnetized in one direction and away from the extinction position for domains oppositely magnetized. Thus one set of domains appear brighter and one darker, although whether the domains magnetized into or out of the surface are light is determined by how the analyzer is arranged.

Strictly speaking the Kerr effect converts plane polarized incident light into light that is elliptically polarized. Most of the light is reflected with its plane of polarization unchanged, but a small component appears with its plane of polarization turned through a rightangle. The intensity of the unchanged component predominates, and the effect is equivalent to rotating the plane of polarization through a small angle, usually less than ½°, determined by the ratio of the amplitudes. A process known as blooming, which consists in coating the surface of the specimen with a substance such as zinc sulphide, to absorb some of the unchanged reflected light, and therefore effectively increase the rotation of the plane of polarization, increases the contrast between the domains. A quarter wave plate may also be used to improve the contrast.

Examples of domain patterns obtained by the Kerr effect are given in the next chapter on materials. Incidently the Kerr effect can also provide information about phases present and crystal orientation in permanent magnet materials.

The subject of the experimental investigation of domains is described in two specialist books, Craik and Tebble (1965) and Carey and Isaac (1966).

At the combined Intermag MMM Conference in Pittsburgh in 1976 Wells and Ratnam reported that grooves appear in $RCo_5$ alloys at the position occupied by domain walls, when the specimens are mechanically, chemically or electrolytically polished for microscopic examination. The grooving is apparently caused by a difference of energy in the wall. How far this phenomenon may be generally useful for the study of domains remains to be seen.

### 3.8.2 Stress theories of wall pinning

The theory of coercivity and other features of the hysteresis loop were developed at considerable length by Becker and Döring (1939), who assumed that 90° and 180° Bloch walls exist, but that their movement is hindered by fluctuating internal stresses. In modern terms they could be described as pinned. The basic assumption is that the energy of the Block wall discussed in Section 3.7.2, is a function of position as shown in Fig.3.15. In the absence of any applied field the wall comes to rest in one of the energy minima. The wall energy varies as a result of the variation of internal stress with position.

Suppose a magnetizing force $H$ is applied in a direction $\theta$. If the wall moves a distance $\Delta x$, the change in magnetostatic energy of the boundary wall is $(\mathrm{d}\gamma/\mathrm{d}x)\Delta x$. The change in the magnetostatic energy of the domain is $2J_s H \Delta x \cos\theta$. Equating these values for unit area of wall, shows that the wall comes to rest in a position such that

$$H = \frac{1}{2J_s \cos\theta} \frac{\mathrm{d}\gamma}{\mathrm{d}x} \tag{3.31}$$

As an increasing $H$ is applied, the domain wall moves to positions where the slope of the energy curve $\mathrm{d}\gamma/\mathrm{d}x$ in Fig.3.15 is greater and

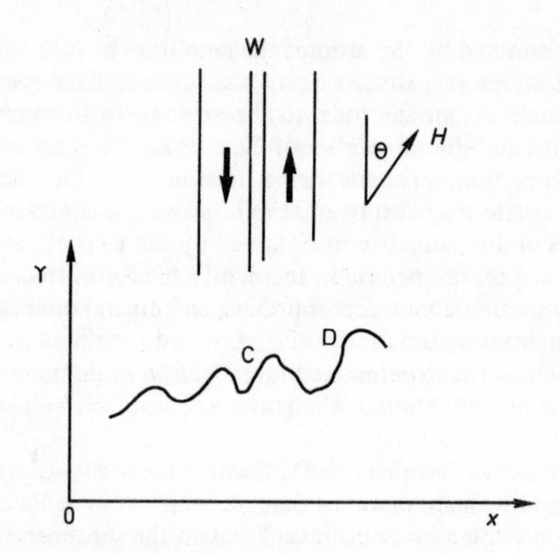


*Fig.3.15. Variation of energy γ of boundary wall with position. Wall moves under influence of magnetizing force H*

greater. The movement is reversible until a position of maximum slope such as c is reached, when the wall jumps some distance to a point D, where the slope is greater and it is pinned more strongly. Such sudden jumps in magnetization actually occur, and are known as the Barkhausen effect. They were originally detected by means of a many turn coil wound round the sample. The coil was connected through an amplifier to a loudspeaker in which the Barkhausen jumps produced clicks.

To calculate the coercivity produced by stresses requires assumptions not only about the magnitude of the internal stresses, but also about the rate at which they change.

Becker and Döring assumed that Bloch walls are plane, and that the stress varies sinusoidally in a direction perpendicular to these planes with mean amplitude $Z_i$. They considered mainly 180° boundaries, and obtained the following equation for the intrinsic coercivity $H_{ci}$:

$$H_{ci} = \frac{\lambda Z_i}{J_s} p \tag{3.32}$$

In this equation λ is the magnetostriction constant, $J_s$ the saturation polarization, and $p$ is a constant of the order unity, when the wavelength of the fluctuations is of the same order as the thickness of a domain wall, but otherwise is somewhat smaller.

There is some X-ray evidence for the existence of stresses in permanent magnet materials, but it is scarcely possible to ascertain their magnitude and the pattern of their fluctuations. The value of $Z_i$ required to account for the coercivities of the older Al–Ni–Co alloys (about 50 $kAm^{-1}$, $\mu_0 H_c = 0.06$ T) is about $2 \times 10^9$ $Nm^{-2}$ and is rather greater than the probable tensile strength of the materials. The difficulty in explaining coercivity in terms of strains becomes even greater in more modern materials.

Becker and Döring suggested formulae for calculating random stresses from other magnetic measurements such as initial permeability, reversible permeability at the remanence point and the shape of the magnetization curve as it approaches saturation. They then inserted these values of $Z_i$ into Equation 3.32 to calculate the coercivity. The objection to this type of calculation is that it may only reveal some connection between the various parts of the hysteresis loop, that may arise from many other hypotheses besides internal stresses. An apparent agreement is not therefore evidence for internal stresses playing a primary role.

### 3.8.3 Kersten inclusion theory

Another hypothetical barrier to the movement of domain walls is the presence of non-magnetic or less magnetic inclusions. This theory was developed by Kersten and an account of it is given by Hoselitz (1952) and Stewart (1954).

The essential assumptions in this theory are again that the domain walls are plane, and also that the inclusions are arranged in a regular manner. Let us consider for example a regular array of spherical non-magnetic inclusions arranged at the corners of a cubic lattice as shown in Fig.3.16. Let there be $n$ such inclusions in unit volume, and let the radius of each inclusion be $r$. If the Bloch wall $B$ cuts any of these inclusions, its energy is reduced by the area of the inclusions cut. In the position shown in Fig.3.16, unit area of the wall cuts $n^{2/3}$ inclusions, and the area of each inclusion intersecting the wall is $\pi(r^2 - x^2)$, $x$ being measured as shown from the centre of one of the spheres. If the wall moves a distance $\mathrm{d}x$ under the influence of a magnetizing force $H$, the work done by the magnetizing force is $2HJ_s\,\mathrm{d}x$ per unit area, and the increase in the wall energy is the differential

$$-\mathrm{d}[n^{2/3}\pi\gamma(r^2 - x^2)] = 2\pi\gamma n^{2/3} x\,\mathrm{d}x \tag{3.33}$$

Equating the energy changes we have:

$$HJ_s = \pi\gamma n^{2/3} x \tag{3.34}$$

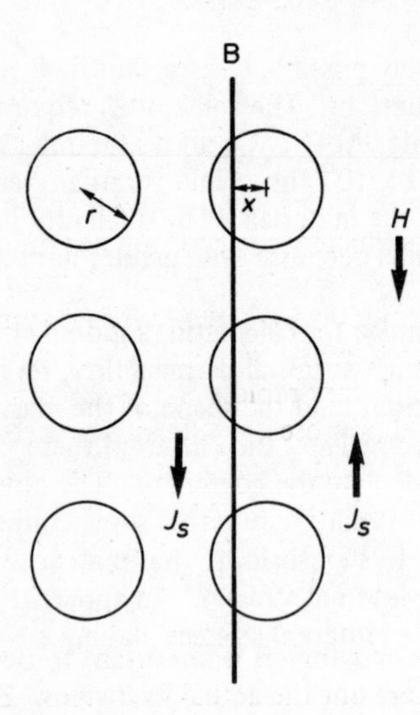


*Fig.3.16. Bloch wall intersecting regular array of inclusions: Kersten inclusion theory.*

If we equate the coercivity to the magnetizing force just necessary to separate the wall from the plane of inclusions:

$$H_{ci} = \pi\gamma n^{2/3} r / J_s \tag{3.35}$$

The X-ray work on Al–Ni–Co alloys suggested a precipitate as well as stresses, and is therefore equally evidence for the inclusion and the stress theory. Equation 3.35 was capable of accounting for the coercivities of the permanent magnet alloys that were known at the time when it was proposed, although it required about 50% of the volume to consist of inclusions.

### 3.8.4 Néel disperse field theory

Néel (1946) criticized both the stress and inclusion theories of coercivity as described above, on the grounds that the simplifying assumptions are improbable and greatly exaggerate the coercivity. Thus in Fig.3.16 a plane wall is shown cutting an array of inclusions also lying in a plane. Thus in one position the wall cuts a large number of inclusions, and if it is displaced through a distance $r$, the radius of the inclusions, it cuts none at all. These assumptions predict a large change in the energy of the wall for a small displacement. If the boundary wall on the contrary is able to bend it can be displaced without ever passing through the state of no inclusions at all. Alternatively if the wall remains plane, but the inclusions are distributed at random instead of in planar arrays, the variation of the area of inclusions cut by the wall is much reduced. Either or both of the above possibilities greatly reduce the variation of the wall energies with position, and according to Néel the large coercivities found in permanent magnets are not predicted.

Néel's objection is explained above in relation to Kersten's inclusion theory. Although it may not be quite so obvious, the criticism applies equally to Becker's stress theory, which is also founded on the hypothesis of plane walls, and stresses which vary rapidly in one direction.

To overcome these objections Néel proposed an alternative theory. If a magnetic medium contains inclusions of a non-magnetic or less magnetic material, magnetic poles appear around these inclusions. The disperse fields connected with these poles involve considerable magnetostatic energies. If the inclusions are cut by 180° walls, so that the sign of half the poles is changed, the energy of the disperse field is reduced. This process is illustrated in Fig.3.17, in which it is important to note that the areas with N and S polarity are not the actual inclusions. Even in the simplest case, when an inclusion is wholly within one domain, it will produce one area of N and one of S polarity. This theory shows that the magnetostatic energy is less if the domain wall is in the position shown in Fig.3.17(b), than in the position of Fig.3.17(a), but the difference in energy lies not within the walls so much as within the domains themselves. For this reason Néel considered this theory was better able to account for high coercivities. Random stresses can also produce disperse fields, and contribute to the coercivity according to Néel's theory.

In his original paper Néel derives complicated formulae for the coercivity by even more complicated mathematical methods. Summaries of the final formulae and Néel's attempts to find experimental justification for them are given in Hadfield (1962), Hoselitz and McCaig (1952) and in Carey and Isaac (1966).

The formulae are however difficult to handle and contain quantities that are difficult to estimate. Stoner (1950) wrote that the deduction process of the boundary energy theories was clear and satisfying, but the premises were wrong or inadequate. The premises of the disperse field theory could tentatively be accepted as correct, but a clearer exposition of the deductive arguments as well as a more searching comparison of the theoretical conclusions with experiment was desirable.

The clearer exposition and searching comparison have not materialized. The coercivities of Al–Ni–Co alloys, the ESD materials made by electrolysis into mercury and probably the Cr–Fe–Co alloys can easily be accounted for by shape anisotropy in single domain particles. The problem is rather to explain why the coercivities are less than predicted,

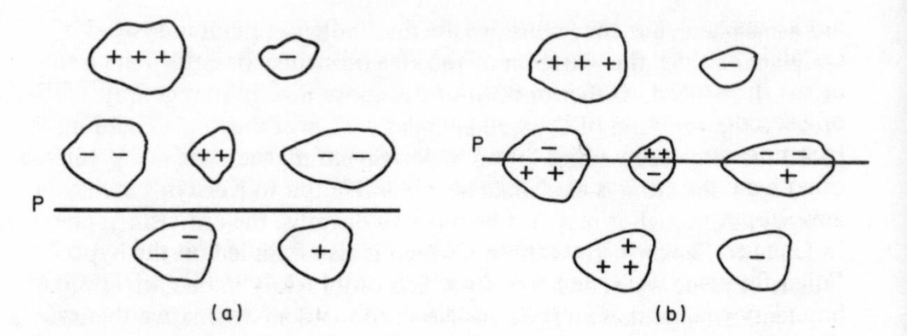


*Fig. 3.17. Magnetostatic energy is reduced when boundary wall P cuts regions of free poles as in (b). (After Néel, 1946)*

but as we have seen imperfect particles, particle interactions, curling and buckling, or the presence of some larger particles can all be invoked to account for the short-fall in coercivity. Bitter-figures are sometimes obtained on these materials, but the concentration of powder is less than it is for normal Bloch walls, and it is thought that the fine particles or precipitates form themselves into domains. Obviously when a fine particle material is demagnetized, the particles cannot reverse their magnetization in a random manner, but are likely to reverse in groups separated by planes, but with no gradual transition of the magnetization as in a normal Bloch wall.

The grain size of barium ferrite is usually close to some of the estimates for the critical size for single domain behaviour. There are some reports of domain walls being observed in such materials. The coercivity is never more than one-third of the anisotropy field, and usually considerably less in commercial materials. There is thus scope for departures from single domain behaviour.

On the whole it seems that the disperse field theories have been replaced either by single domain theories, or more modern pinning and nucleation theories.

### 3.8.5 Modern nucleation and pinning theories

In the modern rare-earth-cobalt permanent magnets, the optimum grain size is considerably greater than that for single domains, and there is abundant evidence for the existence of domain walls. The anisotropy fields $H_a$ are fantastically high, reaching $2.5 \times 10^7$ Am$^{-1}$ ($\mu_0 H_a$ = 30 tesla), but the coercivity is never greater than one fifth of $H_a$.

The theoretician is challenged to explain coercivities far higher than have previously been encountered, but in materials that are certainly not acting as single domain particles. Magnetostriction and stress are certainly inadequate, and some of the materials possess more than 95% of the theoretical density, and contain at the most only a small amount of a second phase, so that any explanation can be rejected, if it requires inclusions to occupy more than a small fraction of the volume. On the other hand the materials do possess a fantastically high crystal anisotropy.

One idea that deserves mentioning is the so-called 'Brown paradox'. Fuller-Brown (1945) suggested that there is no way in which a domain boundary can form in a perfect single crystal, unless a field greater than the anisotropy field is applied. Although the state with a domain boundary has a lower energy than the state in which the crystal is a single domain, the anisotropy field is necessary to surmount the energy barrier between the two states.

Single crystals of most materials contain domain boundaries and it is generally supposed that domain walls are nucleated at imperfections.

Brown's paradox seems to be at work in some modern permanent magnets such as the rare-earth-cobalt alloys. These materials contain grains that are appreciably but not enormously larger than the critical size in which a domain wall can persist. It may be imagined that such grains contain a relatively small number of defects, capable of nucleating domain walls. Some of these defects may only nucleate domain walls, if there is an appreciable demagnetizing force. This demagnetizing force may be much less than the anisotropy field but still quite considerable. If all domain walls are removed by a saturating field, they will reappear only if this nucleation field is applied. An alternative suggestion is that even if walls are present there may be some mechanism that pins them.

In the last few years the processes occurring during the magnetization and demagnetization of $SmCo_5$ have been intensively studied. Accounts of this work are given by den Broeder and Zijlstra (1974) and Livingston (1973). It seems probable that both nucleation and pinning processes operate, and that the relative importance of these two processes depends on the precise composition of the samples and the heat treatment they have received. Certainly in these materials Néel's objection to Bloch walls being pinned by relatively small volumes of defects seems difficult to uphold. Possibly the pinning forces are much greater in these materials of high anisotropy. Néel's objections may also not hold in particles, which although larger than the pinning faults are by no means infinitely large. The compositions and treatment for optimum properties are discussed in Section 4.5 and the following sub-sections that deal with rare-earth-cobalt materials.

A matter that arises in several contexts is the presence of two magnetic phases. When the two phases have large crystal anisotropies, in contrast to shape anisotropy, the coexistence of two such phases may or may not be compatible with a large coercivity. McCurrie and Gaunt (1964) attributed the large coercivity of PtCo alloy to the coexistence of a fcc and tetragonal phase, both of which are magnetic. The present opinion on rare-earth-cobalt magnets is that if $SmCo_5$ coexists with a little $Sm_2Co_{17}$, the coercivity is low, whereas a little $Sm_2Co_7$ can be tolerated and quite probably is beneficial. The conditions may be different for other rare-earth compounds.

Experimentally it is now claimed to be possible to recognize from magnetic measurements, supplemented by Kerr effect domain studies, whether nucleation or pinning predominates, but at the time of writing the precise processes that prevent nucleation or promote pinning are not fully established. If a small quantity of a second phase is involved, the relationship between the crystal structures, the existence or not of coherency at the boundary and the nature of the exchange interaction at this boundary are likely to be important factors. It may be that by the time this book is in the hands of the reader some of these problems will have been solved.

## 3.9 METALLURGICAL CONSIDERATIONS

To understand permanent magnets it is frequently necessary to refer to the metallurgical processes that have led to the structure and are responsible for their properties. Consequently it is necessary to explain the meaning of a few metallurgical terms, and to describe briefly a few metallurgical processes and methods used in experimental metallurgy.

### 3.9.1 Preferred texture

From the emphasis in earlier sections on magnetocrystalline anisotropy, the reader should not be surprised that it is often desirable to produce magnets with some special form of texture or crystal orientation. The best known methods of producing single crystals or materials with some special form of crystal texture are: (1) mechanical work particularly cold drawing or rolling, and (2) allowing the material to solidify from the melt with the heat extracted in one direction only. Cold work is used to produce the desired crystal orientation in two minority permanent magnet materials, namely Cu–Ni–Fe and Vicalloy, but most permanent magnets are not workable. There is a possibility that the properties of a new permanent magnet alloy Cr–Fe–Co may be improved by working.

Certain alloys of the Al–Ni–Co group can be improved by giving them a columnar structure (Section 4.2.5), although the improvement is due to the orientation of a precipitate in certain crystal planes, rather than being a direct consequence of the crystal structure, and crystal anisotropy.

Single crystals of many materials can be grown from the melt, provided the solidification takes place very slowly. Such methods are unsuitable for the commercial production of permanent magnets, but fortunately several of the Al–Ni–Co alloys form columnar crystals even though they cool at normal speeds, provided the heat is extracted in one direction only.

The presence of small amounts of certain elements such as Ti, S, Te and C may help or hinder the formation of columnar crystals. Several hypotheses have been proposed to explain these differences, which are naturally very important practically. One of these explanations supposed that under certain conditions determined by the difference between the solidus and liquidus temperatures, latent heat, thermal conductivities and rate of cooling, constitutional supercooling may occur. By this it is meant that conditions may arise which permit crystals of a different composition from that of the alloy as a whole to start forming in advance of the main solid surface. Any such crystals are expected to have a random orientation and therefore to spoil the columnar structure. A columnar structure is formed if the solid surface advances steadily by the crystals already in the solid growing, rather than new crystals being formed. A completely satisfactory answer to the question of what conditions permit or inhibit columnar growth is not yet agreed, but the fact that very small quantities of impurities make large differences to columnar growth, seems to favour explanations in terms of the presence or absence of nuclei on which new and therefore possibly misoriented crystals can start to grow.

There is a third method of obtaining a preferred crystal structure, which is applicable only to magnetic materials. The material is formed into a powder consisting of single crystals; this powder is aligned in a magnetic field, and then pressed and sintered or possibly frozen into a binder. (In certain materials mechanical methods may replace the magnetic field.) The magnetic field should not exceed the anisotropy field, and in the rare-earth-cobalt alloys much smaller fields suffice. Care is necessary to ensure that the pressing operation does not destroy the orientation. In ceramic magnets the particles of the powdered material are disc shaped and it is desirable for the pressure to be applied in the preferred direction. Rare-earth cobalt particles tend to be more nearly spherical, and at first it was thought that the pressing direction was immaterial. It has now been found that the pressure should preferably be in a transverse direction, since in this way there is less risk of breaking up chains of aligned particles.

The field is often maintained during the pressing, although this is unnecessary if isostatic pressing which does not destroy the orientation is employed. The sintering operation may sometimes actually improve the orientation. The explanation is that because of friction fine particles are less likely to be aligned than large ones, but during sintering the misaligned fine particles are absorbed by the well aligned large ones. The sintering of aligned powder is employed particularly with barium and strontium ferrite, and with rare-earth-cobalt alloys.

An aligned structure benefits the permanent magnet properties in several obvious ways. When the magnetizing force is removed the polarization of a crystal reverts to the nearest preferred direction. If this direction is inclined at an angle $\theta$ to that of the magnetizing force, the component of the polarization in the direction of magnetization is $J_r = J_s \cos\theta$. If the angle $\theta$ can be made zero for all crystals in a material, the remanence $J_r$ reaches the maximum value $J_s$. A square demagnetization curve with a high value of $(BH)_{max}$ also usually results. The influence of alignment on the coercivity is more variable. Columnar Al–Ni–Co alloys usually have rather higher coercivities than their randomly oriented counterparts, but ferrites tend to have slightly lower coercivities, when they are well aligned with square curves. Theory does actually suggest that the influence of orientation on coercivity may differ for various types of domain process.

### 3.9.2 Precipitation processes

Although many permanent magnet alloys are very complex, and contain as many as five or six elements in significant amounts, they can usually be obtained in a homogeneous single phase at a temperature just below the melting point. It is remarkable in that many permanent magnet materials, there is some temperature at which an undesirable phase can be precipitated. Such an unwanted change can occur in the steels, the Al–Ni–Co and Cr–Fe–Co alloys and in rare-earth cobalt, although the nature of the change is different in each case. It is a constraint on the production process that the material must be cooled sufficiently quickly through the temperature range in which this phase is produced to avoid this happening. If the phase is accidently allowed to form, it can usually be dissolved by homogenization just below the melting point.

Frequently some form of two phase structure of a different kind must be produced, and in Al–Ni–Co, Cr–Fe–Co and possibly some other alloys, what is known as a spinodal dissociation is involved. The meaning of this term is explained in the next section.

### 3.9.3 Spinodal dissociation

The idea of spinodal processes is sometimes traced back to Gibbs, although he may have been thinking more of fluids. In the late 1950s interest was focused on spinodal dissociation from solid solutions in metals. Nicholson and Tufton (1966) were prominent in relating these ideas to permanent magnets. Earlier references are contained in this paper and also in de Vos (1966) and Schüler and Brinkmann (1970).

Figure 3.18(a) shows a simple binary phase diagram with a miscibility gap at low temperatures. Outside the miscibility gap shown by the full line A and below the solidus–liquidus line, the system is single phase. Figure 3.18(b) shows the free-energy curve for the system at the temperature marked $T$ in the upper figure.

The important points on the energy curve in Fig.3.18(b) are firstly the minima $M_1$ and $M_2$, which correspond approximately to the intersection of the phase boundary with the isothermal $T$ in Fig.3.18(a), and secondly the points of inflection $F_1$, $F_2$. The latter fix the position of the dotted line known as the spinodal in Fig.3.18(a). The significance of the spinodal is as follows; alloys which are cooled from the single phase region to the two phase region within the miscibility gap become supersaturated. Precipitates are formed whose morphology and distribution depend markedly on whether the alloy composition lies inside or outside the dotted spinodal curve. For alloys that lie between the phase boundary and the spinodal curve, the second derivative of the free energy $d^2F/dC^2$ is positive, the energy curve is concave upwards and there is an energy barrier against small fluctuations in composition as explained in Fig.3.19. For alloys, which lie inside the spinodal curve $d^2F/dC^2$ is positive, the energy curve is convex upwards and there is

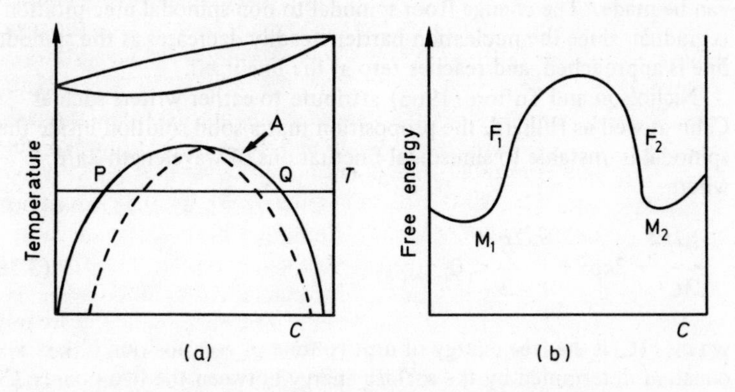


*Fig.3.18. (a) Phase diagram, (b) free energy curve for binary system*

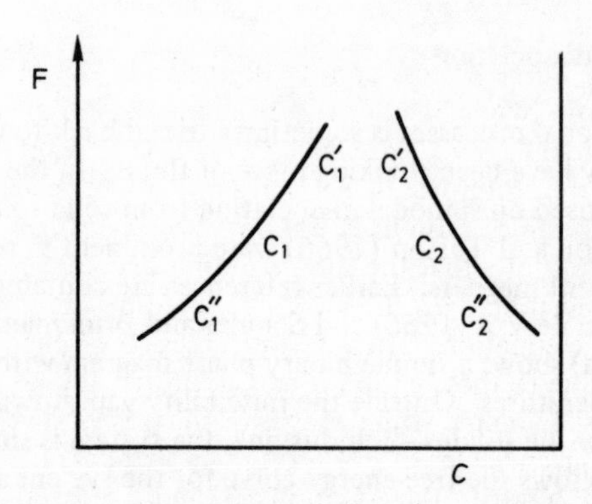


*Fig.3.19. Free energy F plotted against composition C for $d^2F/dC^2$ positive. If alloy of composition $C_1$ segregates into $C_1'$ and $C_1''$ or $C_2$ segregates into $C_2'$ and $C_2''$ there is a net increase in energy, so alloy is stable against small composition changes*

no barrier to precipitation. The alloy is unstable to small composition fluctuations as explained in Fig.3.20.

In the first case when the alloy is stable against small fluctuations, precipitation at temperature $T$ can start only by the nucleation of a precipitate with one of the compositions P or Q in Fig.3.18(a). Such nucleation usually originates at a few imperfections such as dislocations or grain boundaries. Once nucleated such precipitates may grow very large until a coarse structure with compositions P and Q is formed.

In the second case the precipitates are uniformly distributed and are little affected by grain boundaries. At least in the early stages, the composition differences between precipitate and matrix are small. It is not even certain that a sharp distinction between precipitate and matrix can be made. The change from spinodal to non-spinodal precipitation is gradual, since the nucleation barrier steadily decreases as the spinodal line is approached, and reaches zero at the line itself.

Nicholson and Tufton (1966) attribute to earlier writers such as Cahn as well as Hilliard, the proposition that a solid solution inside the spinodal is unstable to sinusoidal fluctuations of wavelength $2\pi/\beta$, where

$$\frac{\partial^2 F}{\partial C^2} + 2k\beta^2 + \frac{2\eta^2 E}{1-\nu} < 0 \tag{3.36}$$

where $F(C)$ is the free energy of unit volume of composition $C$, $k$ is a constant determined by the surface energy between the two phases, $E$ is Young's modulus, $\eta$ is the linear expansion of the lattice per unit composition change. Some writers omit to define $\nu$; de Vos (1966) refers to it as Poisson's ratio, while Schüler and Brinkmann (1970) call it the reciprocal of Poisson's ratio. Possibly they define Poisson's ratio differently. If $\nu$ is the reciprocal of Poisson's ratio as defined in most English books, the third term in Equation 3.36 is negative.

Equation 3.36 shows that the simple thermodynamic condition for spinodal decomposition $\partial^2F/\partial C^2 < 0$ is modified by the second and third terms in the equation. These terms represent respectively the energy barriers due to the creation of an additional area of interface in the lattice, and the volume strain energy of the misfitting phases. We can see that for given values of $\partial^2F/\partial C^2$ and $\eta$ there are certain maximum values of $\beta$ to satisfy the equation and hence a certain minimum of the wavelength of composition fluctuations into which the alloy can decompose. All wavelengths above the critical value $\lambda_c$ are stable, but Cahn suggested that the most likely wavelength to be observed is $\lambda_c\sqrt{2}$. Since $E$ is a minimum along the [100] axes in most cubic metals, orthogonal fluctuations in the three cube edge directions of a cubic crystal are preferred. The result is a microstructure with a distribution of second phase particles, consisting of an ordered array of approximately spherical particles for small volume fractions of precipitate, and an array of rods along [100] directions, which Nicholson and Tufton

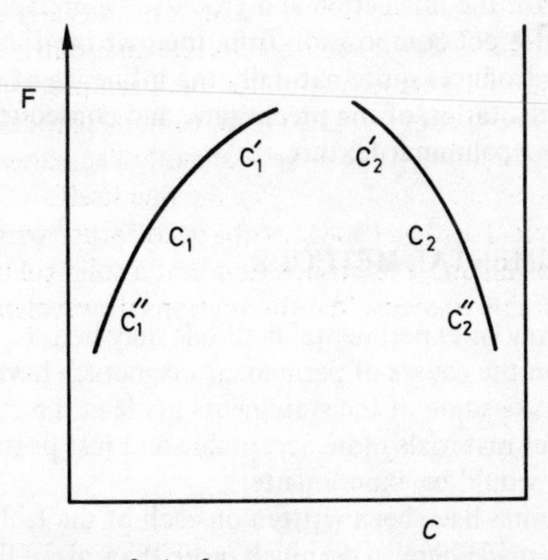


*Fig.3.20. As Fig.3.19 but $d^2F/dC^2$ negative. A small segregation such as $C_1$ into $C_1'$ and $C_1''$ causes a net decrease in free energy, so an alloy is unstable against small composition changes*

compare to 'scaffolding' for larger volume fractions of precipitate. The spacing of the spheres or rods is initially $\lambda_c\sqrt{2}$, but increases steadily during ageing by normal coarsening processes.

The concept of spinodal precipitation has become extremely important for permanent magnets. The spinodal process leads to a very fine precipitate, capable of producing high coercivities dependent upon shape anisotropy. The nucleation process gives rise to a precipitate, which is much too coarse for this purpose.

Until the theory of spinodal dissociation was introduced, I always felt unhappy about the ability of a magnetic field to influence the orientation of precipitate particles. The conventional view was that the difference in magnetostatic energy would be sufficient to cause elongated particles to grow parallel to the applied field in a non-magnetic matrix. My difficulty was that at the important time when the particles were first nucleated, their volume would be very small. The magnetic energy available to influence the process would be limited by the product of the applied field, the polarization of the particles (not very large at a temperature just below the Curie point) and a vanishingly small volume. It seemed to me that it would be easier to account for the precipitation of an oriented non-magnetic particle in a magnetic matrix, since the field would then be acting on practically the whole volume of the material.

The spinodal process avoids these difficulties. The process starts as a simultaneous fluctuation throughout a material which is already magnetized in the direction of the applied field. Furthermore the energy involved in starting a spinodal dissociation is clearly much smaller than that required for the nucleation and growth of a precipitate with a completely different composition from the matrix. Finally the spinodal explanation introduces quite naturally the influence of the crystal axes on the orientation of the precipitate, and consequently the importance of using a columnar texture.

## 3.10 EXPERIMENTAL METHODS

This brief survey of experimental methods may help to show how theoretical ideas on the causes of permanent magnetism have been reached. It may also make some of the statements made in the chapter on permanent magnet materials more acceptable and just possibly may give ideas to some would be experimenters.

Whole volumes have been written on each of the techniques used. No attempt is made here to do much more than name the method and indicate the particular type of information each can yield about permanent magnets.

The use of measurements of various parts of the hysteresis loop is an obvious method of investigating magnetic materials, but the experimental details are described in the chapter on testing. Such experiments are usually combined with domain observations, which have already been discussed.

### 3.10.1 Optical microscopy

The structures responsible for producing coercivity in magnet steels and the Al–Ni–Co alloys are much too fine to be seen in the optical microscope. It happens that in certain Al–Ni–Co alloys this structure coarsens after a period of two or three weeks at a temperature a little below 800°C, and although this coarsening destroys the permanent magnet properties, it enables the precipitate to be viewed in the optical microscope. By means of such microscopic observations, forecasts about the nature of the precipitate were made, that later proved to be substantially correct. This use of the optical microscope is now only of historical importance. Nevertheless the optical microscope is a useful tool for revealing unwanted spoiling phases, and also for investigating textures such as columnar crystals.

A rather similar situation occurs in cast rare-earth-cobalt-copper alloys in which a precipitate is formed but can only be seen in an optical microscope if it has grown too large to produce good permanent magnet properties. There are also small amounts of second phases in some of the sintered rare-earth magnets, which have considerable influence on the magnetic properties. Some of these precipitates are harmful, but some may actually be beneficial, and the optical microscope seems to be a useful tool for observing these structures.

### 3.10.2 Electron microscopy

The electron microscope is an ideal tool for revealing the structures responsible for the properties of the Al–Ni–Co alloys and a number of other permanent magnet materials in which the patterns to be recognized have dimensions in the range less than 0.1 μm. Most of the early experiments made use of replica techniques, but similar results have now been obtained by transmission electron microscopy on thin specimens. As mentioned in Section 3.8.1 the Lorentz method of transmission electron microscopy can be used to investigate domain patterns.

Thick samples can be investigated by means of the scattered electrons produced from their surface layers, with the aid of the scanning electron microscope. There are now also transmission scanning electron microscopes. Attempts have been made to discover the composition of precipitates by the secondary X-rays emitted and electron probe analysis. The results are not yet completely convincing, but the method is probably the most promising for tackling this very difficult problem. A number of earlier experimenters tried to analyse the materials removed by etching or removed by an adhesive film, but the results obtained were obviously highly improbable.

### 3.10.3 X-rays

Conventional methods of X-ray diffraction enable crystal types and lattice constants to be identified. Spinodal dissociation often leads to phases differing very slightly in lattice constants, so careful measurements may be necessary. These difficulties were overcome many years ago, and the crystal structures and lattice constants of the fully treated Al–Ni–Co alloys were identified long before the hypothesis of spinodal dissociation was introduced.

Another important and more difficult problem is the determination of the angular distribution of crystal axes in columnar or semicolumnar alloys. We at the Central Research Laboratory of the Permanent Magnet Association became interested in this problem in the late 1950s, in connection with a study of the degree of perfection of columnar textures. At this time we did not possess our own X-ray equipment, and so we sought the assistance of other laboratories. The only method suggested to us was to select a few crystals on the surface of a sample, and determine the orientation of each one separately. Each crystal required an exposure of several hours, so the experiment was not carried out on many samples. Fortunately more convenient methods of solving this problem are now available. Goniometers which give a distribution of crystal axes with angle have been used on ferrite magnets, and more recently on rare-earth cobalt. See for example Swift, et al (1976).

### 3.10.4 Metallographic methods for crystal orientation

Crystal orientation in Al–Ni–Co alloys is often observed by simple inspection of a fracture, or by polishing and etching a surface and observing it with a low power microscope.

An elegant method of combining crystal orientation with domain techniques was described by Wells and Ratnam (1974). A full-wave plate is inserted between the specimen and analyzer in a polarizing microscope. Colour variations then indicate differences in crystallographic orientation from grain to grain, although the optical anisotropy which produces these variations arises from different directions of magnetization and the Kerr effect. The basal plane is consequently optically isotropic and does not change colour as the specimen stage is rotated. The axial planes do change colour, and the orientation of the crystals can be deduced from the colours.

Martin et al (1975) studied the alignment in $RCo_5$ alloy by deliberately promoting the eutectoid reaction by holding the specimen for 10 days at 750°C. This treatment produces a precipitate in the form of lamellae perpendicular to the alignment direction. This structure is clearly visible in the microscope, as shown in Fig.3.21. Martin considers the method could be applied to $RCo_5$ alloys in which R is Ce, La, Pr, Sm or Gd.

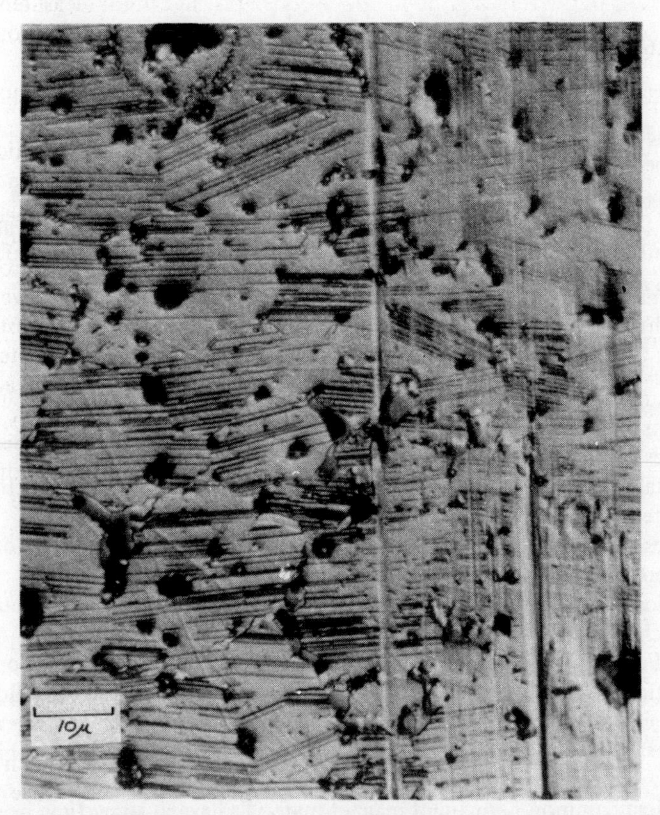


*Fig.3.21. Lamellar structure in alloy after eutectoid transformation. Lamellar planes are perpendicular to hexagonal axes of grains. (Courtesy D. L. Martin, General Electric, Schenectady, N.Y.)*

### 3.10.5 Miscellaneous metallurgical investigations

The various constitutional changes that determine the structure of a permanent magnet, as well as the disappearance of magnetic order when the Curie point of a substance or one of its constituent phases is reached, alter many of the characteristics of the material. The characteristics concerned include linear expansion, specific heat and electrical resistivity. Specific heat is studied by differential thermal analysis, and expansion measurements are often referred to as dilatometry. These methods will nearly always give some indication of a phase change or a Curie point, but do not give very precise information about the nature of the change, and are most profitably used in conjunction with other investigations.

### 3.10.6 Curie points

Much information can be obtained by measuring the variation of $\sigma$ with temperature. In CGS units $\sigma$ was the intensity of magnetization divided by the density, although many writers used the quantity without stating the units. If units were stated they were usually described as EMU per gramme. No decision appears to have been confirmed to define an SI unit of $\sigma$, but it so happens that $J/\mu_0\rho = M/\rho$ has the same numerical value as the CGS unit. I have seen few papers that claimed to quote $\sigma$ in SI units, and shall therefore use the CGS values, with the understanding that they can be regarded as $J/\mu_0\rho$ in SI; $\rho$ being the density.

A material with two or more magnetic phases has two or more Curie points as shown schematically by the $\sigma$, $T$ curve in Fig.3.22. The actual values of these Curie points are functions of the compositions and crystal structures of the phases. If the upper part of the curve is extrapolated back to room temperature, or better still to 0 K, $\sigma_1$ and $\sigma_2$ represent contributions from the two phases. These values depend on the concentration of each phase and its own value of $\sigma$.

Hoselitz (1952) describes this method of investigation in some detail in a chapter headed 'Magnetic analysis'. In binary alloys such as Fe–Ni and Fe–Si this method of analysis permits the concentration and composition of the phases present to be determined. The Al–Ni–Co alloys contain too many components to permit such a precise analysis. Nevertheless as is shown in Section 4.2.6 valuable insight is afforded by this method even in these complicated alloys.

Most common permanent magnet materials have at some time been examined by this method. In spite of difficulties presented by the high field necessary to achieve saturation, the method appears particularly

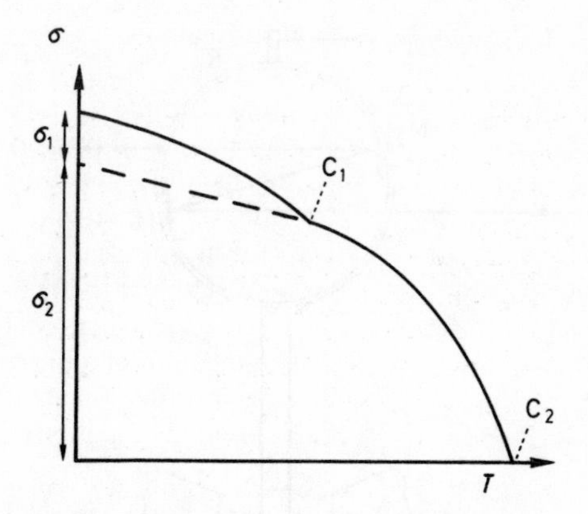


*Fig.3.22. $C_1$ and $C_2$ are the Curie points and $\sigma_1$ and $\sigma_2$ the extrapolated saturation magnetizations of the two different phases 1 and 2*

promising for investigating rare-earth-cobalt alloys with their many compounds with distinctive Curie points.

Most of this work has been carried out with the Sucksmith ring balance. In this balance the force experienced by a small sample of substance is made to distort a ring carrying two mirrors as shown in Fig.3.23. The deflection of a beam of light reflected by two mirrors is followed in a travelling microscope. The force is exerted on the sample by a magnetic field sufficient to produce saturation, on which is superposed a field gradient. The field gradient may be obtained by shaping the polepieces of the electromagnet in the manner indicated in the figure. The departure of the faces of the polepieces from being parallel has been exaggerated in this sketch, the actual value of $\theta$ being only 2 or $3^\circ$. A drawback of this method is that the field and field gradient cannot be varied independently, and therefore the variation of $\sigma$ with field cannot be investigated. Materials such as $SmCo_5$, with large crystal anisotropy that makes saturation of a polycrystalline sample difficult, may be examined in the form of a loose powder of single crystals, that become oriented with the easy direction of each grain parallel to the field.

It is now possible to replace the optical method of taking readings manually by an electronic balance in which the $\sigma$, $T$ curve is drawn automatically by an XY recorder. The sample and sample holder are on the end of a rod and supported by some form of spring as in the ring balance. This moving part carried a shield with a slit that allows a beam of light to fall equally on two photocells, when the system is in the rest position. When a force is applied to the sample more light falls on one photocell and less on the other. The signal from these photocells is

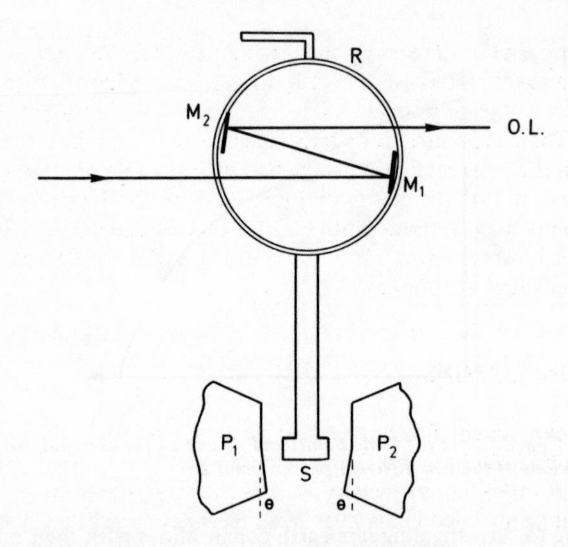


*Fig.3.23. Sucksmith balance. R: Ring. S: Sample holder. $P_1$, $P_2$: Pole-pieces of electromagnet shaped to produce field gradients. $M_1$, $M_2$: Mirrors. OL: Optical lever*

amplified and by a process of negative feedback used to restore the shield to its original position. What is measured is the amplified current that prevents the shield and sample system moving. A negative feedback system can also be used to promote electronic damping.

The advantages of an electronic balance lies not mainly in the saving of labour, but in the greater accuracy of a continuous record of $\sigma$ with temperature. The curve obtained with the ring balance was often drawn by plotting points at temperature intervals of say 10 or 20°C., and one or two incorrect points could give a false impression of a subsidiary Curie point.

For work at high temperatures a good vacuum or protective atmosphere must be used. It is necessary to be on guard against confusing true Curie points with inflections in the $\sigma$, $T$ curves produced by phase changes. True Curie points are reversible and can be obtained with the temperature rising or falling.

It is useful to provide for measurements at temperatures below as well as above room temperature. Some permanent magnets have phases with Curie points just below room temperature.

It has been known for a long time that Curie points can be deduced from peaks in the curves of initial permeability plotted against temperature. Care must be taken in interpreting such curves, because peaks can also arise from other causes, such as minima in the value of the crystal anisotropy, and until recently the method has been little used on permanent magnet materials. Recently this technique has been used rather effectively at the University of Dayton on various alloys of rare-earth cobalt. There is no need for a large magnetizing field; substantial peaks appear in the permeability curve just below the Curie point of each compound, so that the method is more sensitive for detecting subsidiary Curie points than the Sucksmith balance type of experiment. No information is, however, given about the quantities of the different phases or compounds that are present.

## 3.11 CONCLUSIONS

Modern experimental techniques enable us to determine the microstructure of a permanent magnet. In a good magnet there always appears to be some form of inhomogeneity such as grains separated by pores or a precipitate embedded in a matrix of different composition. Such structures may be reached by a precipitation process such as a spinodal decomposition of a solid solution, or by compacting and possibly sintering a powder.

Although some of the grains or precipitate particles may behave as single domain particles, the distinction between single domain particles and larger particles is now somewhat blurred. A particle that is normally too small to contain a domain boundary, may yet change its direction of magnetization under the influence of an applied field by an incoherent rotation process that approximates to the temporary formation of a domain wall. There is also a partial truth in Brown's paradox, and it is often difficult to nucleate a domain wall, once it has been removed from a particle large enough to contain one. Even when a domain wall is nucleated within a grain its free movement may be prevented by pinning at imperfections.

No material can be a good permanent magnet without some form of anisotropy to impede the rotation of the magnetization vector. The most important sources of this anisotropy are believed to be the shape anisotropy of elongated particles and the magnetocrystalline anisotropy that directs the magnetization along certain crystal axes.

In most permanent magnets it is now established which form of anisotropy predominates, for example shape anisotropy in Al–Ni–Co and ESD magnets and crystal anisotropy in ferrites and $RCo_5$. With either form of anisotropy it is an advantage to make the preferred direction of magnetization parallel throughout the material, so as to increase the remanence of the material in one direction at the expense of that in other directions.

The structural requirements for magnets with shape anisotropy differ from those with crystal anisotropy. The coercivity of magnets based on shape anisotropy falls rapidly if the fraction of volume occupied by the magnetic phase much exceeds 50%. It is also important that the matrix should be non-magnetic or nearly so. With crystal anisotropy the magnetic phase may occupy over 95% of the volume. Sometimes a second magnetic phase proves disastrous, but in other cases it may even be an advantage. The reasons for these differences are only partially understood.

While experimental methods now enable nucleation and pinning mechanisms to be distinguished, the problems of the precise nature of the defects that promote nucleation and pinning are still not resolved. It is not impossible that the same defect may prevent a very high coercivity by promoting nucleation, but produce some coercivity by causing pinning. The solutions of these problems might point the way to better magnets. The practical aim must be to combine high remanence with high coercivity. There is no theoretically known reason why this desirable combination should not be achieved, and a search for a material with these desirable properties is more likely to be successful if guided by theory, than if it is carried out by the 'cut and try' methods of the past.

On the other hand it is always necessary to be on guard against the risk of pure research descending to the trivial. Pure research often requires the painstaking carrying out of many different observations on a few samples. The investigator must be alert to avoid being trapped into following up some trivial peculiarity of a particular sample, when what is required is insight with some general relevance.

# Chapter 4

# Permanent magnet materials

This chapter contains a description of the principal commercial permanent magnet materials, and a brief account of a number of other materials that have been reported experimentally, although they are not made commercially. During the last thirty years many new materials have been discovered in the laboratory. If these materials had been discovered 20 or 30 years previously, many of them would have been hailed as providing the best ever permanent magnet. When they were actually found they were already inferior in properties and probably more expensive than existing materials.

Figure 4.1 shows how the $(BH)_{max}$ obtainable from permanent magnet materials has improved during the last 100 years. When a new class of materials has been discovered, there has been for a few years a very rapid improvement in $(BH)_{max}$, followed by an asymptotic approach to a maximum, until a break through has been made with a new class of material.

The important classes of material are the old carbon, chromium, tungsten and cobalt steels, the Al–Ni–Co alloys, ferrites and rare-earth cobalt. This chapter contains sections on each of these types of material. The ferrites are possibly the most important commercial permanent magnet at the present time, yet they have never enjoyed the title of having the best $(BH)_{max}$. Their success lies partly in their cheapness and partly in their high coercivity. Figure 4.2 shows the advance in $H_c$ with time, and the ferrites show up to greater advantage in this figure than in Fig.4.1. A similar graph of the intrinsic coercivity $H_{ci}$ would be even more impressive, but it would be difficult to compile such a graph, as many authors and manufacturers have not given this information.

Some indications of the other physical properties and of the methods of manufacture are also given for the more important materials, and there is an attempt to show how the properties of each material fit into the theoretical framework of the previous chapter.

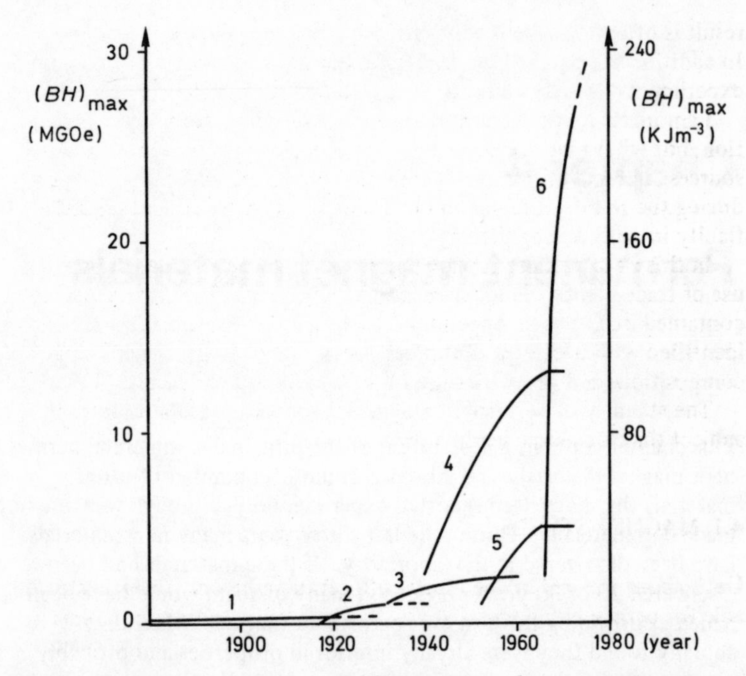


*Fig.4.1. Improvement of $(BH)_{max}$ of different classes of permanent magnets during the last hundred years. (1) Pre-cobalt steels. (2) Cobalt steels. (3) Isotropic Al–Ni–Co. (4) Anisotropic and columnar Al–Ni–Co. (5) Barium and strontium ferrite. (6) Rare-earth cobalt*

Unfortunately the details of manufacturing processes are rarely sufficient to enable you to produce magnets successfully yourself. Even when a process for making permanent magnets is fully and honestly described, it may take several months for someone skilled in the art to reproduce it successfully in a different environment.

Perhaps understandably manufacturers are reluctant to publish complete information about their most successful processes, although experiments that have not led to a practical method of production are often described in great detail. Patents are supposed to give sufficient information to enable someone skilled in the art to repeat the process, but patent examiners usually lack the facilities to check whether this information has been given. Their main function is to ensure that there has been no prior publication.

Patentees and their agents even when publishing a viable process often describe additional possibilities that have not been tried or that give inferior results. One justifiable object is to prevent anyone circumventing the patent by means of a trivial variation in the procedure, but the result is often to make it very difficult to discern the optimum process. In addition the patent literature contains many examples of 'thought experiments', which would be very unlikely to work.

I am under no obligation to any employer to withold any information, but what I write cannot be more complete and accurate than my sources, and in addition our Members often found difficulty in reproducing the results obtained in our Laboratory, as we often found difficulty in reproducing theirs.

In this chapter materials are described as far as possible without the use of trade names. Trade names and the names of manufacturers are contained in Tables in Appendix 2. For simplicity materials have been identified with a letter and number, which may cover a certain range of compositions and properties used by different manufacturers.

The stability of permanent magnets is such a large and important subject that it is discussed in a separate chapter.

## 4.1 MAGNET STEELS

Until about the end of the nineteenth century the only materials available for permanent magnets apart from natural lodestone were the

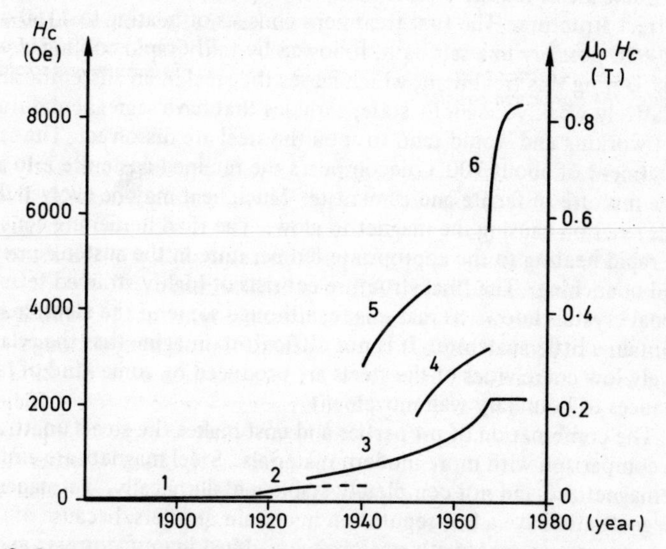


*Fig.4.2. Improvement of coercivity of different classes of permanent magnets during the last hundred years. (1) Pre-cobalt steels. (2) Cobalt steels. (3) All Al–Ni–Co alloys. (4) Barium and strontium ferrite. (5) Platinum-cobalt. (6) Rare-earth-cobalt.*

hardenable carbon steels. A 1% carbon steel 'silver steel' was probably the best, although the much higher carbon cast iron also appears to have been used. The first steels made specially for permanent magnets contained up to 6% of W or Cr. About 1917 cobalt magnet steels were discovered in Japan. Any quantity of Co up to about 40% may be used, but in 1930 some English manufacturers combined to form the Cobalt Magnet Association, and one of the first actions of this association was to standardize six grades of cobalt steel. Steel magnets are usually designated by the percentage of the most important alloying element. Compositions and properties of some of these steels are given in Table A2.1, in Appendix 2. Many of the cobalt steel magnets contained considerable quantities of other alloying elements such as Cr or W. In other countries many other formulae were used.

In an annealed state all the steels can be hot worked and machined, although machining is somewhat difficult in 35% Co. Steel magnets can be cast, but forging and hot pressing are typical processes. The final treatment of all the permanent magnet steels is to harden by quenching from a high temperature, the actual temperature and quenching medium varying with the steel. In this state the steel has the optimum high coercivity necessary for a permanent magnet, but is hard, brittle and can no longer be machined.

Some steels require a more complex triple treatment to obtain the correct structure. The first treatment consists of heating to 1150 or 1200°C possibly in a salt bath, followed by fairly rapid cooling in air or oil. During this treatment, which leaves the steel in an austenitic and relatively weakly magnetic state, carbides that have segregated during hot working and would tend to spoil the steel are dissolved. The second treatment of about 700°C decomposes the retained austenite into a fine mixture of ferrite and cementite. Much heat may be evolved during this reaction causing the magnet to glow. The final hardening consists of rapid heating to the appropriate temperature in the austenitic region and quenching. The final structure consists of highly strained tetragonal crystals known as martensite, although some of the steels may contain a little austenite. It is not difficult to imagine that the relatively low coercivities of the steels are produced by some kind of hindrances to boundary wall movement.

The combination of properties and cost makes the steels unattractive in comparison with more modern materials. Steel magnets are easily demagnetized and not completely stable metallurgically. Permanent magnets still have a bad reputation in certain quarters, because of unhappy experiences with steel magnets. Most manufacturers are now trying to phase out their production. The last publication of the Permanent Magnet Association, before it was dissolved, listed only 35% and 3% Co steel.

The only justification today for making steel magnets is for applications such as hysteresis motors for which a semi-hard rather than a genuine permanent magnet material is required. It is not certain that they are the best materials even for such applications.

This account of the various steel magnets has been kept deliberately short. Fuller accounts are given in more than one chapter of Hadfield (1962).

## 4.2 Al–Ni–Co ALLOYS

In this book the term Al–Ni–Co is used to describe all permanent magnets based on the Al–Ni–Fe system. The first alloy discovered by Mishima (1931) did not actually contain any Co. Its metallurgy resembles that of the Co containing alloys, so that it does not seem worth treating it as belonging to a separate system. In the UK the name Alnico without hyphens is used only for isotropic alloys, containing about 12% of Co; field treated anisotropic alloys with 20 to 25% Co and moderate coercivities of about 45 to 60 $kAm^{-1}$ (580 to 780 Oe) are known by most manufacturers as Alcomax with various distinguishing numbers, while high coercivity alloys with 30% or more Co and some Ti are designated Hycomax. In the USA practically all the manufacturers use a common system of designation, calling all the alloys Alnico with distinguishing numbers, whether they are isotropic or anisotropic and whether they contain Co or not. The Philips organization based in Holland, but with manufacturing facilities or licencees in many other countries uses the name Ticonal, but has recently changed from a system of letters to one of numbers to distinguish grades. More details of these trade names and their meanings are given in Appendix 2.

In this chapter Section 4.2.1 deals with isotropic alloys, containing 12% Co or less, Section 4.2.2 with the moderate coercivity field treated alloys, containing 20 to 25% Co, Section 4.2.3 with high coercivity alloys containing more than 30% Co and Section 4.2.4 with columnar varieties of either of the last two groups. Section 4.2.5 describes how the ideas of Chapter 3 have been applied to explain the properties of the Al–Ni–Co alloys.

In general the Al–Ni–Co alloys are all more or less brittle. They are made either by casting or sintering. Sintering is recommended for manufacturing large numbers of small magnets, and usually leads to slightly inferior magnetic properties, but slightly superior mechanical ones. Normally the only satisfactory method of machining is grinding, although drilling is occasionally tried but is not recommended. Electrolytic machining and spark erosion are also possible. There have been laboratory reports of limited success in hot forging certain Al–Ni–Co alloys, but the author is not aware of the process being used commercially.

In the 1950s Al–Ni–Co alloys certainly comprised the majority of permanent magnets manufactured throughout the world, whether measured by tonnage or value. They have now been overtaken by ferrites in terms of weight and possibly also in terms of value, but this is less certain.

Some procedures are common to the manufacture of all the Al–Ni–Co alloys. Raw materials must be of good quality free from impurities. Carbon is particularly injurious; it must almost always be kept below 0.05%, and a maximum of 0.02% is recommended for some alloys. A little sulphur, say 0.25%, can usually be tolerated, and is sometimes added in the belief that it reduces brittleness. Sulphur also helps to promote columnar growth in Ti containing alloys.

Aluminium contents are rather critical, and the values quoted are those desired in the final alloy. More or less Al is always lost in melting, and an excess is charged to allow for this loss. The excess required depends on the skill of the operator, the procedure he adopts and the type of furnace he is using.

Melting nowadays is almost invariably carried out in high frequency furnaces; the turbulence promoted by the eddy currents induces mixing, but mechanical stirring may also be employed. The size of furnace used industrially typically ranges from 50 to 500 kg. Smaller furnaces used be used in the laboratory. The diversity of alloys and applications means that the manufacturer has difficulty in fulfilling some small orders unless he possesses some furnaces that are not too large.

Air melting is normally quite satisfactory, but occasionally an apron of inert gas may be blown over the crucible. Vacuum melting is unnecessary and actually increases difficulties in controlling the Al content, because of evaporation. If for some reason a vacuum furnace is used to obtain a particularly pure alloy for experimental purposes, the furnace should be filled with a pure inert gas after evacuation.

Green sand moulds (sand bound with a little clay and water) may be used for casting. Most of the more modern moulding techniques are applicable, and shell moulding which gives a better surface finish and more accurate dimensions is now very common. In this process the sand impregnated with a little thermosetting resin is dumped on to a hot pattern plate. After a few seconds a hard shell with the shape of the pattern is formed. Two shells are stuck or clamped together to form a mould. The mould should hold together just long enough for the metal to solidify and then disintegrate. A number of moulds may be placed together and filled through one common runner. Special methods of making hot moulds for casting alloys with columnar texture are described in Section 4.2.4.

Although theoretically it is desirable to fill the moulds directly from the furnace, particularly when casting columnar alloys, some at least of the manufacturers, who have tried this method, have encountered difficulties, and have reverted to the use of ladles manipulated by hand.

When Al–Ni–Co alloys are produced by sintering, the raw materials must be in the form of powders of the constituent metals or suitable prealloys. Al is always added in the form of a prealloy with other required metals such as Fe, Ni or Co, because of its low melting point, and some manufacturers use other prealloys. The use of crushed scrap of the final alloy is obviously desirable, but presents difficulties.

The powders are thoroughly mixed and then pressed in a suitable die, designed to give the required shape and size of magnet, after allowing for shrinkage during sintering. The linear shrinkage is of the order 10%, but depends on the reactivity of the powders, the pressure applied to the die and the sintering time and temperature. A little binder may be added to strengthen the green compacts, provided it is volatile and leaves no impurities. The pressure applied to the die is usually between 5 and 10 kbar, and the sintering temperature a little below the melting point of any of the constituents. Although liquid phase sintering in which one constituent is liquid is now a well-known industrial process, it is believed that the methods used today in the sintering of Al–Ni–Co alloys involve mainly or wholly solid diffusion processes.

Sintering may be carried out as a batch process in a good vacuum. A disadvantage of this method is that although the actual sintering requires only 2 or 3 hr, heating and cooling are so slow in a vacuum, that the time required for the whole process may be as long as 24 hr. An alternative method is to use a protective atmosphere of pure hydrogen, which must be particularly free from oxygen or water vapour. With a slight excess pressure of hydrogen, the ends of a long tubular furnace may be open, and the operation can be carried out as a continuous process.

In order to have good magnetic properties magnets must be sintered until they have a density as close to the theoretical maximum density as possible. The $(BH)_{max}$ of sintered isotropic Al–Ni–Co is about 5% less than that of the cast variety, but the difference tends to be greater in the anisotropic alloys for which it may be as great as 20%.

On the other hand sintered magnets are rather stronger mechanically than their cast counterparts.

In Britain about 10% by value of Al–Ni–Co magnets are produced by sintering. BOC Magnets specialize exclusively in the sintered variety of magnets, and a few other firms do some sintering. For some possibly accidental historical reason concerned with the availability of equipment, the proportion of magnets made by sintering is greater in Germany than elsewhere, and one of the best descriptions of the sintering of magnets is given in the German book, Schüler and Brinkmann (1970).

The initial cost of the dies for sintering is much greater than that of the patterns for casting. The subsequent cost of grinding small sintered magnets and the associated wastage of material is less than that of the cast variety. Thus there is a size below which it is more economical to sinter and above which it is more economical to cast magnets. The German manufacturers seem to find this critical size rather larger than other people.

To prepare a small number of sintered magnets for trial purposes a large sample can be pressed, and small samples can be carefully cut from this green sample before sintering. The method may be compared with the way in which a cook cuts intricate shapes from a sheet of pastry. It is a useful method of enabling a customer to try out a design without incurring the cost of making expensive tools.

Magnets can be sintered with soft iron pole pieces in one piece, making other methods of assembly such as the use of adhesives, brazing or clamping unnecessary.

After the appropriate heat treatment, which varies considerably in detail from one Al–Ni–Co alloy to another, the final finishing processes are performed. Magnets that are required to have precise dimensions must be ground, and any surface through which flux passes in a magnetic assembly requires grinding, but grinding of other surfaces just to make them look nice is a wasteful process. Even the rough outside scale on a cast Al–Ni–Co magnet is no disadvantage magnetically. In having this outside scale ground away, one is paying to dispose of material that supplies some useful flux, even though it may not be quite so efficient as the rest of the magnet. Grinding may appear to improve the test results on the remaining material, but is justified only if the outer scale is really in the way, precise dimensions are essential, or an unsightly unground surface will really detract from the saleability of the final article. Shot or sand blasting can sometimes be used to remove scale instead of grinding.

Some customers ask for magnets to be plated or painted. All normal finishes are possible. They are unnecessary to protect the magnets in normal environments, and the end surfaces of magnets in assemblies with pole-pieces must not be plated, as such plating acts as an undesirable gap in the magnetic circuit.

### 4.2.1 Isotropic Al–Ni–Co alloys

The original alloy discovered by Mishima contained 30% Ni, 12% Al balance Fe, but considerable variation is possible. Fairly good properties can be obtained with as little as 22% Ni and 10% Al.

The repercussions of this discovery on the permanent magnet industry were very great. In England the newly formed Cobalt Magnet Association decided to negotiate jointly for a licence. It was then found difficult to reproduce Mishima's results. At the time it was not realized that the magnetic properties of the alloy were very sensitive to the precise cooling rate and subsequent heat treatment. Consequently a procedure that worked on one size of magnet or arrangement of castings might fail on a different magnet or arrangement that cooled at a different speed. To surmount these difficulties the English manufacturers agreed to share technical information. They renamed themselves the Permanent Magnet Association and embarked on 50 years of co-operation as described in the Dedication.

It was ultimately found beneficial to include some copper and cobalt in the alloy. Typical compositions and properties are given in Table A2 of Appendix 2. In general an increase in Ni or Cu tends to increase $H_c$ but reduce $B_r$ with little change in $(BH)_{max}$.

The heat treatment of the Co free Al–Ni–Fe magnets may consist merely of cooling in air, possibly with a gentle air blast, from a temperature of about 1200°C, at which the alloy is a single phase solid solution. Sometimes the alloy is cooled more quickly and then given a short heat treatment at a temperature between 650 and 700°C. Variations in this treatment as well as of composition can be used to produce either a high $H_c$ or a high $B_r$.

The alloy containing about 12% Co has a $(BH)_{max}$ about 25 to 30% better than the alloy without Co, and requires a similar heat treatment.

A feature common to a varying extent to most of the Al–Ni–Co alloys is that a spoiling, $\gamma$ or fcc phase is liable to segregate out at temperatures between 1000 and 1100°C. This phase may change to a bcc structure at lower temperatures, but whether this happens or not the alloy is left with a coarse two phase structure easily visible in an ordinary optical microscope. Neither of the two phases has the correct composition for the production of good permanent magnet properties. Once the spoiling reaction has taken place the alloy must be reheated to the solution temperature to homogenize the alloy again. This reheating is conveniently made the start of the heat treatment proper. Heating to a temperature of 900–950°C, just below the spoiling range does not produce the spoiling reaction, but does not redissolve the phases if they have already formed. Thus the heat treatment may start from a temperature just above 900°C, if spoiling has not occurred during the initial cooling of the cast.

Between 1935 and 1939 there was considerable cooperation between industrial and academic scientists interested in magnetic materials. In July 1937 the Manchester and District Branch of the Institute of

Physics organized a two day conference on magnetism, at which half the papers were given by industrial and the other half by academic scientists. The lectures given at this conference were published by the Institute of Physics and distributed free to all its Fellows and Associates. One of the lectures was given by Oliver (1938) and is a very good account of permanent magnet technology at that time. Another was given by Bradley and Taylor (1937, 1938) and is a remarkable account of a thorough X-ray investigation of the Fe–Ni–Al system. The ternary diagram they produced was also published in the Proceedings of the Royal Society, and was thought worth reproducing by Hoselitz (1952) and by Hadfield (1962), that is 14 and 24 years later. The reader is warned that Bradley and Taylor labelled phases differently from the present convention. What they called $\beta$ is now called $\alpha$, and what they called $\alpha$ is now called $\gamma$. They found that in the permanent magnet state Fe atoms had congregated into islands with no sharp boundary. Although they attributed coercivity to stress their description of the state of the material is not a bad example of what is now known as spinodal decomposition (Section 3.9.3.).

### 4.2.2 Anisotropic Al–Ni–Co alloys

The discovery of anisotropic Al–Ni–Co alloys by Oliver and Sheddon (1938) was another example of the cross-fertilization between academic and industrial scientists. The circumstances were described by Oliver at a meeting of the British Association in Brighton in 1948. Improved properties had been found in soft magnetic materials such as silicon-iron and nickel-iron alloys by making them anisotropic with directional properties. These anisotropic properties had been obtained by hot or cold work and also by cooling in a magnetic field. It appears that research on magnetic materials was being discussed at an informal meeting in Manchester that included such people as W. L. Bragg, A. J. Bradley A. Taylor and D. A. Oliver, and the suggestion emerged that an attempt should be made to obtain similar improvements in permanent magnet materials, by producing directional properties. Oliver agreed to try the effect of hot pressing and field cooling on the 12% cobalt Al–Ni–Co alloy then being made in Sheffield.

As a result no advantage was obtained by hot pressing, but Oliver and Sheddon (1938) reported a small improvement in magnetic properties resulting from cooling in a magnetic field. Unfortunately as is now known, the improvement that can be obtained in an alloy with only about 12% Co is quite small, and so the result was dismissed as having little practical value.

Philips in Holland were using alloys with higher Co content and some

Ti, possibly to avoid the range of British patents. Their workers discovered that a much larger improvement in $(BH)_{max}$ could be obtained by the field cooling of alloys with more than 20% Co. News of this discovery reached England just before the wartime occupation of Holland, and production of field treated permanent magnets was quickly started in England for military purposes, particularly for magnetron valves used in radar. During the war the British Government set aside patent priorities, but at the end of the war, the Permanent Magnet Association appeared in danger of requiring a licence to manufacture magnets by a process which one of its members had discovered but without appreciating its full significance.

However, Philips had a patent which specified a Ni content of 12 to 20%, and the PMA developed alloys with less than 12% Ni (Alcomax II) and more than 20% Ni (Hycomax I). Later they found that the addition of Nb increased the coercivity, and two more alloys Alcomax III and IV were introduced. The patent position was still unclear, but the PMA decided to make these alloys and wait for Philips to act. No action was taken until the Philips Ticonal patent had nearly expired and then an infringement action was brought against an agent selling PMA magnets in Italy. The PMA acting collectively successfully defended this action.

This account has been included because although all the original patents have now expired these efforts to circumvent patents, for better or worse, motivated a considerable amount of permanent magnet research and still have some influence on the trade names and compositions in use. One may also wonder whether any individual British magnet manufacturer would have been sufficiently confident of its case to risk defending a patent action in a foreign country.

Hadfield (1962) quotes a large number of nominal compositions and properties. With the expiry of patents it is probable that some of these have been modified over the years. In Appendix 2, alloys with a $(BH)_{max}$ of about 43 kJ $m^{-3}$ (5.4 MGOe) are classed together as material A2; for this material $B_r$ can vary between 1.25 and 1.3 tesla while $H_c$ varies in the opposite sense to $B_r$ between 46 and 54 $kAm^{-1}$ (580 and 680 Oe.). The composition is likely to be about 22 to 25% Co, 12 to 14% Ni, 8% Al, 2 to 4% Cu and 0 to 1% Nb. In England Nb is added to obtain the higher coercivities, i.e. Alcomax III rather than Alcomax II. The Al content is very critical; there is a value somewhere between 7 and 8% below which $\gamma$ phase is precipitated and the alloy spoilt, but the very best properties are obtained very close to this limit.

The heat treatment consists in cooling from the solution temperature of about 1250°C to black at an average speed of 1.2°C $sec^{-1}$. A magnetic field is applied in what is to be the preferred direction of magnetization. At one time it was thought that the field operated as the magnet cooled through its Curie point. Actually, what is important is that the field should operate, while the spinodal decomposition takes place. By experiment it was found that the field need only be switched on while the magnets cool from 850 to 750°C. It is of course necessary that the alloy is magnetic while the spinodal dissociation occurs, and field cooling is ineffective in alloys with a lower cobalt content, because the Curie point is lower than the temperature at which the spinodal decomposition begins. Provided $\gamma$ phase has not been precipitated the treatment can be started at a temperature between 900 and 950°C. Whenever possible this lower temperature is chosen as it produces less scale and is less costly in fuel and furnace elements.

A brief description of some experiments carried out in our laboratory by P. L. Burkinshaw may illustrate the nature of the $\gamma$ phase problem. Samples of size 10 × 10 × 10 mm with a thermocouple attached were heated to 1250°C for 15 min, and then blasted down to a predetermined temperature at which they were held for various lengths of time and were finally blasted to room temperature. The samples were polished and examined by means of an optical microscope, a spot counting technique being used to assess the proportion of $\gamma$ phase present. This proportion of $\gamma$ phase is shown by means of time, temperature, transformation curves for four different alloys in Fig.4.3. The rate at which the transformation occurs, and also the temperature at which it occurs most quickly is sensitive to small changes in the alloy composition. Some of the changes are rather confusing. Thus Ti increases the rate of precipitation at temperatures below 1000°C, but reduces the precipitation at higher temperatures. This complication was discovered because of an apparent contradiction between Turelli and Leopardi (1971), who reported that Ti increased the tendency for $\gamma$ formation, whereas we believed it to be a $\gamma$ suppressor.

The best advice is to avoid composition combinations that are found to induce the $\gamma$ phase precipitation, and employ procedures that permit quick cooling of the castings. The castings should for instance be removed quickly from the moulding material.

The magnetic field may be applied by means of an electromagnet, a solenoid or sometimes even a large permanent magnet. A minimum value of 80 $kAm^{-1}$ (1000 Oe) is recommended, and as the magnet being treated is not usually in a closed circuit, some allowance for its self-demagnetizing field should be made. Occasionally horseshoe or C-shaped magnets are treated by allowing them to cool, while placed on a single bar carrying a current of several thousand ampere. In this method AC has been tried; it works but is not quite so effective as DC. Presumably the direction of the field is unimportant, but with AC there is a short dead time while the current is reversing.

Following the field treatment the magnets require a fairly long treatment at temperatures around 600°C. Formerly we used a two stage treatment in the laboratory of 48 hr at 590°C, followed by 48 hr at 560°C. This procedure was satisfactory, but although it was occasionally used industrially, most manufacturers preferred something shorter. Various programmes can be adopted, and it is probable now that most magnet makers have developed their own method. We found in the laboratory that 2 hr at 625°C, 6 hr continuous cooling from 625°C to 560°C followed by 16 hr at 560°C gave satisfactory results. This treatment can be carried out in one working day followed by an overnight treatment that requires no supervision.

Three methods have been used to increase the coercivity of anisotropic alloys without increasing the Co content. The resulting alloys are classified in Appendix 2 as A3. Merely by increasing the Nb content to 2 to 3%, the coercivity can be raised to 60 $kAm^{-1}$ (720 Oe) but with a $(BH)_{max}$ reduced to 36 kJ $m^{-3}$ (4.5 MGOe). This is the alloy that the PMA listed as Alcomax IV. Before the use of Nb was discovered the PMA, as already mentioned, developed an alloy with 9% Al, 21% Ni, 20% Co and 2% Cu. This alloy had a coercivity of 67 $kAm^{-1}$ (830 Oe), but a $(BH)_{max}$ of only 26.2 kJ $m^{-3}$ (3.3 MGOe).

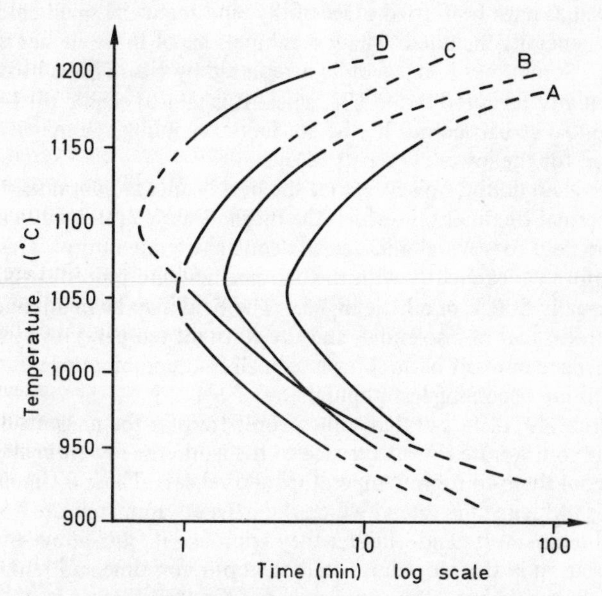


*Fig. 4.3. TTT curves for 1% γ precipitation in Al–Ni–Co alloys.*

| | *A* | *B* | *C* | *D* |
|---|---|---|---|---|
| *Ni/wt %* | *13.8* | *13.6* | *15.25* | *14.75* |
| *Ti/wt %* | *0.56* | *–* | *0.56* | *–* |

This alloy was called Hycomax before Hycomax II with a better $(BH)_{max}$ and $H_c$ was developed. It may still be made occasionally as Hycomax I. In the USA alloys with coercivities between 60 and 80 $kAm^{-1}$ (750 and 1000 Oe) are classified as Alnico VI. The increased coercivity is obtained by adding Ti and increasing the Ni content.

### 4.2.3 High coercivity Al–Ni–Co alloys

The attempts to increase the coercivity of Al–Ni–Co alloys by various additions without increasing the Co content were all accompanied by a reduced $(BH)_{max}$. By increasing the Co content at the same time as adding Ti and possibly Nb, it is possible to obtain alloys with coercivities ranging from 80 to over 160 $kAm^{-1}$ (1000 to 2000 Oe) and even obtain a slightly increased $(BH)_{max}$. In Appendix 2 these alloys are placed in two classes; A4 with not more than about 35% Co and 5% Ti and coercivities up to about 128 $kAm^{-1}$ (1000 to 1600 Oe), and A5 with up to 40% Co and 8% Ti and higher coercivities. Typical combinations of Co and Ti are: 30% Co, 5% Ti, 35% Co and 5% Ti and 38 to 40% Co and 8% Ti. Ni is usually kept at about 13 to 14%, although higher values have been tried successfully, and the usual small amount of Cu is generally included. Many combinations of these elements are possible. Sometimes 1 or 2% of Ti is replaced by Nb. This substitution is particularly favoured in the UK, where it is said to enable the heat treatment to be carried out by the continuous cooling in a magnetic field used for the lower coercivity alloys.

There is no doubt, however, that the best results are obtained with an isothermal treatment, in which the magnets are maintained in a magnetic field for several minutes at a constant temperature. This temperature varies slightly with the composition and procedure adopted, but is usually 800°C or a little higher. The field may be maintained by an electromagnet or a solenoid, and the constant temperature by an air filled furnace or a salt bath. I have seen all four combinations used but salt baths are becoming less popular.

Practice also differs in that some people transfer the magnets directly from the solution treatment furnace to the isothermal equipment, while others cool them to room temperature and reheat. These different methods influence the rate at which the magnets approach the isothermal temperature and whether they approach it from above or below. Any variation in the procedure alters the optimum time and temperature of the treatment. The actual value of the temperature is very critical, and a departure of as little as 10 or 20°C is sufficient to produce extremely poor results. Too low a temperature often produces demagnetization curves that are concave upwards.

A tempering treatment is necessary as with the standard alloys of moderate coercivity, although the proportion of the coercivity that accrues during this stage appears to be rather less than with the standard A2 class of alloys. A typical tempering cycle is 4 hr at 640°C plus 16 hr at 570°C. It is possible that a minimal advantage may be obtained by an additional period at a lower temperature such as 520°C.

The precise cycle of operations has to be worked out by the manufacturer himself to suit the composition used and the equipment available.

A council of perfection is that the magnetic field should be applied at any time that the magnet is at a temperature between 750 and 850°C, that it should be as high as possible, certainly three or four times greater than was recommended for normal alloys in the last section, and that one should avoid if possible cooling the magnets to room temperature between different stages of the treatment.

The high coercivity alloys are certainly very brittle and liable to breakage, which sometimes occurs spontaneously. These difficulties are particularly serious with the 40% Co, 8% Ti alloys with coercivities of 160 $kAm^{-1}$ (2000 Oe) or more, and in consequence not many manufacturers will tackle this grade. There is some evidence, although it is perhaps not conclusive, that breakages are reduced by avoiding intermediate cooling to room temperature.

The high coercivity alloys like the other Al–Ni–Co alloys may be sintered instead of cast. Some manufacturers favour giving small sintered magnets of these expensive alloys a solution treatment in a protective atmosphere. They claim that the better finish means that less material is lost in grinding, and saves more than the cost of the atmosphere.

### 4.2.4 Columnar Al–Ni–Co alloys

It occurred to Dr K. Hoselitz, the first Director of Research of the Permanent Magnet Association, that an improvement in the magnetic properties of the anisotropic Al–Ni–Co alloys might be obtained by producing them with a suitable crystal texture. What happened bore some resemblance to the fate of the earlier discovery of the benefit of cooling in a magnetic field by Oliver and Sheddon. An experiment was carried out which showed a marginal benefit which was considered insufficient to justify further work. A worth-while crystal texture was then achieved by Ebeling (1948) of the US General Electric Co. The desirable structure requires each cubic crystal of the alloy to have one ⟨100⟩ cube edge parallel to a given direction. This structure can be induced in the Ti free alloys during solidification by ensuring that heat flows from the molten alloy in this direction only. It is only practical to attempt this operation with simple shapes such as cylinders or cuboids.

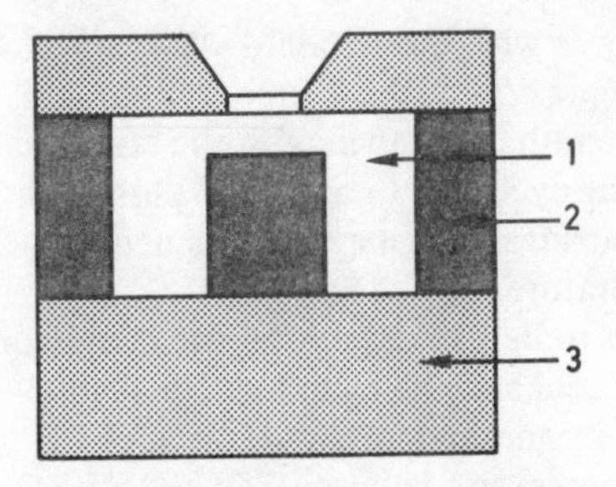


*Fig. 4.4. Mould pre-heated in furnace or by exothermic reaction for casting columnar magnets. (1) Mould cavity. (2) Hot mould. (3) Cold chill plate*

Radial heat flow may be prevented by casting into a hot mould, while heat is extracted from one end surface by a cold chill. A substantial block of steel makes an adequate chill. The method is illustrated schematically in Fig.4.4. Two main methods are in use for heating the mould. British manufacturers favour the use of a special mould composition heated by an exothermic reaction. Initially a material containing iron oxide and aluminium powder rather like thermite or an incendiary bomb was used. This material was already being supplied to foundries to prevent metal solidifying in the feeding heads and runners of castings, and in the first experiments it was ignited by contact with the hot metal. This method gave some good results but was a little erratic in operation. The recipe now preferred contains nitrate to oxidize the aluminium powder, and it is pre-ignited by a gas flame, an electric heater or something resembling a fuse, before filling the mould. This procedure is much more reliable, and also gives magnets with a better surface.

In the other method, which is certainly extensively used in America, the moulding sand is impregnated with binders based on silicates and $CO_2$, which enables it to be preheated in a furnace from which it is removed just before pouring in the hot metal.

There is little to choose between the magnetic properties obtained with the two methods, so the decision between them is made on the basis of cost and convenience. The exothermic moulds can be quickly made and used in any foundry without any very specialized equipment, while the other method requires the juxtaposition of a large furnace for heating moulds, close to the melting unit. Such an installation may be preferable for the large scale production of columnar magnets that justifies a special manufacturing unit. It is desirable to use rather hotter metal for columnar than ordinary magnets, and this suggests that it is desirable to arrange if possible to fill the moulds directly from the melting furnace.

Philips in Holland developed a method of making long bars 2 or 3 cm diameter of columnar material by continuous casting. Figure 4.5 is a sketch of a possible method of operating this process. Philips used this method for several years, but have now deleted columnar materials from their catalogues.

A method of making an equiaxed bar of an Al–Ni–Co alloy columnar by passing it through a zone, where a short length is melted, was developed in Japan. The zone melting can be performed with a high frequency coil and possibly a graphite susceptor. A sketch of this method is shown in Fig.4.6.

We succeeded in making columnar magnets experimentally by both zone melting and continuous casting, but it is not certain whether either method is now in use on an industrial scale.

Columnar magnets can be made from the 25% Co Ti free alloys by any of the above methods with or without Nb, although the relative improvement from making the alloys columnar seems to be greater in the alloys containing an appreciable amount of Nb. We have obtained in the laboratory much the same $(BH)_{max}$; about 68 kJ $m^{-3}$ or 8.5 MGOe with over 2% Nb and with less than 1% Nb, although in the non-columnar form the higher Nb alloy has about a 20% lower $(BH)_{max}$. The properties of commercial columnar materials, listed as material A2 col in Appendix 2 are of course somewhat lower than the above values.

Besides fully columnar materials, there is a logical justification for making semicolumnar material. Semicolumnar magnets are cast with a cold chill plate, but a normal mould material that is not preheated. Heat loss from the sides of the castings can be reduced slightly if the partitions of mould material between the castings are made as thin as possible without risk of them breaking down and allowing the individual castings to join. The process gives only a small improvement, but also costs little

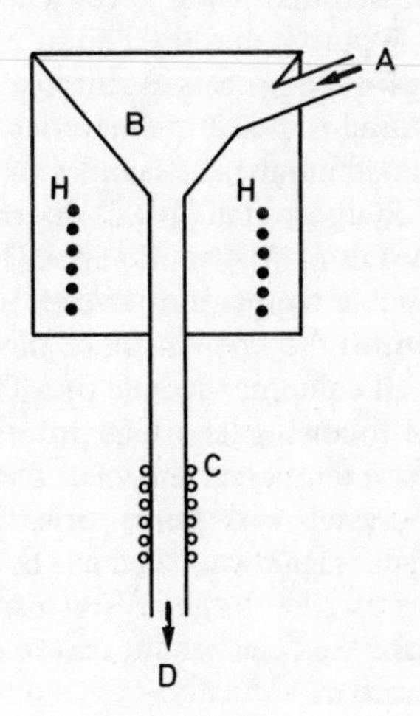


*Fig.4.5. Simplified sketch of continuous casting process. A: Raw materials fed in. B: Space in which ingredients are arc-melted in protective atmosphere. H: Molybdenum heater in refractory brick. C: Water cooled copper coil. D: Cast rod with columnar crystals withdrawn here*

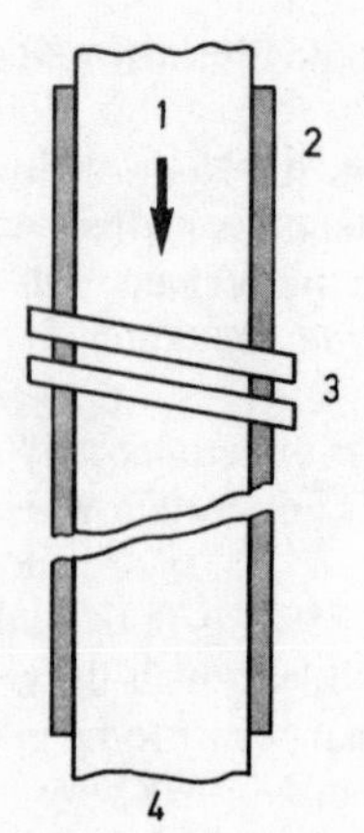


*Fig.4.6. Zone melting. (1) Bar to be zone melted. (2) Refractory tube. (3) HF heating coil. (4) Emerging bar cooled by air or water*

more than normal casting. English manufacturers designate this class of material by adding the letters SC after the name of the alloy, while American manufacturers use the letters DG (directed grains) for the same purpose.

The high coercivity alloys, containing appreciable quantities of Ti, solidify with a fine-grained structure, and present special difficulties in obtaining a columnar texture. At a meeting of the British Association in Sheffield in 1956, Gorter of the Philips Laboratories at Eindhoeven, orally announced success in producing a sample of such material with columnar properties. Shortly afterwards Luteijn and de Vos (1956) published the details. The values they obtained were: $B_r$ 1.18 T, $H_c$ 104.5 $kAm^{-1}$ (1315 Oe) and $(BH)_{max}$ = 87.5 kJ $m^{-3}$ (11 MGOe); they stressed that it was necessary to use raw materials of the greatest possible purity, and the Philips organization made it clear that they did not regard the process as suitable for commercial exploitation. They refused to make the material for sale, although they occasionally gave a small number of samples to academic research workers.

Steinort et al (1962) discovered a method of growing large single crystals of Al–Ni–Co alloys by a solid state reaction, induced by a suitable temperature cycle. It is, however, practically impossible to control the orientation of the crystals. We managed to produce a very small columnar sample of a Ti containing alloy in our Laboratory by the following laborious process. A sample was subjected to the appropriate temperature cycle. The orientation of a crystal or a small group of crystals with similar orientation was determined by X-rays, and a small magnet was then cut by means of a slitting wheel so that its length was parallel to the crystal axis. The labour costs of such a process would make the final product more costly than if it had been made of platinum.

Harrison (1962, 1966), working for one of the Members of the PMA solved the problem of making the high Ti alloys columnar. In view of the advice given by Luteijn and de Vos all the laboratory efforts to achieve this result had been made with high purity metals. Harrison's firm were making ordinary columnar alloys and also some non-columnar high coercivity Ti containing alloys. One day an experiment was tried in the factory of pouring some of the high coercivity alloy into an exothermic mould, and quite surprisingly good columnar material was obtained.

Harrison found that he was unable to repeat this result in the laboratory, and therefore investigated the difference between the commercial and laboratory material. He found that 0.2% of S was added to the commercial material with the object of reducing brittleness. How successful the sulphur is in reducing brittleness is open to question, but certainly 0.2% S is sufficient to enable good columnar material to be made with alloys containing up to 5% Ti. Any of the recognized methods that have been described for growing columnar crystals may be used.

Sulphur additions are not sufficient to enable columnar crystals to be grown in material with 8% Ti, but later Harrison discovered that even this more difficult task can be performed by using Te or Se. These materials do have some deleterious effects on the magnetic properties of the alloys, and in addition are rather unpleasant to use.

The Laboratory of the International Nickel Co developed a somewhat complicated procedure, which they named the Magnicol process for producing columnar structures in high Ti alloys. The process is described with a statement of the magnetic properties obtained by Palmer and Shaw (1969) and Dean and Mason (1969). The base charge of Fe, Co, Ni, and Cu is melted and deoxidized by the addition of about 0.05% of Si until a test sample solidifies with a flat top. Then 0.05% of carbon is added as ferrocarbon, and after 2 min Al, Ti, and S are added, the last named as ferrous sulphide. The melt is maintained at 1650°C for 10 min, and then poured into a mould pre-heated to 1150°C, but placed on a water cooled copper chill.

The procedure must be carried out with meticulous care, and there are also critical factors in the subsequent heat treatment. Difficulties must be expected in introducing the process in different foundry conditions. We were rather unsuccessful in our own laboratory in operating the process according to the written instructions, and succeeded only when one of the inventors came to instruct us in person. There is no doubt that the method can work, when it is properly operated, and that it eliminates the need to use Te.

The 5% Ti columnar alloys are listed in Appendix 2 as A4col, and are the best commercial Al–Ni–Co magnets available. It is difficult to find a manufacturer of columnar 8% Ti alloys. The best laboratory properties for any Al–Ni–Co alloy of which I am aware are quoted by de Vos (1966). Using a composition 35% Fe, 34.8% Co, 14.9% Ni, 7.5% Al, 5.4% Ti and 2.4% Cu, he obtained after an isothermal treatment of 10 min at 800°C, and tempering for 2 hr at 650°C + 20 hr at 585°C: $B_r = 1.15$ T, $H_c = 121.5$ kAm$^{-1}$ (1525 Oe), and $(BH)_{max}$ = 106.5 kJ m$^{-3}$ (13.4 MGOe).

### 4.2.5 Structure of anisotropic Al–Ni–Co alloys

Two body centred cubic phases with slightly differing lattice parameters were observed by X-ray diffraction in an anisotropic Al–Ni–Co alloy by Oliver and Goldschmidt as long ago as (1946). This experiment has been repeated by many different workers, but although they have measured the lattice parameters with much greater accuracy, X-ray measurements on their own cannot take our understanding of how permanent magnetism arises much further than was done by Oliver and Goldschmidt.

At that time the electron microscope was in its infancy, and no precipitate could be seen in the optical microscope, unless the alloy was spoilt magnetically by holding at a temperature between 750 and 800°C for much too long. It was therefore natural to suppose that the permanent magnets were in a state of incipient precipitation.

The Stoner and Wohlfarth shape anisotropy hypothesis requires the existence of two distinct phases with sharply differing magnetic properties. Gradually electron microscopy was improved and the existence of the two phases was confirmed. The first pictures were rather blurred and obtained only on material that had been at least a little over-treated. Ultimately quite clear electron micrographs were obtained on practically all the Al–Ni–Co alloys in the optimum permanent magnet state. Figure 4.7 is an example.

My own views on the processes occurring during the heat treatment of anisotropic Al–Ni–Co magnets are largely based on work done by Clegg and published by him and myself (1957). In this investigation saturation magnetization $\sigma$ and coercivity $H_c$ were measured as a function of temperature on a material of class A2 after various prior heat treatments. The work was partly inspired by two earlier unexplained observations on Al–Ni–Co alloys:

(1) Jellinghaus (1943) had reported that after cooling the alloy in a magnetic field but before final tempering there is a large positive temperature coefficient of coercivity.

(2) Hansen (1955) and van der Steeg and de Vos (1956) had observed that if a fully tempered alloy is reheated to a temperature in the range 650 to 750°C, the coercivity at room temperature is

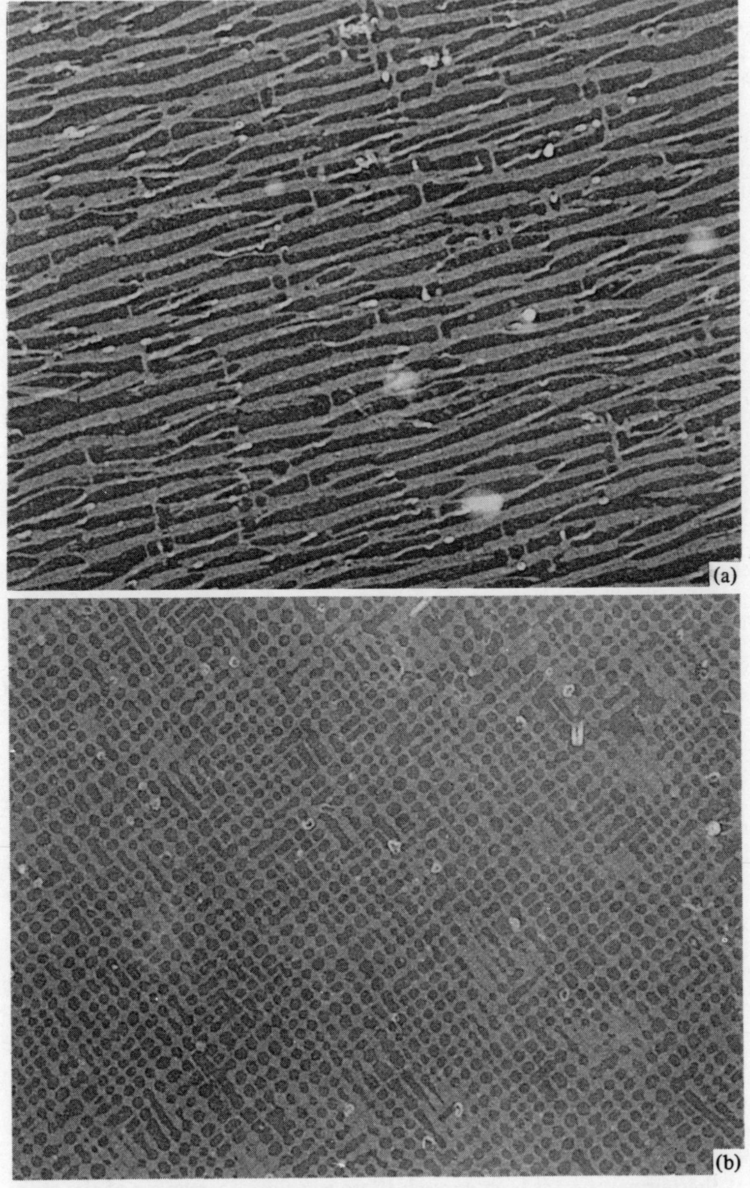


*Fig.4.7. Electron micrographs of the $\alpha + \alpha'$ structure of Alnico 8 (×50 000) after 9 min at 800°C. Refined replica technique gives greyish white oxide layer on 'Ni–Al' phase. (a) Parallel, (b) perpendicular to field direction. (Courtesy K. J. de Vos, Philips, Eindhoeven)*

reduced, but if the alloy is then retempered at a temperature below 600°C, the coercivity is partly or wholly restored.

Measurements of $\sigma$ as a function of temperature were made with a Sucksmith balance as described in Section 3.10.6, and coercivity temperature curves were measured by an extraction method, which involved having a fixed search coil inside a furnace, both being inside a solenoid. The results are shown in Figs.4.8 and 4.9.

In Fig.4.8. curve (a) is a $\sigma$, $T$ curve for a sample quenched from 1250°C. There is no sign of any subsidiary Curie point, the coercivity is very low and the sample is assumed to be single phase. Sample (b) was cooled at an average speed of 1.2°C $s^{-1}$ from 1250°C to 600°C in a magnetic field. As found by Jellinghaus the coercivity more than doubles as the temperature is raised from room temperature to 500°C. There are no subsidiary Curie points but the slope of the $\sigma$, $T$ curve between room temperature and 500°C has become distinctly steeper. At the time we had never heard of spinodal decomposition, but we said that the sample might contain a continuous variation of compositions with differing Curie points. This suggestion can now be identified with the sinusoidal fluctuation of compositions that is supposed to occur in the early stages of spinodal decomposition. This state can reasonably be described as incipient precipitation.

The completely tempered sample (c) has a $\sigma$, $T$ curve with no subsidiary Curie point and a coercivity that falls slightly with increasing temperature. Samples (d) and (e) have had treatments at 650 and 700°C respectively. Subsidiary Curie points now appear, and the coercivity temperature curves display maxima at roughly the same temperatures as these Curie points. The reduction in room temperature coercivity that results from treatment at 650°C, occurs almost wholly in the first hour.

Figure 4.9 shows that the spoiling that is induced by holding at 700°C can be largely restored by tempering at 590°C, but that the alloy can be spoilt again by heating again to 700°C. After long periods at temperatures in excess of 700°C, the magnetic properties can no longer be restored by retempering.

The above results are important in connection with the temperature stability of Al–Ni–Co alloys to be considered in Section 5.5.3, but here they are used to explain the metallurgical processes that control the magnetic properties of the alloy. After a magnet has been cooled in a magnetic field and the spinodal dissociation has been initiated, two processes can occur during a subsequent isothermal treatment of the alloy. If the temperature is high enough there is a slow coarsening of the structure. At 780°C this coarsening makes the structure visible in an optical microscope after two or three weeks. Probably this coarsening occurs also at lower temperatures but even more slowly. This coarsening process is irreversible, and once it has taken place the properties of the magnets cannot be restored without reheating them to the solution temperature and repeating the whole treatment.

Much more quickly than the coarsening process, the material acquires a two phase equilibrium composition, which is characteristic of the temperature of the treatment. One phase is rich in Fe and Co and strongly magnetic, the other is rich in Ni and Al and weakly magnetic.

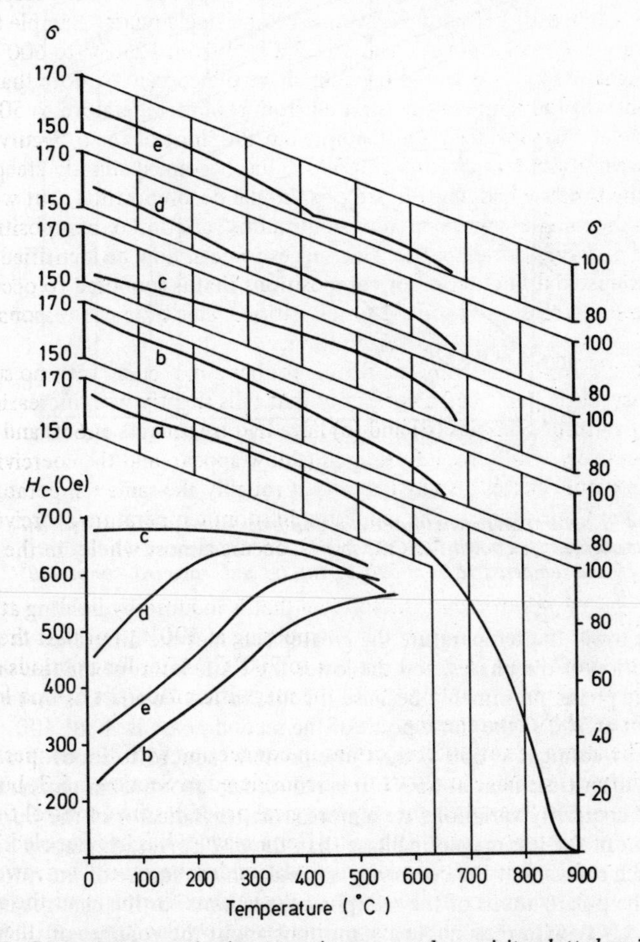


*Fig.4.8. Saturation magnetization per unit mass σ and coercivity plotted against temperature. (a) Quenched 1250°C. (b) Cooled 1.2°Cs$^{-1}$ from 1250°C. (c) Tempered 48 hr at 590°C + 48 hr at 560°C. (d) Tempered 48 hr at 650°C. (e) Tempered 4 hr at 700°C.*

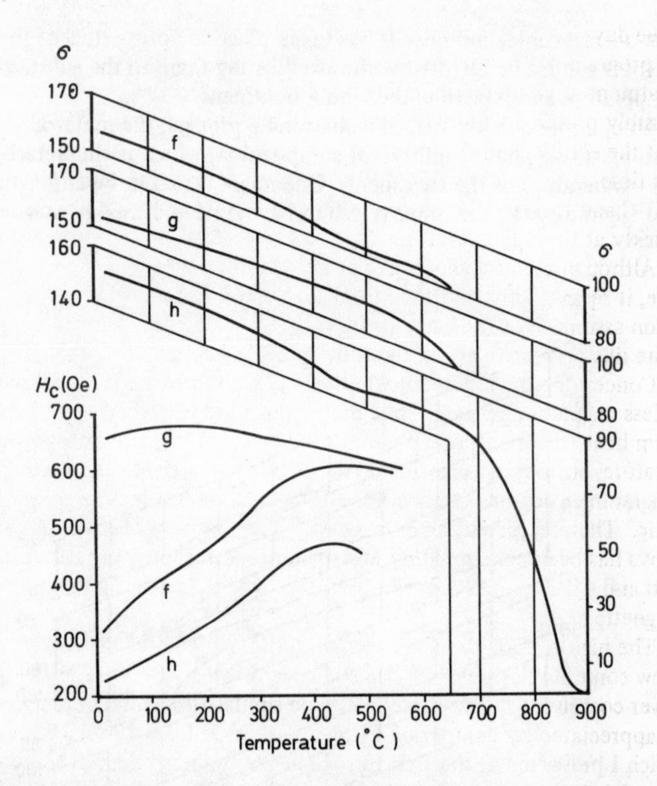


*Fig.4.9. Saturation magnetization for unit mass σ and coercivity plotted against temperature. (f) Cooled 1.2°Cs$^{-1}$ from 1250°C, tempered 4 hr at 700°C. (g), (f) and tempered 16 hr at 590°C. (h), (g) and tempered 4 hr at 700°C*

The lower the temperature the greater becomes the difference in composition of the phases, and the lower the Curie point of the less magnetic phase, presumably because it contains less Fe and Co. After treatment at 700°C the Curie point of the second phase is about 400°C, after treatment at 650°C the Curie point is reduced to about 200°C, and after treatment at 560°C it is reduced to about room temperature. The coercivity variations are a more sensitive indicator of the Curie point of the less magnetic phase than the $\sigma$, $T$ curves. Equation 3.29, which shows that the coercivity depends on the square of the difference in the polarizations of the two phases, accounts for the changes in coercivity with reasonable assumptions about the volumes of the phases.

At temperatures of 650°C and higher the equilibrium compositions are reached within an hour or so, whether the material has been treated at lower temperatures or not. At temperatures below 600°C two or three days may be required to reach this equilibrium state. Most practical programmes for the tempering of these alloys involve a step-wise treatment at gradually reducing temperatures. Such programmes may possibly produce a more complicated structure, but it is quite probable that the equilibrium composition at say 560°C can be reached in overall less time, if the alloy is first put into the state corresponding to say 620°C and then 590°C, since the diffusion processes take place more quickly at these temperatures.

Although these changes take place with little coarsening of the structure, it seems likely that the initial 'sine wave' fluctuations in composition assumed for the early stages of spinodal decomposition, become more like a square wave in the fully tempered magnets.

Concerning the argument whether the second phase is non-magnetic or less magnetic, my opinion is that if the alloy has received the optimum heat treatment, one phase should be non-magnetic at room temperature, although it is probable that some industrial heat treatment programmes designed to save time and costs may leave it slightly magnetic. The presence of a paramagnetic phase in fully treated Al–Ni–Co alloys has been confirmed by Mössbauer spectroscopy; see for example Asti and Criscouli (1970), who find that this phase has become ferromagnetic at 85 K.

The high coercivity alloys containing Ti and an increased Co content show some similarities, but also some signfiicant differences from the lower coercivity Ti free materials. The similarities and differences can be appreciated by comparing Fig.4.10, which is taken from a paper which I presented at the first European Conference on Hard Magnetic Materials in 1965 (McCaig 1966a) with Figs.4.8 and 4.9. Figure 4.10 refers to an alloy with 39.6% Co and 7.5% Ti. The quenched alloy (a) still has a low coercivity but the $\sigma$, $T$ curve has an inflection at about 300°C. The isothermally treated alloy (b) also has an inflection at about this temperature. The significance of these inflections is very different, as shown in Fig.4.11, which shows $\sigma$, $T$ curves for heating to 500°C and cooling again. For the isothermally treated material (b) the heating and cooling curves coincide, and the inflection clearly indicates a Curie point. For the quenched material (a) the heating and cooling curves diverge, and the inflection clearly shows a phase change the nature of which has not been elucidated.

In absolute terms the increase in coercivity as the Curie point is passed in material (b) is similar to that in field treated 25% Co alloys, but the relative increase is much less, because even the material in which both phases are magnetic at room temperature has already a coercivity of 80 $kAm^{-1}$ (1000 Oe).

Work on these and other Al–Ni–Co alloys is reviewed by de Vos (1966), who includes many beautiful electronmicrographs, and also by Gould (1971). Both authors give many references.

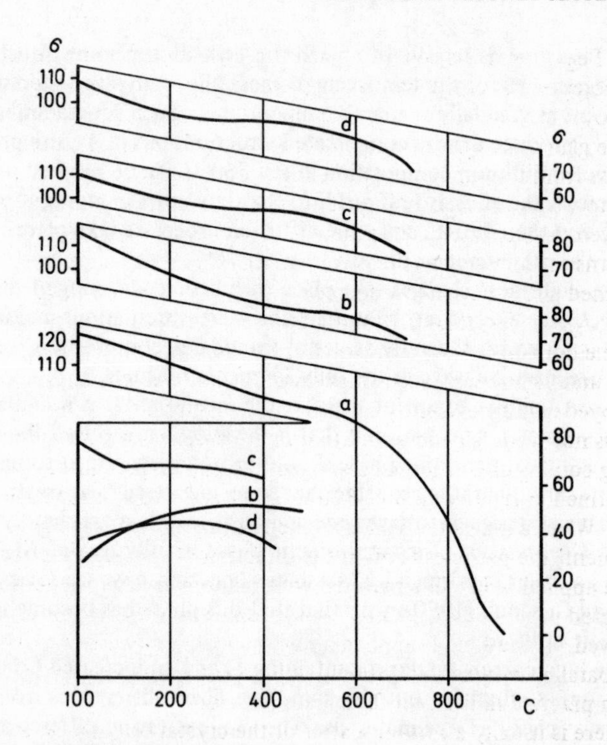


*Fig.4.10. σ, T and $H_C$, T curves for high coercivity Al–Ni–Co. (a) 1250°C quenched. (b) 825°C, 12 min. (c) 48 hr at 590 + 48 hr at 560°C. (d) 1250°C quenched + 700°C quenched*

Reasons that have been put forward to explain the higher coercivity of the Ti alloys are numerous. Experimentally torque curves show the apparent anisotropy constants are about 150% higher in these alloys than in those without Ti. Electron micrographs suggest that the magnetic particles may be more elongated in the Ti alloys, and also that the volume of the non-magnetic phase may be greater, but Gould (1971) quotes Bulygina and Seregev (1969) as regarding these causes as inadequate to account for the difference in coercivities and Granovoskii et al (1967) as suggesting a contribution to the coercivity from stresses. Paine and Luborsky (1960) suggested that the coercivity of the 25% Co alloy is reduced by an excessive number of cross-ties in the structure of the alloys in the permanent magnet state. It is relevant to remark that we are seeking to explain why the 25% Co alloys have coercivities so much lower than predicted, rather than why the high Co and Ti alloys have such high coercivities.

Higuchi (1966) suggested that Ti produces a third ϵ phase of composition $Fe_2Ti$, which is coherent with the less magnetic α phase, and that this increases the coercivity, but he was adding Ti to the standard 25% Co alloy.

The glamour of the rare-earth-cobalt alloys has rather distracted interest from the fundamentals of the Al–Ni–Co system, and until I came to write this section I must admit that I had forgotten how many unsolved problems remain in the Al–Ni–Co system. One such problem concerns the alloys quenched from 1250°C. It is not surprising that the quenched alloys have a low coercivity, but they also have a low remanence; $J_r/J_s$ is less than 0.1 (McCaig 1957). In this paper I suggested that the domains must form closed rings, but I am only half satisfied that I understand why such rings should form, and why they should be destroyed as they evidently must be by tempering.

It is now generally accepted that in the absence of a magnetic field during cooling or isothermal treatment, elongated precipitate particles are formed at random in the [100], [010] and [001] cube edge directions. When a magnetic field is applied during the initial stage of heat treatment, the particles form in the direction of the nearest cube edge to the applied field. In a material with random crystal orientation the predicted value of $J_r/J_s$ is 0.837, but often rather higher values are observed. If the field is applied along the columnar axis of a material with parallel crystals, $J_r/J_s$ should with a perfect structure reach 1.0, and in practice values over 0.95 are obtained.

There is usually a certain scatter of the crystal orientation in columnar magnets, and this produces small reductions in $B_r$ and $H_c$ and rather larger reductions in $(BH)_{max}$ (McCaig and Wright 1960).

The reason why it is necessary to add sulphur to the Ti containing alloys in order to grow columnar crystals has led to considerable speculation. One explanation involved the solidus liquidus temperature gap and constitutional super-cooling. Under certain conditions it is supposed that solid crystals, probably with a composition different from that of the alloy as a whole, may be nucleated in front of the advancing solid surface, and the orientation of such crystals is expected to be random. In view of the importance of small amounts of oxygen and carbon in the Magnicol process, it seems more likely that the important criterion for columnar growth is the absence of nuclei capable of initiating the growth of misaligned crystals.

In normal casting procedure some columnar growth occurs inwards from the sides as well as the ends of any specimen. It can be shown mathematically that the columnar crystals growing in from the sides have their cube edges on the average slightly nearer the direction of magnetization than would be the case with completely random crystals. Without going into these mathematical details it may be pointed out that a completely random direction may be up to $\cos^{-1}(1/\sqrt{3})$ or about 53° from the nearest cube edge. For the crystals grown in from a side surface the direction of magnetization lies in a {100} plane and cannot be more than 45° from a cube edge.

### 4.2.6 Pressed Al–Ni–Co magnets

Al–Ni–Co alloys can be pulverized and the powder hot pressed with some plastic binding material. The pressure required is between 1 and 4 kbar, and the pressing is carried out at a temperature at which the binding material is soft. There are firms operating the process in Germany, France and Japan. Various grades of material are made, but $(BH)_{max}$ does not appear to exceed 8 kJ $m^{-3}$ (1.0 MGOe) in isotropic material. The process is justified as a cheap method of producing magnets that would be expensive or fragile if they were cast or sintered. The commercial materials were originally isotropic, although it is now claimed that anisotropic samples can be made, by crushing magnets that have been field treated, and pressing the resulting powder in a magnetic field. Further details of these pressed magnets are given in Schüler and Brinkmann (1970).

## 4.3 Cr–Fe–Co ALLOYS

Alloys of Cr–Fe–Co are considered at this stage because of their similarity to Al–Ni–Co and because of their possible importance in the future, rather than their actual importance at present. Kaneko, Homma and Nakamura announced the discovery of this permanent magnet alloy system at the Chicago Conference on Magnetism and Magnetic materials (1972). They had already given papers on the subject at conferences in Japan in October 1970 and April 1971. They attributed the permanent magnet properties to a spinodal decomposition as in the Al–Ni–Co alloys. The composition suggested was 30% Cr, 25% Co and balance Fe with optional additions of Mn or Si, so that the alloy may be compared with Al–Ni–Co with the Al, Ni and Cu replaced by Cr. The heat treatment recommended was solution treatment at 1300 to 1350°C in argon, followed by a rapid quench, 30 min at 630 to 640°C in a magnetic field, and further treatments without a field for up to 6 hr at 600 and 560°C. The best properties claimed were $B_r$ 1.3 T, $H_c$ = 46 kAm$^{-1}$ (580 Oe) and $(BH)_{max}$ = 42 kJ $m^{-3}$ (5.3 MGOe).

There are a number of important differences between Cr–Fe–Co and Al–Ni–Co alloys. In favour of Cr–Fe–Co the raw materials required are cheaper, the magnetic field treatment is carried out at a lower temperature, and after solution treatment it may be forged or machined. Against the use of Cr–Fe–Co, the temperature of the solution treatment is high, and the quenching from this temperature must be very rapid to prevent the alloy being spoilt by the formation of an injurious $\gamma$ phase. In order to permit this rapid quenching much of the development work on Cr–Fe–Co has been carried out on samples with sections of diameter less than 10 mm.

A claim to have improved $(BH)_{max}$ up to 49 kJ $m^{-3}$ (6.2 MGOe) by a 50% reduction in cross-section by working after the field treatment, but before the final tempering has been made. This claim is made by Inoue Japax Research Inc in a French patent (1972), but a German patent by the same organization (1971), which is otherwise identical, does not include this claim.

Higuchi, Kamiya, and Suzuki (1974) claim to have made a 500 kg cast with $(BH)_{max}$ values in the range 40 to 46 kJ $m^{-3}$ (5.0 to 5.8 MGOe). The magnets were small, 10 mm long and 2.8 mm in diameter. These samples were swaged between field treating and ageing. Perhaps most significantly the composition was given as 27.5% Cr, 17.5% Co, 1% Si and balance Fe. With such a low Co content the alloy is even more competitive. Kaneko, Homma, Fukunaga and Okada (1975) have reduced the Co content even further to 15%, by adding 1% Nb and 1% Al. At the same time the solution temperature is reduced to 900°C. The magnets are, however, cooled quickly after casting. $(BH)_{max}$ values up to 40 kJ $m^{-3}$ are claimed.

Experiments in the P.M.A. laboratory, sometimes in collaboration with Members, confirmed many of the Japanese results, improved on some but failed to reproduce others. Magnets with $(BH)_{max}$ over 40 kJ $m^{-3}$ (5 MGOe) were obtained, but not as consistently as was

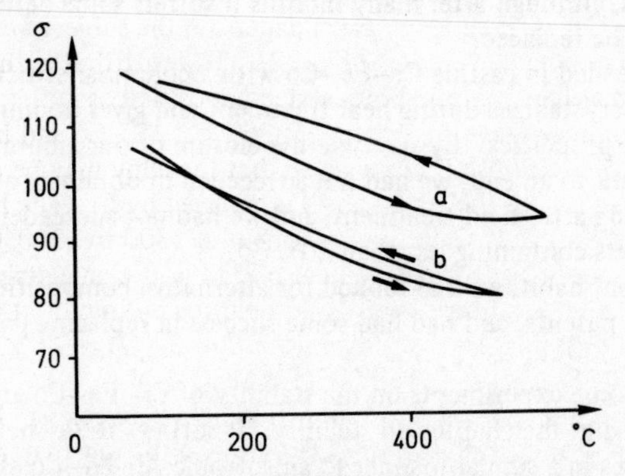


*Fig.4.11. σ, T curves, heating and cooling on same material as for Fig.4.10. (a) 1250°C not reversible, (b) 825°C reversible.*

desired. Sometimes two magnets from the same cast were treated together, but finished with quite different properties.

One discovery owed something to serendipity. It was desired to use ferrochrome, which is cheaper than Cr metal. The samples obtained were not only not permanent magnets, they were not even magnetic. Investigation revealed that the ferrochrome used had been purchased for a quite different purpose, and contained a deliberately large amount of nitride.

From this accident Wright, Johnson and Burkinshaw (1974) were led to investigate the influence of nitrogen that might be accidently picked up during melting. They found that nitrogen can easily be picked up in this way and greatly increases the formation of $\gamma$ phase. Conversely if precautions are taken to exclude nitrogen, the $\gamma$ phase is less likely to form, quenching does not need to be so rapid and larger magnets can be made. Rather than using costly methods such as vacuum melting, they recommend adding a nitride forming element; 0.5% of Ti is effective.

An interesting method of applying the magnetic field during heat treatment is to use Al–Ni–Co permanent magnets, preferably Columax. A magnetic circuit is built up of the magnets to be treated, blocks of Columax and steel. The circuit is magnetized after assembly, and is kept together for the whole of the heat treatment cycle from 640 to 560°C. There is certainly no harm to the Cr–Fe–Co as a result of maintaining the field at the lower temperature, although the method cannot be used with an intermediate working of the alloy. In consequence of the reversible processes explained in Section 4.2.5, the deterioration of the Columax suffered at 630 or 640°C is repaired by the treatment at the lower temperature. The Columax may be used many times, although after many months it suffers some deterioration and should be replaced.

We succeeded in casting Cr–Fe–Co with a columnar structure, but the alloy recrystallizes during heat treatment and gives no improvement in magnetic properties. By the time the closure of our laboratory brought work to an end, we had not succeeded in obtaining any benefit from working after field treatment, and we had not succeeded in making good magnets containing less than 20% Co.

As was our habit, we had looked for alternative compositions outside the original patents, and had had some success in replacing part of the Cr by V.

We did some experiments on the stability of Cr–Fe–Co and these are described in the chapter on stability. It suffices to say here that the alloy behaves in a similar manner to anisotropic Al–Ni–Co alloys and should be suitable for applications such as instruments that require good stability.

We found that isotropic Cr–Fe–Co alloys can be made with $(BH)_{max}$ over 20 kJ $m^{-3}$ (2.5 MGOe) quite consistently. This value is better than can be achieved with Al–Ni–Co.

It is rather surprising that so little interest has been shown in this alloy, which has already been shown to equal the normal Co alloys in $(BH)_{max}$. The raw materials are cheaper and the alloy is workable. Already by the addition of Mo, the coercivity can be increased to 80 $kAm^{-1}$ (1000 Oe). Admittedly there are certain difficulties in production, but it seems to the author that a tiny fraction of the enormous expenditure that is being incurred in research on rare-earth cobalt would solve these difficulties, and the potential economic reward is much greater. If Cr–Fe–Co had been developed before Al–Ni–Co, it is doubtful whether anyone would have wished to make the latter.

Very recently a single crystal with composition 30% Cr, 23% Co, 1% Si and balance Fe was prepared by Kaneko et al (1976). The properties claimed were: $B_r = 1.2$ T, $H_c = 83$ $kAm^{-1}$ (1040 Oe) and $(BH)_{max}$ 64 kJ $m^{-3}$ (8 MGOe). The method of preparation, which appears to involve some solid recrystallization, may not be immediately suitable for commercial use, but the result shows that the full potential of this alloy has not yet been realized.

## 4.4 FERRITES

In the present context the word ferrite implies a material, containing the ferrite group $Fe_2O_3$, and must not be confused with the steel making term, which describes a bcc phase. Natural lodestone has the formula $Fe_3O_4$, but can be regarded as $FeOFe_2O_3$ and is therefore ferrous ferrite. Ferrite magnets are also referred to as ceramic or oxide magnets.

The first artificially produced permanent magnet ferrite was cobalt ferrite. Its chemical formula is approximately $CoFe_2O_4$. Compared with metallic magnets it has a low $B_r$ and a high $H_c$, but compared with modern ferrites, its $B_r$ is rather high and its $H_c$ low. Sintered cobalt ferrite was made for a time in the USA, and resin bonded cobalt ferrite magnets were made in England just after the second world war. So far as I am aware cobalt ferrite is no longer manufactured for use in permanent magnets. There is some scientific interest in cobalt ferrite, because its properties can be influenced by treatment in a magnetic field.

During the 1940s the Philips organization developed many soft magnetic ferrites. These materials have a cubic structure and are sold as Ferroxcube for use in radio frequency transformers and other electronic applications. Besides Fe and O they contain metals such as Mn, Mg, Zn, Ni and Cu singly or in various combinations.

Probably as a result of their research into soft ferrites Philips in the early 1950s discovered hard ferrites based on Ba, Sr or Pb. These materials have the nominal formula $MFe_{12}O_{19}$ where M is Ba, Sr or Pb. This formula can be regarded as $MO.6(Fe_2O_3)$, but in practice better results can often be obtained with rather less iron oxide, say 5.5 $Fe_2O_3$, and some permanent magnet properties can be obtained over a fairly wide range of compositions, some involving different compounds.

It is possible to find some much earlier scientific papers that mention the magnetic properties of barium ferrite, but it is fair to say that the Philips discovery was a surprise to most magnet manufacturers. An exception is Krupp's in Germany who discovered the material independently.

### 4.4.1 Production of ferrite magnets

Ferrite permanent magnets are made in isotropic and anisotropic forms. The anisotropic ferrites are now made in grades with a high $B_r$ and $(BH)_{max}$ or with a high $H_c$.

The raw materials for making ferrite magnets are normally ferric oxide and the carbonate of Ba, Sr or Pb. Particle sizes and the amount and nature of impurities are important. The raw materials may be mixed in a wet or dry mixer. They are then calcined at a temperature between 1000°C and 1350°C. The actual temperature chosen depends on the raw materials and influences the subsequent treatment of the material. Some manufacturers have succeeded in making isotropic magnets with only one sintering operation, but although this method is cheap it is difficult to control, and the properties of the magnets are inferior. In the single sintering process, the raw materials must obviously be pressed in dies of the correct shape, but when it is intended to remill the powder, the raw materials are often formed into small pellets. The initial calcining is then economically performed as a continuous process in large kilns.

The calcined material is next crushed and milled, usually in a ball mill with water. The size of powder depends on the grade of magnets being made. Although of course many manufacturers carry out the whole process, there is now a considerable trade in the calcined and milled powder, which smaller manufacturers buy and fabricate into magnets. The powder manufacturers usually supply several different grades of powder for making different grades of magnets, and should supply the producer with advice on its subsequent treatment.

To make isotropic magnets the powder is dried and pressed. Pressing of the dried powder for isotropic magnets is a quick process that can be carried out in an automatic press.

To make anisotropic magnets, the powder is aligned by a magnetic field while it is in the die. The field should be applied in the same direction as that in which the pressure is applied. The grains tend to have the shape of discs, with their planes perpendicular to the hexagonal axis; there is consequently with this arrangement a small pressure effect that assists the magnetic alignment. The field can be applied by making the die walls of a non-magnetic material, and the plungers of a magnetic steel, and placing a large coil round the die. If a flux density of 0.5 to 1.0 T, can be measured in the die, when the gap length is the same as at the final size of the pressing, the field is probably adequate.

Anisotropic magnets can be made with the dried powder, and this is the most convenient method. Better results are obtained by filling the die with a slurry of powder and water. The water acts as a lubricant and permits better alignment of the powder in the magnetic field. Unfortunately this is a rather slow and messy process. It is necessary to provide grooves or holes for the water to escape and filter it through paper or other material. Consequently the pressing cycle may take between half a minute and a minute. Most manufacturers use a quite fluid slurry and feed the die automatically with a pump, but I have seen a stiffer slurry used, and in this case the die was filled by hand with a tool that looked like a builder's trowel.

Next the compacts are sintered at a temperature within the range 1100 to 1300°C. The heating and cooling must be carried out slowly to avoid cracking. These temperatures as are those for the first sintering are the ones quoted by Ireland (1968). The wide limits cover different raw materials, milling procedures and end products. In any particular production process a much narrower temperature control is necessary. Polgreen (1966) suggests that the initial firing should be at 1200°C and the final firing at a slightly higher temperature. These temperatures may form part of a successful process, but our own laboratory experiments show that many other combinations may be used. The higher the initial calcining temperature, the less reactive is the powder, and the higher is the necessary final sintering temperature.

Oxygen should be maintained at approximately the partial pressure at which it occurs in air, during the sintering operations. If the atmosphere becomes impoverished in oxygen, as may happen if there is any oxydizable material in an enclosed furnace, the magnets tend to lose oxygen with deleterious consequences to their properties. The evolution of $CO_2$ or water vapour may also drive out oxygen. The furnaces should therefore be well ventilated.

A wide range of magnetic properties can be produced by varying the heat treatment and milling sequence. The character of the magnets and the way in which they may be assembled, depend very much on whether $\mu_0 H_{ci}$ is greater than $B_r$ or not.

High sintering temperatures and light milling tend to produce high values of $B_r$ and $(BH)_{max}$ but rather low coercivities. Lower sintering temperatures and excessive milling tend to produce low densities, low $B_r$ and $(BH)_{max}$ but high coercivities.

In order to obtain good alignment in a magnetic field, milling must proceed until every grain is a single crystal. Statements are sometimes made that the grains must be single domain particles. This condition may be desirable for good coercivity, but is completely irrelevant for good alignment. Actually the finer the powder, the greater is the frictional or viscous resistance to alignment. Thus to make good anisotropic magnets the initial calcining must be carried on until crystal growth is sufficient to permit the production of single crystal powder that is not too fine to be aligned. In the final sintering process alignment tends to be improved by small misaligned grains being assimilated by larger aligned ones.

In the opposite sense a large coercivity requires the final material to have a fine grain structure. The grains should be less than or at least not too much greater than the critical size for single domain particles. This fine structure cannot be obtained if the powder from which the compacts are pressed is too coarse, or if the sintering is continued too long. The milled powder often has a rather low coercivity, which increases considerably during sintering. Possibly milling the powder to the required fineness produces defects in the crystal lattice, which are healed during sintering.

These considerations show why a high $B_r$ and a high $H_c$ are to some extent mutually exclusive. Many small additions have been claimed to produce a better combination of magnetic properties. In the early days these additions were made on a purely empirical basis. Some of them may have been beneficial in the context of the particular experiment, but only because they corrected some other fault in the process. The subject of additions is better discussed in connection with recent attempts to understand better the kinetics of the reactions.

Ferrite magnets are hard and as with alloy magnets the only practical method of machining is grinding. Unlike the alloys, ferrites can be successfully ground with diamond wheels. Although diamond wheels are initially expensive, they have a very long life and the overall cost is less than for the soft wheels necessary for Al–Ni–Co. The wear on diamond wheels is so small that the concave surface of segments can be ground with a wheel profiled to have the correct radius. Alternatively a cylindrical wheel with radius equal to the internal radius of the segments may be used.

### 4.4.2 Sr and Pb ferrites

Pb or Sr may be used instead of Ba to make permanent magnets. With Pb ferrite it is claimed that it is possible to obtain a certain amount of anisotropy by pressure alone. Presumably the platelet shape of the grains is more pronounced than in barium ferrite. It is also claimed in favour of Pb ferrite that the sintering temperature can be lower (900 to 950°C for the first and 1050 to 1100°C for the final sintering). The magnetic properties appear to be somewhat inferior with Pb ferrite, and there is also more public apprehension of the dangers of Pb poisoning. There does not appear to be any commercial manufacture of Pb ferrite, although it is possible that small amounts of Pb are sometimes added to other ferrites.

Philips in their original patents claimed that Sr could be used instead of Ba, but there is no evidence that they thought Sr was any better than Ba. Cochardt, who worked partly as a private entrepreneur in Germany and partly as an employee of the Westinghouse Corporation in Pittsburgh, carried out an enormous number of experiments and showed that Sr ferrite is in some respects superior to Ba ferrite; in particular it seems possible to obtain a better combination of a good $(BH)_{max}$ and a high coercivity with Sr.

The fact that many manufacturers are now using Sr is sufficient acknowledgement that Cochardt had made a genuine discovery. In support of his patent claims he stated that to obtain the best results from Sr ferrite some sulphate was desirable and that this could be economically achieved by using celestite, a cheap source of Sr that is mainly $SrSO_4$.

Krijtenburg (1965) of the Philips Laboratories, Eindhoeven, published a report of a Sr ferrite magnet that conatined no sulphate, and which had a $(BH)_{max}$ of 36 kJ m$^{-3}$ (4.8 MGOe). This still seems to be the highest $(BH)_{max}$ claimed for any ferrite magnet.

In Appendix 2, ferrite magnets are classified as follows: F1, sintered isotropic, F2, sintered anisotropic, high $B_r$, F3, sintered anisotropic, high $H_c$, F4, bonded isotropic, and F5, bonded anisotropic. Manufacturers rarely state whether Ba, Sr, or a mixture of the two is used. Recently some manufacturers have advertised strontium ferrite.

Good isotropic ferrite has a coercivity such that $\mu_0 H_{ci}$ is greater than $B_r$ and $\mu_0 H_c$ is only a little less. When $H_{ci}$ is measured or is stated it is usually such that $\mu_0 H_{ci}$ is greater than 0.3 T.

The ferrite magnets with the highest $(BH)_{max}$ have a very square intrinsic hysteresis loop. The coercivity is not usually greater than about 160 kAm$^{-1}$ and $\mu_0 H_{ci}$ at about 0.2 T is little more than half $B_r$. This material is normally built into assemblies and magnetized in situ like alloy magnets. Grades with higher coercivity can be used alone in quite short pieces and if built into magnetic circuits can be magnetized before assembly. The higher coercivity materials are made with a considerable range of properties. The last PMA Bulletin quoted $H_c$ as 240 $kAm^{-1}$ ($\mu_0 H_c = 0.3$ T). With this material $\mu_0 H_{ci}$ may not be greater than $B_r$, which is quoted as 0.37 T, but in America ceramic magnets with $\mu_0 H_{ci}$ greater than 0.4 T, but a considerably lower $B_r$ are made. Such properties are useful in motors and generators, in which the magnets are subjected to demagnetizing forces from armature reaction. With this range of properties and even with prices based on the use of fairly expensive synthetic iron oxide, ferrites became equal in importance to Al–Ni–Co alloys in the world market, although in Britain there are still more firms making alloy magnets. Recent advances in ferrite magnets have produced better combinations of $(BH)_{max}$ and $H_{ci}$ rather than higher values of either alone.

### 4.4.3 Temperature variation of anisotropy and coercivity of ferrites

Figure 4.12 shows curves of (i) $K_1/J_s$ the theoretical value of the coercivity of randomly aligned barium ferrite particles (ii) the experimental value for a fine grained unsintered compact (iii) $J_s(N_b - N_a)/2$ for platelets, and (iv) the difference between curves (i) and (iii), all plotted

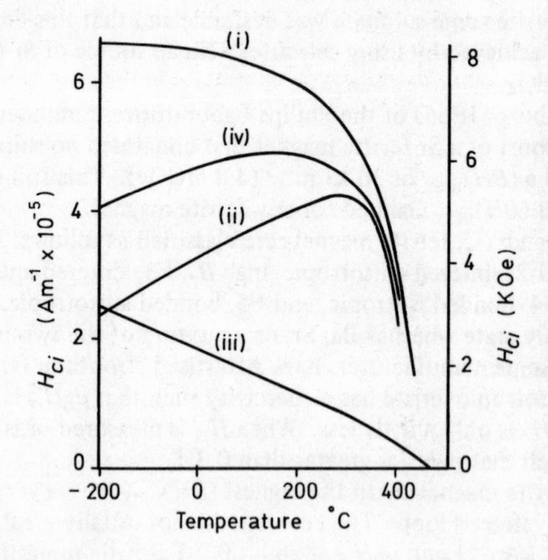


*Fig.4.12. Temperature variation of $H_c$ for $BaFe_{12}O_{19}$. (i) $K_1/J_s$ theory. (ii) Experimental value for fine powder. (iii) ½$(N_b - N_a)J_s$ or $2\pi(N_b - N_a)J_s$ in CGS units. (iv) Difference between curves (i) and (iii).*

against temperature. Two suggestions for the difference between curves (i) and (ii) have been made. Sixtus, Kronenberg and Tenzer (1956) suggested that since the barium ferrite powder consists of flat platelets, these platelets should make a negative shape anisotropy contribution plotted in curve (iii). This contribution is opposed to the crystal anisotropy contribution (i), and is therefore subtracted from it to obtain curve (iv), the shape of which is somewhat more like that of the experimental curve.

The alternative explanation suggested by Went et al (1952) is that the critical size for the formation of Bloch walls is proportional to $\sqrt{K_1}/J_s$. This quantity is plotted against temperature in Fig.4.13, and it is thus possible that Bloch walls form more easily in barium ferrite at low temperatures and account for the lower coercivity. This reduced coercivity, whatever its cause, is a fairly serious practical drawback in some grades of ferrite magnet, as will be seen in Section 5.5.5.

The coercivity of barium ferrite is well-maintained up to 95% of the theoretical density, as is characteristic of most materials that have high coercivities based on crystal anisotropy. Although the coercivities are less than the anisotropy fields, the differences are less than in many other materials.

### 4.4.4 Recent trends in ferrite manufacture

At the Third European Conference on Hard Magnetic Materials held in Amsterdam in 1974, 15 out of 67 papers could be classified under the title of this section. On listening to these papers being delivered, and even on reading them for the first time, I was inclined to think they were rather humdrum, and dealt with rather minor practical or scientific details of the process by which ferrite magnets are made. For example no magnetic properties as good as those reported at the first of these European Conferences in Vienna in 1965 were mentioned. On reading the papers a second time, I began to collate the papers with one another and with the existing practice described in Section 4.4.1, and I began to realize that some of these papers may disclose facts of considerable commercial importance. Just how important they are is difficult to assess because it is not easy to distinguish in papers of this nature processes that are suitable for factory production from those that can only be performed in experimental conditions in the laboratory.

One of the main objects of this work has been to reduce production costs. Steinort (1974) suggests that raw materials account for nearly a third of these costs. In the past, high quality synthetic iron oxide made for the paint industry, has been used both commercially and experimentally. Various additions have been tried with the object of improv-

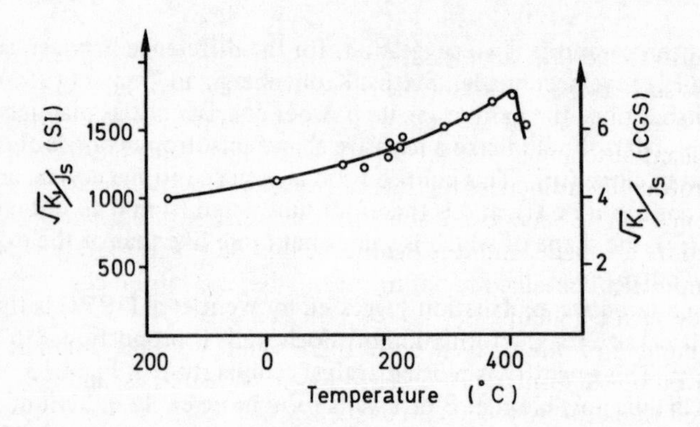


*Fig.4.13.* $\sqrt{K_1}/J_s$ *plotted against temperature (Went, et al 1952)*

ing the final product. Recently the accent has been on finding alternative sources of iron oxide, considering what impurities they contain, and modifying the sintering and milling procedure to correct for these impurities.

Ruthner iron oxide, reclaimed from HCl pickling baths costs about 40% of the price of synthetic oxide. Natural haematite contains about 0.5% of $SiO_2$ and 0.5% of $Al_2O_3$ and costs only 14% of the price of synthetic oxide. Steinort attributes to Alexander Cochardt 'about five years ago' the invention of the H and C process, using haematite and celestite. He gives no references but describes the process in some detail. The raw materials are haematite, celestite ($SrSO_4$) and soda ash ($Na_2CO_3$), Soluble $Na_2SO_4$ is formed and removed. Presumably this process presents difficulties, because Steinort went on to state that he was using 80% Ruthner iron oxide, but he is now hopeful of using a mixture of this and haematite.

As a background to this practical work there has been a more scientific study of the reactions by which barium or strontium hexaferrite is formed. Gadalla and Hennicke (1974) discuss the formation of various intermediate compounds such as the monoferrite. Work on these lines was started by Howard (1969) and Bye and Howard (1971) at Sheffield University in collaboration with the Permanent Magnet Association. Their methods included magnetic, gravimetric, infra-red spectroscopy, scanning electron microscope, X-ray and electron-probe analysis techniques. They found that the reaction proceeds in stages:

$$BaCO_3 + Fe_2O_3 \rightarrow BaOFe_2O_3 + CO_2$$

$$BaOFe_2O_3 + 5Fe_2O_3 \rightarrow Ba0.6(Fe_2O_3)$$

they also found that the rate of the reaction was greatly influenced by factors such as the texture of the iron oxide, degree of mixing and compaction and the size of the compacted agglomerates. Perhaps the most significant outcome is that the presence of a liquid phase during sintering profoundly influences the rate of reaction; Howard used LiF. The influence can be beneficial if the correct modifications in the sintering procedure are made, and it is significant that $SiO_2$ and $Al_2O_3$, which are impurities contained in natural haematite, are substances that may provide such a liquid phase.

A number of other papers on ferrite magnets were presented at the Third European Conference on Hard Magnetic Materials, and are referred to below by the names of the authors and page number only. The remainder of the reference is as for the papers by Steinort or Gadella and Hennicke above.

Krijtenburg and Stuijts (p.83) find that the best results are not obtained when sulphate is present, although they are not catastrophically bad. Additions of $SiO_2$ are on the other hand beneficial.

Van den Broek (p.53) gives a useful chart, showing the effect on $B_r$, $H_c$ and shrinkage of trying to correct errors in the initial presintering by subsequent modifications in the milling and final sintering procedure. He concludes that complete correction is not possible, so it is most important to use the optimum pre-sintering programme, which of course depends on the raw materials used.

Kools (p.98) compares methods of measuring particle size. Direct measurements of the particle size of fine magnetic powders are difficult because of their tendency to agglomerate. Control of particle size is clearly very important. The Fisher 'sub-sieve-sizer', which uses the rate of flow of air through a standard compact of the powder is commonly used. This air permeability method really messures the surface area of the particles, but the results are usually converted to a Fisher number, which is the calculated diameter of the particles assumed to be monosize spheres. Another instrument is the Areatron, in which gas adsorption is measured and the diameter given is obtained by assuming monosize platelets with diameter to thickness ratio of 3:1. These values are very different as shown by Fig.4.14. This figure is derived from Kools' paper and shows the particle size of powders as given by the Fisher and Areatron method plotted against the actual size measured by a scanning electron microscope. Kools concludes that the Fisher method is preferable.

Processes described in Section 4.4.1 included a method of making an inferior grade of isotropic barium ferrite in a single sintering process, and a method of making a slightly anisotropic ferrite (usually Pb ferrite) by pressure alone. The latter method is possible because of the disc shape of the presintered grains. It is rather surprising to find two papers that describe different methods of producing anisotropy in a single sintering process. These methods make use of topotactical reactions in which the crystal axes of the product are determined by those of one of the reagents.

Esper and Kaiser (p.106) pressed needle shaped iron hydroxide particles and strontium carbonate. Although the needle shaped particles take up a position at right angles to the pressing direction, the hexagonal axis of the sintered ferrite is parallel to the pressing direction. It appears that $B_r$ greater than 0.3 T can be obtained, but with a rather low coercivity. There is a longish discussion of difficulties, and it is improbable that the process is at present considered ripe for industrial development.

Stablein (p.110) achieved anisotropy, starting with a very differently shaped raw material. He used haematite with platelet shaped grains and rather high impurity contents: 1.6% $SiO_2$ and 0.9% $Al_2O_3$. He used pressures varying from 0.9 to 26.3 kbar, which he described as quasi-isostatic. $H_{ci}$ increases with sintering temperature up to 300 $kAm^{-1}$ ($\mu_0 H_{ci}$ = 0.375 T) at 1200°C. The best values of $(BH)_{max}$ quoted are 10.4 kJ $m^{-3}$ (1.3 MGOe) with a pressure of 0.9 kbar and 12.8 kJ $m^{-3}$ (1.6 MGOe) with a pressure of 26 kbar. Obviously 26 kbar can only be used in a laboratory process, but pressures well in excess of 0.9 kbar can be used commercially. It is an attractive proposition to make a material superior to normal isotropic barium ferrite in a single sintering operation from rather impure raw materials, provided there are no undisclosed snags.

If ferrites can be made appreciably cheaper, they are likely to make further progress relative to Al–Ni–Co magnets. The latter still have the advantage of a low temperature coefficient, but a ferrite with a reduced

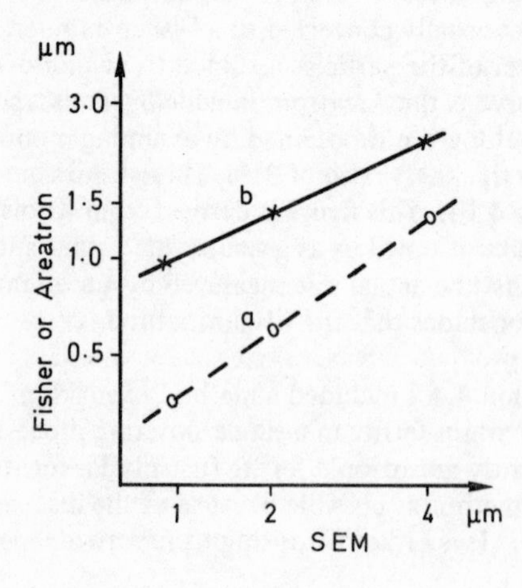


*Fig.4.14. (a) Fisher (b) Areatron particle size plotted against SEM value (Kools, 1974)*

temperature coefficient is described by Esper and Kaiser. This material, however, contains arsenic, and may be regarded as too dangerous for widespread use.

### 4.4.5 Bonded ferrite

Useful magnets can be made by bonding barium or strontium ferrite powder in various resins, plastics or natural rubber. The resulting products may be hard, but for many purposes it is useful to make them flexible.

If no special steps are taken the material is isotropic, and $(BH)_{max}$ is unlikely to exceed 5.5 kJ $m^{-3}$ (0.7 MGOe). If good flexibility is required the proportion of ferrite powder must not be too large, and $(BH)_{max}$ may be considerably lower.

A method of making anisotropic bonded magnets was developed by the Leyman Corporation (now taken over by the Minnesota Metallurgical and Mining Co.). It is understood that the anisotropy is produced by rolling the mixture of ferrite powder and natural or synthetic rubber into very thin sheets. A large number of sheets may be placed on top of one another and rolled again. Flexible anisotropic magnet material in sheet or strip form is now available from a number of manufacturers with $(BH)_{max}$ up to about 12 kJ $m^{-3}$ (1.5 MGOe). The preferred direction is normal to the plane of the sheet.

A number of scientific papers have described a method of making anisotropic bonded ferrite by aligning the magnetic particles with a magnetic field. The method is more likely to be used for rigid magnets.

## 4.5 RARE-EARTH MAGNETS

The term 'rare-earth' is misleading. The rare-earth elements were until recently rare and expensive, but the reason was that they were difficult and expensive to separate and refine, not that the earths from which they were obtained were scarce. Recently the difficulties in separating and reducing rare-earth metals have been largely overcome, and as a result quite suddenly and quite unexpectedly to the author, rare-earth magnets appear likely to compete successfully with older types even for some applications in mass-produced articles, where overall cost is a primary consideration.

Of course some rare-earths are more abundant than others. Mischmetal, which is obtained by fusion electrolysis of rare-earth salts without preliminary separation, contains about 50% Ce. Although cerium is usable in permanent magnets, it leads to rather inferior properties.

Samarium, which up to now has proved the most suitable rare-earth metal for making permanent magnets, comprises about 3% of normal misch-metal. It is thus not the most plentiful rare-earth, but it is not particularly scarce. Pr and Nd, which have also proved interesting, are rather more plentiful than Sm.

To emphasize that rare-earths are not rare earths, the words are in the remainder of this chapter hyphenated.

The rare-earth elements, atomic numbers 58 (Ce) to 71 (Lu) form a transition group similar to but more complicated than that containing Fe, Ni and Co (see Table 2.1). All the rare-earth elements have the stable electron core of 54 electrons as in Xe. They all have 2 electrons in the outermost 6*s* shell (quantum numbers $n = 6, l = 0$). These electrons are responsible for the similarities in the chemical bonding and behaviour of the rare-earth elements. In some of the rare-earth elements there is also a single 5*d* ($n = 5, l = 2$) electron that influences the valency, and in the solid state combines with 6*s* electrons to form conduction electrons.

In this transition group of rare-earth elements, the inner 4*f* ($n = 4$, $l = 3$) shell has different numbers of electrons from element to element. The electrons in this inner shell largely determine the magnetic behaviour, which consequently does vary significantly from element to element. The elements Y with atomic number 39 and La with atomic number 57 resemble the rare-earths proper in some respects, and particularly in their behaviour in magnetic alloys. La does of course immediately precede the rare-earth elements.

Orbital magnetic moments contribute significantly to the atomic moments of some rare-earth atoms. In the light rare-earth atoms, Ce to Eu, the orbital magnetic moment is greater than the spin moment, but antiparallel to it. In Gd the moment is due essentially to spin; it is the only rare-earth element that is ferromagnetic at room temperature, although some of the others are strongly magnetic at very low temperatures.

In the heavy rare-earth elements, Tb to Yb, atomic numbers 65 to 71, spin and orbital moments augment, leading to some very high values of $J_s$ at liquid helium temperatures.

The rare-earth elements have very different atomic radii from the transition elements, and consequently when one rare-earth is mixed with one transition metal, they tend to form a complicated range of stoichometric compounds, rather than solid solutions, although there is sometimes a limited range of solubility at high temperatures. Most pairs of one rare-earth and one transition element form at least half a dozen such intermetallic compounds, the compounds actually occurring differing slightly for different binary combinations. At present $RCo_5$ is the most important compound for permanent magnets, but small amounts of $R_2Co_{17}$, $R_2Co_7$, $R_5Co_{19}$ or $RCo_3$ may greatly influence the permanent magnet properties for better or worse. R represents one of the light rare-earth elements, or quite commonly a mixture of two or more such elements. Considerable effort is at present being directed to seeking permanent magnets based on $R_2Co_{17}$ compounds.

In all the $RCo_5$ compounds the cobalt and rare-earth spin moments are antiparallel. With the heavy rare-earths this means that the whole rare-earth moment is antiparallel to the cobalt and the result is to produce ferrimagnetic compounds with rather low saturation polarizations. In ferrimagnetic materials the magnetizations of the two sub-lattices vary independently with temperature, and anomalies in the magnetization temperature curves such as compensation points may occur. Hopes of combining the very high moments of these heavy rare-earths with that of cobalt have been disappointed.

With the light rare-earth elements, although the net moment of the rare-earth element is rather small (orbital minus spin moment) it does supplement the cobalt moment. Consequently useful, although not spectacular, saturation polarizations can be obtained. These polarizations are qualitatively accounted for by the sum of the cobalt and rare-earth moments, although not very accurately, since other sources of moment such as the conduction electrons may exist.

Although the spins of the cobalt and light rare-earth atoms are antiparallel, the moments on the rare-earth atoms act as whole units, and the $RCo_5$ compounds, when R is a light rare-earth element, behave as ferromagnetic rather than ferrimagnetic materials.

This explanation is rather brief, but there are several good accounts of the subject, including a book by Nesbitt and Wernick (1973). In order to leave room for the very considerable amount of work published in the last two or three years, matters that are discussed in this book are dealt with as concisely as possible. In particular, although some of the more important pioneers are named, details of references are not given if they can be found in Nesbitt and Wernick.

Some early work on $RCo_5$ was carried out in the Bell Laboratories by Nesbitt and co-workers, and in the US Naval Ordnance Laboratory by Hubbard and Adams about 1960. Since Gd was the most ferromagnetic rare-earth element, it was natural that $GdCo_5$ should be investigated. Unfortunately the absence of an orbital contribution to the moment of Gd combined with the antiferromagnetic coupling of Co and Gd spins, make $GdCo_5$ rather weakly magnetic. Thus the results did not appear very interesting and were not pursued for some time.

Strnat must be given the credit for drawing attention to the possibilities latent in rare-earth cobalt compounds. He must have been working for several years on rare-earth transition metal compounds at the US Air Force Laboratories. His publications began to appear in 1966, and a significant early one that Nesbitt and Wernick do not mention is

Strnat and Hoffer (1966). Strnat has since continued his work on rare-earth-cobalt magnets at the University of Drayton. Among other measurements Strnat extrapolated the field necessary to produce saturation in a number of $RCo_5$ compounds, and estimated the anisotropy fields and possible $(BH)_{max}$ values as shown in Table 4.1. The value of 23.0 MGOe (184 kJ $m^{-3}$) predicted for $SmCo_5$ has been slightly exceeded in the laboratory, and closely approached in production, probably because $J_s$ was slightly underestimated. The highest value of all for $PrCo_5$ has not been achieved although 26 MGOe (210 mJ $m^{-3}$) has been obtained with a mixture of $PrCo_5$ and $SmCo_5$. Strnat's predictions are remarkably accurate, particularly as at that time he had not made even a reasonably good permanent magnet.

The first real permanent magnets were made by Philips in Eindhoeven initially by bonding $SmCo_5$ powder in resin, and soon afterwards by using very high pressures. This work is described briefly in Section 4.5.1. Sintered magnets were announced by Das in 1969, and Martin and Benz by a slightly different process in the following year. This work together with the description of experiments on many different rare-earth metals is also described in Section 4.5.1. Parallel with the development of sintered magnets it was discovered independently at the Bell Laboratories in the US and by Tawara and Senno in Japan, that permanent magnets can be made by casting and heat treatment, provided they contain some copper as well as rare-earth metal and cobalt. This method is described rather fully by Nesbitt and Wernick, who developed the method, in their book. It does not appear to be very widely used at present, so this branch of the subject is summarized rather briefly in Section 4.5.2.1. Attempts to explain the behaviour of rare-earth cobalt magnets have been the subject of many papers and to assess this work is a rather daunting task attempted in Section 4.5.3. Finally I have attempted to assess the position of rare-earth cobalt magnets relative to other types of permanent magnet in Section 4.5.4.

### 4.5.1 $RCo_5$ powders and compacts

In the years immediately following Strnat's first publication, most experiments on $RCo_5$ magnets were concerned with powders, and unsintered compacts of such powders.

Certain difficulties are met and precautions are necessary in working with $RCo_5$ materials. The rare-earth metals attack most crucible materials. They can be melted in pure alumina crucibles, although there may be a small pick-up of aluminium. These crucibles may be coated with a wash of $Y_2O_3$. It has recently been reported that boron nitride crucibles have been used successfully. A protective atmosphere is necessary during

**Table 4.1** MAGNETIC AND PHYSICAL PROPERTIES OF $RCo_5$ COMPOUNDS AS MEASURED OR FORECAST BY STRNAT ET AL (1967)

| *Compound* | *Curie point* (°C) | *Liquidus* (°C) | *Peritectic* (°C) | *Density* (g cm$^{-3}$) | $J_s$ (T) | $\mu_0 H_a$ (ext) (T) | $K = \frac{1}{2} H_a J_s$ (MJ m$^{-3}$) | *Theoretical* $(BH)_{max}$ (kJ m$^{-3}$) |
|---|---|---|---|---|---|---|---|---|
| $YCo_5$ | 648 | ~1360 | 1352 | ~7.69 | 1.06 | ~13.0 | ~5.5 | 225 |
| $LaCo_5$ | 567 | ~1220 | 1090 | ~8.03 | 0.909 | ~17.5 | ~6.3 | 165 |
| $CeCo_5$ | 374 | ~1205 | 1196 | ~8.55 | 0.77 | 17–21 | 5.2–6.4 | 118 |
| $PrCo_5$ | 612 | ~1245 | 1232 | ~8.34 | 1.20 | 14.5–21 | 6.9–10 | 288 |
| $SmCo_5$ | 724 | ~1325 | 1320 | ~8.6 | 0.965 | 21–29 | 8.1–11.2 | 184 |
| $MCo_5$ | ~520 | – | ~1185 | ~8.35 | ~0.89 | 18–19.5 | 6.4–6.9 | 158 |

melting and heat treatment. The atmosphere may be of pure argon or helium; we found that commercial grade argon is unsatisfactory.

Presumably hydrogen has been tried as a protective atmosphere, because Zijlstra (1972) reported that it can be taken in by the alloy in considerable quantities, and cause a loss of coercivity. In fact with the related alloy $NiCo_5$ and a moderate pressure, this is suggested as an economical method of storing hydrogen.

No-one seems to have reported using nitrogen, and presumably it is expected to be unsuitable. Obviously it would be useless to try any but a high purity grade. As described later, there is a considerable evolution of heat and nitrogen when a fine powder is heated to 150°C.

As an alternative to melting in rather special crucibles, levitation melting can be used for small samples. Arc melting on a water cooled copper hearth is very satisfactory.

Fluid energy milling in an inert gas was described by Benz and Martin. Ball milling in an organic liquid such as toluene, benzene or a similar hydrocarbon is also quite successful. We used a method in which a mixture of the powder, ⅛ in diameter steel balls and a suitable mixture of pure hydrocarbons sold as petroleum ether, was stirred with a rotating paddle.

There is no advantage in excluding air completely during the milling process, unless it can be excluded until the samples are sintered. If air is excluded during milling, the powder is much more reactive, and may even become pyrophoric if it is later handled in air. Nevertheless even the powders that have not been milled without access to air, tend to deteriorate if they are kept in air for more than a few days.

It should be explained that if a rare-earth oxide $R_2O_3$ is formed, 0.1% of oxygen effectively ties up to 0.6% of rare-earth metal. Since it appears necessary to control the rare-earth content to about ±0.5%, the oxygen content needs to be controlled to about ±0.1%. Rare-earth cobalt magnets probably normally contain about 0.4% oxygen, and it seems easier to control the amount of oxygen at this level, than it would be at a lower one.

Many curves were published showing $H_{ci}$ as a function of the fineness of the $RCo_5$ powders. Powders with $\mu_0 H_{ci}$ in the range of 0.2 to 0.5 T were obtained by milling powders of $YCo_5$, $PrCo_5$, $LaCo_5$ and $MCo_5$, where M stands for misch-metal, the relatively cheap mixture of rare-earth metals, mainly Ce, La and Nd, manufactured for lighter flints. In many of these experiments the particle size of the powders is not clearly defined, and in some only the times of milling on some particular type of mill are given. A feature common to nearly all the curves is that $H_{ci}$ passes through a maximum, and then begins to decrease with further milling. This maximum is reached before the powder has reached the fineness predicted for single domain behaviour, by any method of estimation. The most probable cause of this fall in coercivity is some damage to the particles produced by excessive milling.

$SmCo_5$ is an exception; with this material a much higher coercivity ($\mu_0 H_{ci}$ = 1.7 T) can be attained, and this value appears to be approached asymptotically rather than passing through a maximum with further milling. By chemical removal of the damaged outer layer of the particles, $\mu_0 H_{ci}$ can be increased to 2.5 T.

The first $RCo_5$ magnet with properties comparable with those of the Al–Ni–Co alloys was reported at a Conference in London by Velge and Buschow (1967). This magnet was an aligned compact of $SmCo_5$ powder in a binder, and its properties were: $B_r$ = 0.578 T, $\mu_0 H_c$ = 0.513 T, $\mu_0 H_{ci}$ = 0.9 T, $(BH)_{max}$ 65 kJ $m^{-3}$ (8.12 MGOe). There is some doubt whether this particular sample was completely stable, but production of $RCo_5$ magnets in a polymer binder was started by a British firm in 1975. The composition of both the magnetic powder and binder are probably different from those of the original Philips compact. The properties of the bonded and sintered rare-earth magnets known to be offered commercially are included with the classification letter R in Appendix 2.

Soon after making the original bonded magnet, Buschow and co-workers succeeded in making pressed compacts without a binder, with $(BH)_{max}$ over 160 kJ $m^{-3}$ (20 MGOe). The powder was given a long and complicated treatment, and after alignment was subjected to a combination of isostatic and uniaxial pressure of 20 kbar or more. The cost of constructing dies for such pressures and the necessary safety precautions make the process unattractive for industrial exploitation. An interesting scientific point is that the highest $(BH)_{max}$ was obtained with a composition containing more Co than stoichiometric $SmCo_5$.

Das announced at the Intermag conference in Amsterdam in April 1969 that he had produced a sintered magnet with $(BH)_{max}$ equal to 160 kJ $m^{-3}$ (20 MGOe). In the printed version of the paper, he stated that the powder was pressed in a magnetic field, and sintered in an inert atmosphere for 1 hr at a temperature 'around 1000°C'. In a later paper Weihrauch and 6 others including Das (1973) recommend that the sintering should be carried out at a temperature between 1100 and 1130°C.

In the following year Martin and Benz of the General Electric Laboratory, Schenectady described a method of making magnets with similar properties by liquid phase sintering. Among other publications this discovery was included in a paper at the International Conference on Magnetism in Grenoble in 1970. We were told of this process in personal discussion with Martin, who visited our laboratory shortly after the Conference. The rare-earth cobalt is reduced to powder in a fluid energy mill, in which the powder is agitated by a vigorous flow of nitrogen. A feature of this method of milling is that it permits some control not only of the mean particle size, but also of the particle size distribution of the emerging powder. Two powders were prepared and mixed. One had the approximate composition $RCo_5$ and the other a much higher rare-earth content between 60 and 70 wt %. The powder mixture was then sealed in a rubber container, and aligned in the field of a superconducting magnet. The compact was then transferred still in the rubber container to an isostatic press and compacted with a pressure of about 13 kbar. After sintering, which the additive rich in Sm was believed to facilitate by producing a liquid phase, the magnets were magnetized in a 10 T superconducting solenoid.

Perhaps the most valuable information communicated to us by Martin was that the powder should not be milled too fine before making the compacts.

Up to this time all our efforts to sinter $RCo_5$ magnets had failed. Many of the compacts disintegrated in a violent exothermic reaction, as soon as they were heated to about 150°C (McCaig 1970). This reaction occurred whether the sintering was attempted in hydrogen, argon or a vacuum. The explanation is that during milling oxygen is physically adsorbed on the large surface area of the fine particles, and when the powder is heated to 150°C this oxygen reacts with the rare-earth metal, raising the temperature by several hundred degrees almost instantaneously.

In making barium ferrite magnets the powder is milled until the Fisher estimate of the particle size is 1 $\mu$m or less, and the coercivity has fallen well below its maximum value. We had been following a similar procedure with the rare-earth-cobalt powders, until we were advised by Martin that milling should be terminated before the powder had reached its maximum coercivity, and the Fisher number was still in the range 3 to 10 $\mu$m.

Within a few weeks of receiving this advice, Johnson and Fellows (1971) of our laboratory were producing reasonably good rare-earth-cobalt magnets. They used very simple apparatus. The powders were milled to a particle size less than 10 $\mu$m in a simple attrition mill. That is to say a mixture of powder, ⅛ in steel balls and petroleum ether (boiling point 60 to 80°C) was stirred by a stainless steel paddle, rotating at about 250 to 350 rpm in a stainless steel beaker. It is necessary to crush the powder fairly fine and remove the coarser particles by sieving, before using this type of mill. Alignment of the powder and also the final magnetization was carried out by a pulse technique, using one half cycle of the mains supply. An approach to isostatic pressing was imitated by surrounding the compact with a large plug of rubber that just filled the die cavity, the diameter of which was about three times that of the sample. The maximum pressure that could be exerted was about 7 kbar. With this crude equipment $SmCo_5$ magnets with $(BH)_{max}$ values up to about 150 kJ $m^{-3}$ (19 MGOe) were obtained, although it must be admitted that these properties were nor reproduced as consistently as desired.

Johnson and Fellows used a Sm rich additive as recommended by Martin and Benz. It is now considered doubtful whether such an addition produces much liquid phase, and whether this is of great significance. Mixing two powders of different compositions is, however a useful way of controlling and varying the final composition, which should be rather on the rare-earth side of stoichiometric $SmCo_5$.

Numerous workers have now made magnets with various rare-earth metals and cobalt, although frequently some Sm is added. Table 4.2 contains some of the best laboratory results with compositions. Production properties, given in Appendix 2, are of course lower.

Some of the best values have been obtained with magnets based largely on Pr. Johnson and Fellows (1974) used $PrCo_5$ as the base and a 73% Sm alloy as the additive. They produced a magnet with $(BH)_{max}$ 175 kJ m$^{-3}$ (22 MGOe), the highest value that we ever obtained, but found that this and a number of other magnets we made containing Pr were metallurgically and magnetically unstable. Some of these magnets that were not so good at first actually improved magnetically in the first few weeks, but nearly all eventually deteriorated magnetically, and many eventually disintegrated into a powder.

Although at the time of writing I have not seen this instability of $PrCo_5$ mentioned in print by other writers, I have been told that it has been observed. Possibly it is significant that interest in Pr alloys seems to have waned in the last year or so. Schweizer, Strnat and Tsui (1971) found that the sintering behaviour of $PrCo_5$ is complicated; $H_{ci}$ plotted against sintering temperatures shows peaks at 1050 and 1120°C with a minimum in between. They attempt to explain this behaviour in terms of the phase diagram. In spite of these difficulties, $PrCo_5$ has potentially better properties than $SmCo_5$, because of its superior value of $J_s$, and some manufacturers are successfully incorporating some Pr in their magnets.

Obviously great interest rests in the possibility of replacing Sm by the much cheaper Ce misch-metal. Martin and Benz have obtained excellent results with alloys that contained some misch-metal, but over half the rare-earth content was still Sm. Johnson and Fellows (1972) made a magnet that contained no rare-earth metal other than misch-metal, and had a $(BH)_{max}$ of 70 kJ m$^{-3}$ (8.8 MGOe). In unpublished work this value was shortly afterwards marginally surpassed, and for a time the PMA could claim to have made the best $RCo_5$ magnet containing no rare-earth metal other than misch-metal. As can be seen from Table 4.2 our values have now been well overtaken. Other establishments are possibly more interested in substituting as much misch-metal for samarium as possible, without any loss of properties, than in obtain-

Table 4.2 BEST LABORATORY PROPERTIES OF RARE-EARTH PERMANENT MAGNETS

| *Material* | $B_r$ (T) | $(BH)_{max}$ ($kJm^{-3}$) | $(BH)_{max}$ (MGOe) | $H_c$ ($kAm^{-1}$) | $\mu_0 H_c$ T |
|---|---|---|---|---|---|
| Co-rich $SmCo_5$ pressed compact[1] | 0.91 | 160 | 20 | 610 | 0.77 |
| $SmCo_5$ liquid phase sintered[2] | 0.92 | 191 | 24 | 690 | 0.77 |
| Sm–Pr–Co liquid phase sintered[3] | 1.03 | 207 | 26 | 805 | 1.01 |
| Misch-metal[4] $Co_5$ | 0.81 | 116 | 14.5 | 630 | 0.8 |
| $R_2TM_{17}$ (R = rare-earth, TM = transition metal[5] | 1.1 | 239 | 30 | 560 | 0.7 |

1. Buschow, Naastepad and Westendorp (1969)
2. Foner, McNiff jr., Martin and Benz (1972)
3. Charles, Martin, Valentine and Cech (1972)
4. Nagel (1974)
5. Bachmann, and Nagel (1976)

ing the best properties possible without using any samarium. In terms of raw materials cost alone the use of any Sm or other individual rare-earth metals should be avoided, but if production costs are high the choice of the most economical mixture is less clear cut.

Heat treatment following sintering can significantly alter the properties of rare-earth cobalt magnets. There is a certain temperature range that depends on the composition, but is usually between 600 and 800°C in which a drastic depreciation of properties occurs. It appears that in this temperature range a eutectoid decomposition of $RCo_5$ into $R_2Co_{17}$ and a rare-earth rich phase such as $R_2Co_7$ takes place. The rare-earth-rich phase may be beneficial, but the $R_2Co_{17}$ is extremely detrimental to the coercivity. The magnets should therefore be cooled sufficiently quickly through this temperature range to suppress this dissociation. On the other hand treatment at higher temperatures such as 900°C is sometimes beneficial, particularly according to Benz and Martin in magnets containing misch-metal.

Sintered magnets are usually made with a rare-earth content in excess of that in $RCo_5$. This excess guards against the formation of $R_2Co_{17}$ in the event of rare-earth metal being lost by evaporation (the boiling point of Sm is only about 1900°C) or differential oxidation of the rare-earth element. It appears also that just below the sintering temperature there is a certain solubility range of $RCo_5$, but on cooling to 900°C this solubility range contracts. If there is a small excess of rare-earth metal a little $R_5Co_{19}$, $R_2Co_3$ or $RCo_3$ may be precipitated. These phases are usually magnetic and some of them have a high crystal anisotropy. Ways in which a small quantity of such a precipitate may enhance the coercivity are discussed in Section 4.5.3.

A considerable number of firms are now offering rare-earth-cobalt magnets. In 1973 and 1974 we tested a number of such magnets and typical properties were: $B_r$ = 0.8 T, $H_c$ = 630 kAm$^{-1}$ ($\mu_0 H_c$ = 0.8 T), $(BH)_{max}$ = 125 kJ m$^{-3}$ (16 MGOe). Our apparatus did not permit an accurate measurement of $H_{ci}$, which would be much higher than $H_c$. A high value of $H_{ci}$ is important mainly for temperature stability which has been investigated by Mildrum, Hartings and Strnat (1974). This subject is discussed in more detail in Section 5.5.6. It is interesting to observe here that Mildrum and co-workers in connection with their stability project had to measure the properties of several hundred magnets, some made by die pressing and sintering a single powder and others by isostatically pressing two powders before sintering. They presented their results in the form of a histogram, as shown in Fig.4.15. The mean values were better by the second technique, but the spread of results was also greater. These results were obtained on samples supplied in 1972, and since then both production methods and properties have probably changed.

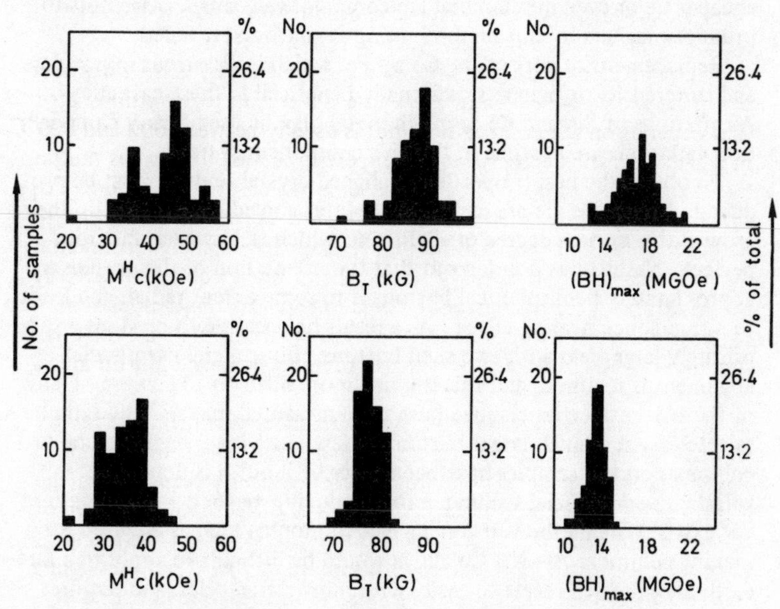


*Fig.4.15. Statistical distribution of magnetic properties of sintered $SmCo_5$. Top pressed isostatically. Bottom diepressed*

### 4.5.2 Cast rare-earth alloys containing copper

Alloys with compositions close to $RCo_5$ are magnetic, but have low coercivities in the as cast state. Permanent magnets can be made by casting and subsequent heat treatment, if copper is also present. That is to say the steps of reducing the material to a fine powder, and then sintering are eliminated.

Nesbitt and Wernick (1973), who discovered this process in 1968, describe it in more detail than is possible here. Fundamentally they consider that $RCo_5$ and $RCu_5$ form a solid solution at high temperatures, but that this solid solution dissociates at lower temperatures, possibly by a spinodal process. In this way a fine structure condusive to permanent magnetism is produced. As with the Al–Ni–Co alloys this precipitate was first observed by means of the optical microscope in material, which had been over-aged to coarsen the precipitate. Its presence was later confirmed in samples that had been correctly treated, an electron microscope then being necessary.

Increasing the amount of Co that is replaced by Cu tends to reduce $J_s$ and consequently $J_r$, but increases $H_c$. As with powders and sintered material, Sm appears to be the best rare-earth metal, but with these cast alloys the difference between the results obtained with Sm and with the cheaper Ce or even misch-metal are comparatively small. Attempts to produce magnets by this method, using Pr, have so far failed.

Replacement of part of the Co by Fe, which is disastrous in powders and sintered $RCo_5$ magnets, is actually beneficial in these cast alloys. As mixtures of Sm and Ce or misch-metal may be used, many composition variations are possible in this five component system.

To obtain the best properties an aligned crystal texture must be produced. If buttons are arc-melted on a water cooled copper hearth, they grow with a certain degree of alignment, which is, however, far from perfect. Nesbitt has pointed out that the orientation of the crystals in approximately hemispherical buttons is to some extent radial, and leads to a flux concentrating effect. As a result he was able to produce a surprisingly large field with two such buttons. For general use parallel alignment is required, and this is much more difficult to achieve. Many of the best properties claimed have been measured on very tiny samples selected from a much larger button. A few quite large single crystal or columnar crystal samples have been described, but it is doubtful whether a commercial technique for producing magnets by this method yet exists. The methods involving heated moulds such as are used for making columnar Al–Ni–Co alloys would be difficult to apply to a rare-earth alloy, which reacts so easily with most refractories, and requires a protective atmosphere.

The optimum heat treatment naturally depends on the atmosphere, but typically involves a quench from 1000°C, followed by a few hours at about 400°C. $(BH)_{max}$ values up to about 100 kJ $m^{-3}$ (just over 12 MGOe) have been obtained. Some more recent papers on rare-earth-cobalt alloys containing copper are mentioned in Section 4.5.4.

### 4.5.3 Theoretical explanation of properties of $RCo_5$

An excellent review entitled 'Present understanding of coercivity in cobalt-rare earth' was published by Livingston (1973). This review contains no less than 83 references. Workers who have contributed greatly to this understanding include: Zijlstra, Strnat, McCurrie, Becker and Buschow all of whom have published numerous papers with various co-workers. Recent papers by den Broeder and Zijlstra (1974) and by McCurrie and Mitchell (1975) may be assumed to summarize their present views. Strnat (1971) presented a review to an AIP Conference and has since contributed numerous papers on particular aspects of the subject. Any references which are omitted in this section can probably be traced through one of the above references.

According to Livingston there are three significant sizes connected with the magnetic behaviour of fine particles. $D_c$ is the diameter of a sphere below which in the absence of a magnetic field the energy of a single domain structure is less than that of a multi-domain structure, $b_c$ is the diameter of a cylinder below which coherent rotation processes require less energy than incoherent curling, while $\delta$ is the width of a domain wall. He gives the following relations for these quantities:

$$D_c = 1.4\ \gamma / J_s^2 \qquad (4.1)$$

$$b_c = 2A^{1/2} / J_s \qquad (4.2)$$

$$\gamma = \pi (A/K_1)^{1/2} \qquad (4.3)$$

$A$ is the exchange constant, while $K_1$ and $J_s$ have their usual meanings. $\gamma$ is the domain wall energy per unit area and is given by:

$$\gamma = 4(AK_1)^{1/2} \qquad (4.4)$$

From these relations:

$$b_c = \tfrac{1}{2}(D_c \delta)^{1/2} \qquad (4.5)$$

and the ratio of $D_c$, the size of the smallest particle in which the presence of a domain wall is energetically possible to $\delta$ the width of such a wall is:

$$D_c/\delta = 2K_1/J_s^2 \tag{4.6}$$

This ratio is also a measure of the relative importance of crystal and shape anisotropies. In $RCo_5$ permanent magnet materials typical values are according to Livingston: $D_c = 1\ \mu m$, $b_c = 0.04\ \mu m$ and $\delta = 0.006\ \mu m$. The idea that the critical size for a single domain particle is of the same order of magnitude as the width of a domain wall is no longer true when $K_1 > J_s^2$. Furthermore it is not correct to suppose that all particles less than $D_c$ in diameter behave as single domain particles, while all larger particles are multidomain. One can imagine a domain wall being produced transiently by a magnetic field in a particle with diameter less than $D_c$, and much of the theory of coercivity in $RCo_5$ magnets is concerned with the difficulty in nucleating domain walls in larger particles once they have been removed.

The fact that curling is easier than coherent rotation when the diameter of a cylinder exceeds 0.4 $\mu m$ is of little significance, since both processes are very difficult when $K_1$ is large. Actually the grain size of most $RCo_5$ magnets exceeds 1 $\mu m$, so walls can exist. The problem is that according to Brown's paradox, a field equal to the anisotropy field is required to replace a domain wall in a perfect crystal. For many $RCo_5$ compounds this involves a magnetizing force of more than 20 $MAm^{-1}$ or a flux density of more than 25 tesla.

The theory of the coercivity of $RCo_5$ is thus concerned with possible processes by which domain walls can be nucleated, and also possible processes by which these walls may be pinned, or in other words by which their motion can be impeded even if they are nucleated. Both nucleation and pinning are believed to be produced by some form of imperfection. The rather complicated and paradoxical situation thus arises that imperfections can reduce coercivity by promoting nucleation and increase coercivity by promoting pinning. The same imperfection may conceivably be capable of promoting both nucleation and pinning.

Shape imperfections such as sharp corners, protuberances, or crevases in the surfaces of grains, although they are important in materials that owe their coercivity to shape anisotropy, seem incapable of nucleating domain boundaries in materials with anisotropy fields a factor of ten or more times greater than any magnetostatic field. Such imperfections could possibly produce moderate coercivities by pinning.

A locality where the magnetocrystalline anisotropy is much reduced or even misoriented could clearly act as a nucleating site for a domain wall. Such a site may arise from a misoriented grain in a compact, by mechanical damage during milling, or by the occurrence of an inclusion with different composition and crystal anisotropy, such as the cobalt rich $R_2Co_{17}$. This alloy may be formed if the alloy is too rich in cobalt, if it becomes depleted in rare-earth atoms by differential oxidation, or as a result of the eutectoid reaction, which it is now generally agreed occurs at intermediate temperatures.

A local failure of the exchange forces could equally well make the nucleation of a domain wall more easy. A very effective nucleation site would be some kind of stacking fault or dislocation, which produced an antiferromagnetic interaction across some surface. Such an antiferromagnetic coupling would account for the nucleation of a reversed domain, while a positive field tending to favour uniform magnetization was still acting. The nucleation of such domains has actually been observed.

A strong pinning site may also promote nucleation by preventing a small volume of reversed magnetization being removed by a magnetizing field. When the magnetizing field is removed or reversed, such a small volume can act as a nucleation site from which reversed domains can grow. Evidence for this type of combined nucleation pinning site is thought to be given by the need to apply very large magnetizing forces to realize the full coercivity of rare-earth-cobalt magnets, some of which appear to have much higher coercivities when magnetized in one direction than the other, although the value of $B_r$ may be much the same.

This behaviour differs in magnitude rather than nature from that observed in other materials. It is recognized as a general rule that reduced coercivities are probable in many materials if the magnetizing force is less than 3 times the coercivity. Asymmetric loops are likely to occur in such circumstances unless the material is put into a cyclic state by reversing the magnetization say half a dozen times. What is so different in $RCo_5$ is that $H_{ci}$ may be 5 or 6 times greater than $H_c$, and peculiarly shaped loops should be expected unless the magnetizing force is several times greater than $H_{ci}$.

Evidence for both nucleation and pinning has been obtained from magnetization curves, and from domain studies. Some of the experiments have been carried out on small samples containing only one or a small number of grains. From the magnetization curves on such samples nucleation and pinning processes can often be distinguished.

If there is pure nucleation there should be no hysteresis until the magnetizing force $H_n$, which sweeps all nuclei of reverse domains out of the sample, is reached. After that there should be a square loop as shown in Fig.4.16. If there is pure pinning there should be some hysteresis even with small magnetizing forces, and the loops are unlikely to be square as shown in Fig.4.17. Sometimes nucleation and pinning may operate in different parts of the sample, and then all manner of complicated curves may arise. Figure 4.18 shows only one of many possible

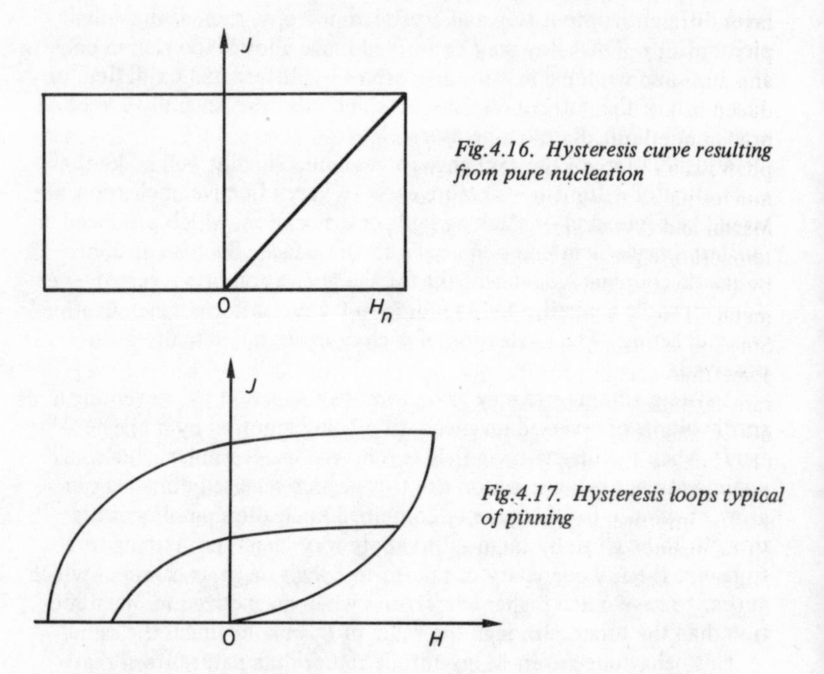


*Fig.4.16. Hysteresis curve resulting from pure nucleation*

*Fig.4.17. Hysteresis loops typical of pinning*

examples. In a bulk sample many processes occur and the loops are smoothed out. Figure 4.19 is a reasonably plausible loop for a bulk sample with both pinning and nucleation. It is also a convenient figure for explaining a parameter, $H_k$, which has recently become popular as a description of the properties of rare-earth-cobalt magnets. $H_k$ is defined as the demagnetizing force that reduces $J$ to 90% of its value $J_r$ at remanence. The curvature in the intrinsic curve shown in Fig.4.19 leads to a low value of $H_k$, and is therefore undesirable.

In the alloys containing copper, the coercivity is produced by a fine precipitate, and it seems reasonably certain that domain wall pinning throughout the bulk of the material is the dominant process (Becker 1976).

The coercivity of powders is very sensitive to the state of the surface, and various types of surface defects are held responsible. The source of coercivity in magnets sintered from such powders also seems to be situated largely at the grain boundaries. There are some differences of opinion as to how far the dominant process in sintered magnets is nucleation or pinning.

Various types of surface imperfections including oxide particles have been suggested as being active on the surfaces. Strnat suggested that a layer of rare-earth-rich phase such as $R_2Co_7$ between the grains could pin domain walls. Livingston expressed doubt about this suggestion and appeared to think that the rare-earth-rich phase should only be produced in the liquid phase sintering process. Ray and Strnat (1975) have now pointed out that there may be a number of possible rare-earth-rich phases such as $R_5Co_{19}$, and that $La_5Co_{19}$, $Ce_5Co_{19}$ and $Nd_5Co_9$ have much smaller anisotropy fields than the corresponding $R_2Co_7$ compounds. Martin and co-workers (1975) investigated the phases produced in the temperature range in which the eutectoid decomposition occurs, but by the reverse process of diffusion between pure Co and rare-earth metal. They found that in addition to the $R_2Co_{17}$ phase $Ce_5Co_{19}$ but not $Sm_2Co_7$ is formed. Indirectly this result suggests that $Ce_5Co_{19}$ may sometimes account for the poor coercivity of $CeCo_5$. These various rare-earth-rich phases may be precipitated at temperatures between about 600°C and 900°C, depending upon the alloy. McCurrie and Mitchell (1975) have also confirmed by electron probe analysis the existence of rare-earth-rich phases localized near the grain boundaries.

The improvement in coercivity sometimes observed after treating $RCo_5$ magnets at a temperature of about 900°C may be attributed either to the solution of $R_2Co_{17}$ precipitated at a lower temperature, or

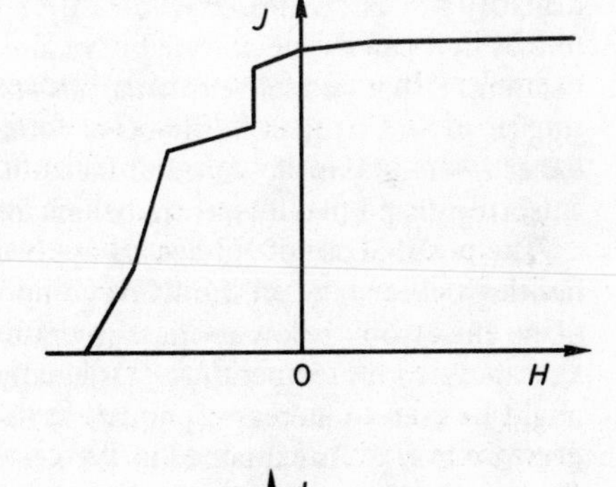


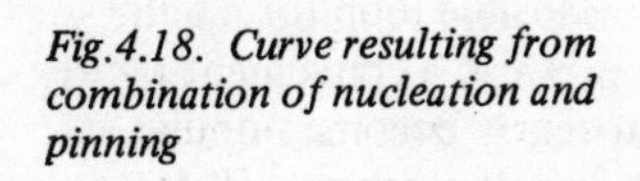
*Fig.4.18. Curve resulting from combination of nucleation and pinning*

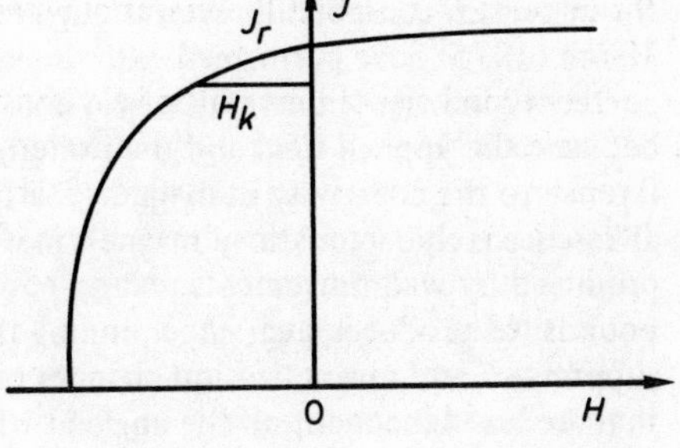


*Fig.4.19. Typical curve for bulk samples*

to the production of pinning sites by the precipitation of $R_2Co_7$ or a similar phase.

Possibly both processes occur, but the first requires that the material has already been spoilt by too slow a cooling rate or holding at a lower temperature. In our laboratory we did not find such large improvements as a result of treating at 900°C as have been reported by other workers. At the end of the sintering treatment our samples were usually removed from the hot zone of the furnace and cooled rapidly by a flow of cold argon, it is therefore unlikely that much $R_2Co_{17}$ was precipitated, and consequently the magnets could not be improved by its removal.

Although the coercivity of $RCo_5$ magnets is very much less than that of the anisotropy field, variations of coercivity with temperature follow fairly closely variations of the anisotropy field, so it seems desirable to seek some mechanism, which would make the coercivity a fairly constant fraction of this field (the anisotropy field is defined by Equation 3.7). A possible explanation is the fraction of the whole volume of material occupied by pinning sites. In different samples this fraction could be different, and the coercivity need not be related closely to the anisotropy field.

Much attention has been given to the important question of why it is so much easier to obtain good coercivities in $SmCo_5$ than in other $RCo_5$ compounds. The anisotropy field is certainly greatest in $SmCo_5$, but this difference is not great enough on its own to account for the disparity in coercivities. Possible differences in the rare-earth-rich phases that can be precipitated have already been mentioned in this connection. Other possible explanations are that the domain wall width $\delta$ is smaller in $SmCo_5$ than in the other compounds, and pinning may be easier when $\delta$ is small. Another factor is that $Sm_2Co_{17}$ has an easy axis anisotropy, and may therefore be less disastrous if it is precipitated.

The possible use of Nd has always been interesting, because $NdCo_5$ has the highest $J_s$ of all the $RCo_5$ compounds. Pure $NdCo_5$ has an easy plane anisotropy below room temperature, which changes to an easy axis above room temperature. Originally it appeared that a little Nd might be used to increase $J_s$ and $J_r$ at the expense of an acceptable decrease in $H_c$. As explained in the next section, Nd actually increases $H_{ci}$ in certain compounds, even though $H_a$ is reduced. Nagel, Klein and Menth (1975) have performed experiments on the angular variation of coercivity in oriented crystals of Nd-containing magnets. The angle between the applied field and the preferred direction makes little difference to the coercivity in magnets starting with low $H_{ci}$. Such a difference is characteristic of magnets in which magnetization changes are produced by wall movements. Nagel et al suggest that in certain compounds Nd produces such large pinning that boundary movements are suppressed, and magnetization changes take place by rotational processes that are less dependent on the angle at which the field is applied.

Wells and Ratnam (1974) also used a number of interesting techniques with the Kerr effect. Figure 4.20(b) shows a Kerr effect photograph of a two phase sample. The compositions of the phases were also determined by electron probe analysis, and the results are shown in Fig.4.20(a). By comparing these two pictures, it can be seen that the domain patterns in the phase marked $SmCo_5$ is coarser than in that marked $Sm_2Co_{17}$. Figure 4.20(c) is a normal bright field photograph in which $Sm_2Co_{17}$ is shown as irregular white inclusions. Roughly the same area of this sample is shown by the Kerr effect in Fig.4.20(d). Wells and Ratnam conclude from their experiments that $Sm_2Co_{17}$ precipitates in platelets normal to the *c* axis of $SmCo_5$.

They also used the Kerr effect with a full-wave retardation plate between the specimen and analyser. Colour variations then show up the difference in the orientation of the grains.

A good understanding of the domain processes operating in rare-earth magnets, could help in the search for better materials, and make possible more consistent production properties. It has also an important bearing on problems of magnetization and demagnetization discussed in Sections 8.1 and 8.2.

Most published experiments on rare-earth-magnets have been made on specimens that have been premagnetized in a high field. Such magnets can be demagnetized by an alternating or direct field, provided its magnitude is a little greater than $H_{ci}$, although it can be seen from Fig.4.15 that some production magnets require a field as great as 4 $MAm^{-1}$ ($\mu_0 H = 5$ T). Such magnets must then be remagnetized in the original direction, otherwise a very lean demagnetization curve is obtained. This behaviour is not in itself different in nature from that obtained on conventional materials. Inner loops on such materials are usually measured in the cyclic state; indeed by the old ballistic methods it was difficult to measure loops without reversing the magnetization. Ferrites are easily obtained in a thermally demagnetized state, and using modern plotting methods, Dietrich (1970) measured inner loops on such ferrite magnets without first applying a reversed field. We also made some similar measurements. Very asymmetric loops are obtained with magnetizing forces that do not greatly exceed $H_{ci}$, but within experimental error the coercivity does not exceed the peak magnetizing force.

According to Bohlmann (1976) the magnetizing force necessary to magnetize virgin magnets, i.e. thermally demagnetized rare-earth magnets, is comparatively small. He gives an example of such a virgin magnet that required only 1.2 $MAm^{-1}$ (15 kOe) to raise $B_r$ to 98 or 99% of its maximum value, although the same magnetizing force would produce only about 60% of the optimum $B_r$ on a non-virgin magnet. Perhaps even more remarkably a virgin magnet magnetized by a magnetizing force of 400 $kAm^{-1}$ (5000 Oe) retained 85% of its magnetization after

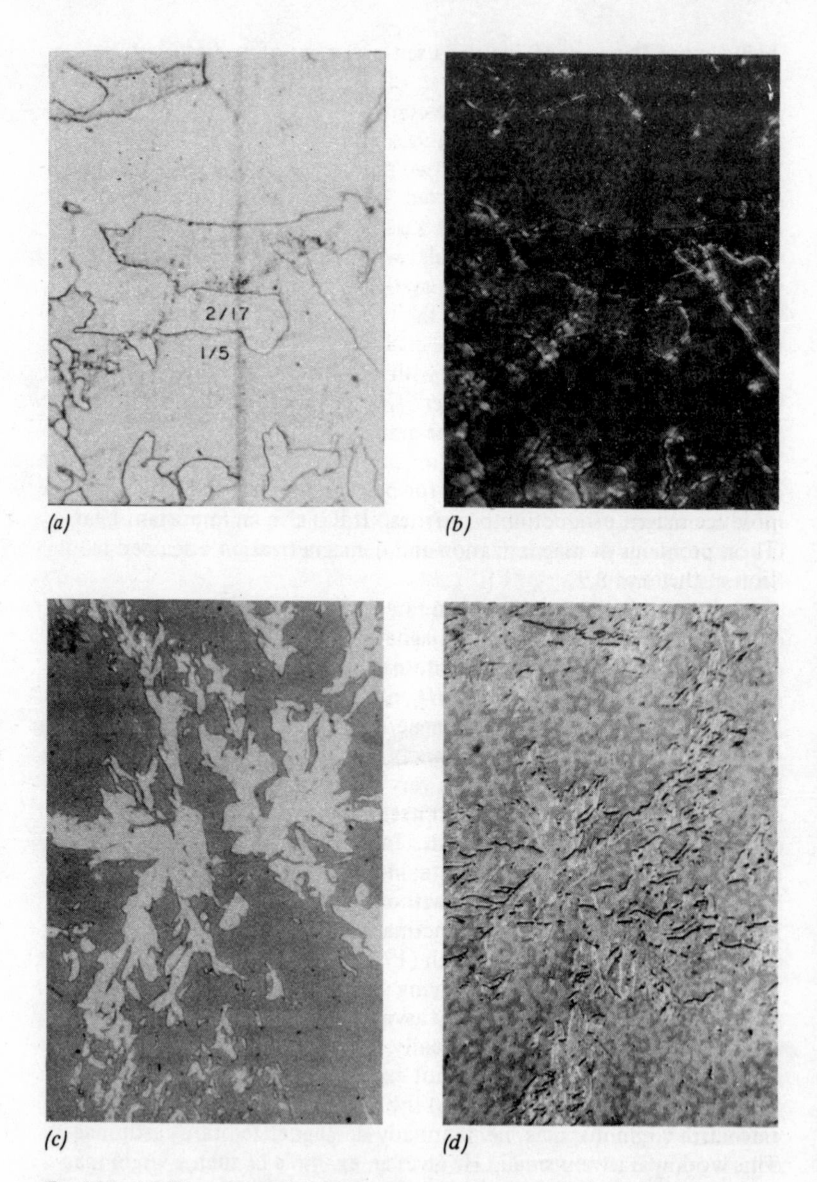


*Fig.4.20. (a) Phases identified by electron probe. (b) Same area as (a) examined by Kerr effect. (c) Normal bright field micrograph of $Sm_2Co_{17}$ in $SmCo_5$ matrix. (d) Roughly same area as (c) examined by Kerr effect. (Wells and Ratnam, 1974, courtesy Colt Ind., Crucible Materials Research Center, Pittsburgh)*

being exposed to a demagnetizing force of more than double this value.

These findings are a challenge to theory to explain, and perhaps also to show how such behaviour can be predicted and controlled. It would be unwise at this stage to assume that all rare-earth magnets will behave in this way. Bohlmann's results are partly but not completely confirmed by Wright and Roberts (1976).

Ease of magnetization, provided it is accompanied by adequate values of $B_r$, $H_c$, $H_k$ and $(BH)_{max}$, is a desirable attribute to permanent magnets. Although values of $\mu_0 H_{ci}$ 2 or 3 times $B_r$ may be necessary to ensure stability; the values of 4 and 5, which this ratio reaches in some production magnets, as shown in Fig.4.15 is for most purposes only a nuisance.

### 4.5.4 Recent developments in rare-earth cobalt magnets

Much effort has been directed during the last few years to an attempt to produce magnets based on the compounds $R_2Co_{17}$ rather than $RCo_5$. These compounds have higher values of $J_s$ and contain a smaller proportion of the expensive rare-earth elements. Unfortunately $R_2Co_{17}$ compounds with the exception of $Sm_2Co_{17}$ have an easy plane of magnetization, which is incompatible with a high coercivity. Even $Sm_2Co_{17}$ has a rather low value of $K_1$ and consequently of $H_a$. Strnat has found that the values of $K_1$ and $H_a$ can be increased to respectable values if part of the cobalt is replaced by iron. The anisotropy field is still much less than in most $RCo_5$ compounds, but several times greater than in ceramic magnets.

Until recently efforts to make permanent magnets based on the formula $Sm_2(Co_{1-x}Fe_x)_{17}$ have been disappointing. The nature of the problem is discussed by Perkins, Gaiffi and Menth (1975), who take a rather cautious but not completely pessimistic view of the prospects. They forecast that to obtain a good combination of $B_r$ and $H_{ci}$ a rather small particle size that presents certain difficulties is necessary. Just as in $RCo_5$ the precipitation of $R_2Co_{17}$ must be avoided, so in $R_2(Co_{1-x}Fe_x)_{17}$ magnets the precipitation of Fe–Co which might result from oxidation would be harmful.

In the Brown-Boveri Laboratories some success has been achieved in producing magnets in which part of the Co is replaced by Cu, and the rare-earth content is intermediate between that of $RCo_5$ and $R_2Co_{17}$. This work was announced at Amsterdam in 1974, and more details were given by Perry and Menth (1975). Permanent magnet properties can be measured in small crystals taken from the material as cast, and then given a suitable heat treatment, but there seems no reason why the material should not be powdered and sintered, if this proves to be an easier method of securing alignment in a magnet of practical size. Optionally about 5% of the Co may be replaced by Fe.

A typical composition was $Sm(Co_{0.84}Cu_{0.16})_{6.9}$, and after 1 hr at a temperature between 1200°C and 1240°C followed by 1 hr at 790°C, this alloy had $B_r$ = 0.92 T and $H_{ci}$ about 350 kAm$^{-1}$ (4300 Oe). By introducing 5% of iron for cobalt, these values were increased to $B_r$ = 0.95 T, and $H_{ci}$ = 400 kAm$^{-1}$ (5000 Oe). For this modified sample the heat treatment was 1 hr at 1230°C air cooled 1 hr 825°C. Further variations in these compositions and improvements in properties were announced at Philadelphia in December 1975.

A remarkable development which was made by Tsui, Strnat and Schweizer (1972) was the advantageous addition of Nd to rare-earth magnets. They were encouraged to persevere with this metal in spite of the low anisotropy field of $NdCo_5$ by their own theory that coercivity results from wall pinning by intergranular material. Thus they argued that if the intergranular material consisted of say $R_2Co_7$ with a high anisotropy field, the anisotropy field of the grains themselves would not be very important. They experimented with sintering $NdCo_5$ or $DiCo_5$ powders with various Pr and Sm rich sintering additives. Di signifies a mixture of Nd and Pr which is obtained at an early stage of the separation of misch-metal into constituent rare-earth metals, and is therefore relatively cheap. They achieved coercivities up to 800 kAm$^{-1}$ ($\mu_0 H_{ci}$ = 1 tesla) by sintering $DiCo_5$ with 60% Sm, 40% Co additive at 1040°C.

Nagel, Klein and Menth (1975) have improved on these results. They tried various combinations of Sm, Nd, and Co, and also M*, Sm, Nd and Co. Keeping approximately to $RCo_5$, they found that replacing Sm or M with Nd reduces $H_a$, but up to about 25% Nd actually increases $H_{ci}$. Although the best results were obtained with Sm, an alloy with only 15% of the rare-earth content in the form of Sm was very interesting. By replacing about a quarter of the M with Nd an $H_{ci}$ of 1.3 MAm$^{-1}$ ($\mu_0 H_{ci}$ = 1.6 T) was obtained. This coercivity could be combined with a $B_r$ of 0.9 T. $(BH)_{max}$ values were not quoted, but the demagnetization curve was described as square.

An important feature of the magnets with Nd is that the coercivity tends to rise with temperature. Problems might arise in low temperature applications, but performance at high temperatures, for which there is already a demand, should be improved.

According to Bachmann and Nagel (1976) the search for a good $R_2Co_{17}$ magnet has at last succeeded. The properties of the material are stated to be: $B_r$ = 1.1 T, $H_c$ = 560 kAm$^{-1}$ ($\mu_0 H_c$ = 0.7 T), $H_{ci}$ = 600 kAm$^{-1}$ ($\mu_0 H_{ci}$ = 0.75 T) and $(BH)_{max}$ = 239 kJ m$^{-3}$ (30 MGOe). The precise composition is not given, but the production process is said to be similar to that for $SmCo_5$.

* cerium misch-metal

Nagel (1974) also reported magnets with $B_r$ = 0.81 T, $(BH)_{max}$ = 110 kJ m$^{-3}$ (14 MGOe) and $H_{ci}$ = 0.9 T, based solely on misch-metal and cobalt. With 20% of the misch-metal replaced by Sm, $H_{ci}$ is doubled but $(BH)_{max}$ is increased only marginally. Since $\mu_0 H_{ci}$ for the $MCo_5$ material exceeds $B_r$ it is as high as is required for most applications. Some of the above topics are also discussed by Nagel and Klein (1975).

Barbara et al (1971) obtained a record $(BH)_{max}$ of 580 kJ m$^{-3}$ (75 MGOe) at liquid helium temperatures in the compound $Dy_3Al_2$. Two factors are responsible for this extraordinary value; one is the very high magnetic moment, which Dy and some of the other heavy rare-earths possess. The other is that in $Dy_3Al_2$ the domain walls are extremely thin. The 180° change in the direction of magnetization may possibly occur in a single jump within one lattice displacement. A thin wall of this type is easily pinned to produce a high coercivity.

Opinions differ about the possibility of producing magnets with this kind of performance at room temperature. At a conference in Paris the Nobel Laureate, Néel expressed optimism, which was obviously not shared by the whole of his audience.

There are actually two separate problems. All the heavy rare-earth metals with very high atomic moments happen to have low Curie points, and so far no way of increasing these Curie points substantially has been found. In the transition metals the FeCo alloy with the highest value of $J_s$ has also a very high Curie point. Thus there is no general principle that couples high $J_s$ with a low Curie point, and this consideration probably prompted Néel's optimistic view.

On the other hand Equation 4.3 shows that the thickness of a domain wall is proportional to $(A/K)^{1/2}$, and the very thin domain walls in $Dy_3Al_2$ are caused by a combination of a small exchange constant $A$ and a large crystal anisotropy $K$. Now if the exchange constant is small, the magnetic order is easily destroyed by thermal agitation, and so the Curie point is low. Consequently it is unlikely that magnets with extremely thin walls will be found at room temperature. These extremely thin walls make a high coercivity more probable, but a more than adequate coercivity is obtained in $RCo_5$ magnets without them. Thus there is a theoretical reason why very thin walls cannot occur at high temperatures, but there is no theoretical reason why high values of $J_s$ should not be found. If such high polarizations can be found there is no reason why it should not be associated with an adequate coercivity. We do not know at present where to seek higher saturation values, but Néel's optimism should certainly be respected.

The most important recent development in rare-earth-cobalt magnets has been a very large reduction in their price. This has been possible largely because of the introduction of the reduction–diffusion method developed by the General Electric Co. at Schenectady, and described in a number of papers, for example Martin and co-workers (1974).

Before the reduction–diffusion process was introduced, Sm was separated from the other rare-earth metals in the metallic state, often by distillation. In the new process the separation is made in the oxide and samarium oxide is reduced by calcium metal in the presence of Co. The resulting product is a samarium-cobalt powder, which can be made with the composition required for the next stage in the process. If the liquid-phase sintering process is used a separate powder for this purpose can be supplied. Thus the reduction–diffusion process, in addition to other advantages, eliminates the step of melting Co and Sm. The process is now operated by the Hitachi Magnetics Corporation, which operates the former General Electric Magnet factory, and the effective price of Sm today is less than a fifth of what it was a few years ago.

The availability of cheap raw materials has enabled magnet manufacturers to plan for larger scale production and reduce their prices even further.

Th. Goldschmidt of Essen operate a similar process, and Domazer and Strnat (1976) described an innovation, which promises even further reductions in cost. A mixed rare-earth oxide can be obtained cheaply as a by-product of a process not concerned with magnetic uses. This oxide contains about 70% samarium oxide, and the remainder is mainly oxides of other rare-earth metals such as Nd, Y, La and Gd. This oxide is treated by a reduction–diffusion process and the material is claimed to be capable of producing permanent magnets with $(BH)_{max}$ values up to nearly 160 kJ $m^{-3}$ (20 MGOe).

Another recent advance is the introduction of magnets made of polymer-bonded rare-earth-cobalt powders. These products were dealt with directly or in passing by several of the papers at the 1976 Second International Workshop on Rare-Earth-Cobalt Magnets in Dayton. The magnets can be made in rigid or flexible forms, and for both types $(BH)_{max}$ values up to about 80 kJ $m^{-3}$ (10 MGOe) have been claimed. At present these materials are only guaranteed to be stable up to temperatures of 60 to 80°C, but it is hoped to increase this range to about 140°C, at least for the rigid variety. There have also been promising experiments with rare-earth-cobalt powders in a soft metal bond.

### 4.5.5 Future prospects for rare-earth-cobalt magnets

The study of the magnetic properties of rare-earth alloys and compounds is an attractive subject for research. There are so many new compositions to be investigated, so many new properties to be discovered and so many interesting phenomena to be explained, that any reasonably competent research worker entering this field is almost assured of finding some results worth publishing. Early in this new era in permanent magnetism, I remarked half jokingly that the subject should provide a thousand PhD's. Surveying the papers that have already appeared, I think this suggestion may not prove to be an exaggeration.

While it is understandable that the subject is very suitable for research in University Departments, because it can give a young research worker experience in many new and difficult techniques, can lead to interesting results and pose problems, which are a challenge to the intelligence to interpret, much work has also been carried out in industrial laboratories. This industrial research was much more of a gamble, and there were numerous voices that argued that the expenditure on research and development could never be repaid, because rare-earth magnets could never be cheap enough to capture more than a small fraction of the market. I have expressed agreement with these pessimistic forecasts, but am glad to have been proved wrong. Rare-earth-cobalt magnets are already being offered at prices that in terms of cost per unit of $(BH)_{max}$ are competitive with some grades of Al–Ni–Co. Forward looking manufacturers of mass produced articles, for whom cost is more important than quality, believe that it will be possible to make articles such as small motors more cheaply by using rare-earth magnets than with even the cheap ferrite magnets. The saving is expected not in the cost of the magnet itself, but in the cost of the rest of the device.

The question of whether it is better to use pure Sm, misch-metal or even additions of other rare-earth metals remains open, and the balance may vary from time to time. The manufacturer, who offers several grades, and can quickly switch his production from one to another, will be in the strongest position. At present the benefit in using misch-metal rather than Sm is marginal, because sufficient Sm can be extracted from rare-earths needed for non-magnetic uses. If the demand for Sm for permanent magnets grows, the use of misch-metal may become more attractive. The problem of how to use the cheapest raw material available at the time is already solved.

Cobalt is actually much more scarce than the rare-earths. Although $SmCo_5$ uses a much higher proportion of Co than the Al–Ni–Co alloys, a $SmCo_5$ magnet uses less Co than an Al–Ni–Co magnet that can perform the same function. A shortage of Co would therefore favour rare-earth magnets relative to Al–Ni–Co, although not of course relative to ceramic magnets.

## 4.6 ESD MAGNETS

The letters ESD stand for elongated single domains. Immediately after the end of the last war the Ugine company in France, acting on the advice of Professor Néel, commenced production of single domain fine powder magnets. Powders of iron or iron cobalt were prepared by the reduction of organic salts such as oxalates or formates in a stream of hydrogen. The particles were nearly spherical with diameters between 0.1 and 0.01 $\mu$m. Iron and iron cobalt particles of this size are pyrophoric, and have to be stored in a liquid such as toluene, benzene or a petroleum type hydrocarbon. As most of the suitable liquids are themselves inflammable the process was hazardous. Magnets were formed by pressing the powder with a small amount of oily substance to protect against oxidation. The properties of the magnets were roughly equivalent to those of isotropic Al–Ni–Co alloys.

The coercivity of the powders probably arose from slight departures of the particles from a true spherical shape, although crystal anisotropy could have contributed. No attempt was made to produce anisotropic magnets by this method.

In the early 1950s the British GEC experimented with the process and suggested among other things that the powders could be stored in specially purified water, or in water that contained certain oxidation inhibitors. The material was launched under the name Gecalloy, but within a few years the process was abandoned in both Britain and France. Meanwhile the Laboratory of the General Electric Co of the USA developed a process for making elongated single domain powder magnets. The process is described by Luborsky, Paine and Mendelsohn (1959).

A solution of iron or a mixture of iron and cobalt salts is electrolysed into a mercury cathode. In order to form good elongated particles the apparatus should be protected from vibration. As electrolysed at room temperature the particles are very fine and may be superparamagnetic. They do not form an amalgam with mercury, but at this stage the mercury is little affected by a magnet. Possibly one may regard the product as a colloidal solution of the metal powder in mercury.

Heating the mixture for a short time at a temperature between 100 and 200°C coarsens the metal powder, and it is now possible to remove a thick slurry of magnetic particles and mercury by means of a magnet. The coercivity of this slurry can be measured if the slurry is first frozen. This coercivity is improved if a little metal such as Sn, Sb or Pb is first added to the mercury. These metals do form amalgams, but apparently some is deposited as a thin coating on the magnetic particles. The above account describes experiments that we confirmed, rather than the General Electric process. By pressing the slurry we managed to make something that could be tested as a fairly good permanent magnet, but which still contained a considerable amount of mercury.

The General Electric Co set up a plant to manufacture magnets by this process at Edmore, Michigan. (This factory has now been sold to Hitachi.) The processes of electrolysis and treatment of the mercury particle mixture are carried out automatically in large enclosed steel vessels. Only two or three people are required to supervise these operations, and among the precautions to protect them from mercury poisoning, the floor is kept wet.

Ultimately the mercury is removed from the magnetic powder by distillation. Members of the staff of the General Electric Co claimed publicly that there were some trade secrets in this stage of the operation, and that these would only be revealed to a purchaser of a licence to operate the process. Certainly we were unable to remove the mercury without spoiling the magnetic properties of the particles.

The magnetic particles, probably in the form of small aligned agglomerates, were then pressed into the shape of the required magnets. Usually anisotropic magnets are required, and for this type a magnetic field is applied during pressing. Since the particles are elongated in contrast to the platelets of ferrite, the field is applied at right-angles to the direction of pressing.

The final magnets consist of iron cobalt powders bound in a matrix of a metal such as Sn, Sb or Pb. Plastic binders were tried, but were abandoned, because although the initial properties of the magnets were acceptable, they were not sufficiently stable.

Although in the laboratory magnets were produced with properties corresponding to most of the anisotropic Al–Ni–Co alloys, the commercial magnets have only 60 to 70% of the $(BH)_{max}$ of material A2 (Alnico V or Alcomax III). They are made with a range of properties shown in Table A2.5 of Appendix 2. The different properties are presumably obtained by varying the concentration of the magnetic powder.

The original stimulus for the work in France was the possibility of making magnets containing no metal more expensive than Fe. The stimulus for the development of ESD magnets was the hope of obtaining greatly enhanced $(BH)_{max}$ values. Neither hope has been realized. Micromagnetics have shown that the improved $(BH)_{max}$ values were never possible by shape anisotropy, and the process is so expensive that it is not worth operating without using the optimum percentage of cobalt.

I was assured on several occasions that the plant at Edmore was working to capacity, making magnets difficult to produce by other methods. On the other hand the process does not seem to have been sufficiently profitable to justify expansion, or to encourage others to take it up.

As a by-product of the research into ESD magnets our understanding of the behaviour of fine particles has been improved. The discovery of exchange anisotropy and displaced loops also arose from this research. There was a suggestion that the properties of ESD magnets could be improved by coating the particles with an oxide: possibly cobalt ferrite. It is not clear whether this process has been operated. In view of discoveries in other fields, further development in ESD magnets seem improbable.

## 4.7 VICALLOY

An interesting series of alloys results from substituting various amounts of Fe by V in the high saturation, 50% Fe 50% Co alloy. If less than 2% V is used the alloy remains a soft magnetic material, known as V Permendur. The purpose of such a small V substitution is to make the alloy more easily workable. An alloy with about 9% V is used as an isotropic permanent magnet slightly better but more expensive than 35% Co steel. The heat treatment of this material is fairly simple, consisting of a quench from 1200°C followed by tempering at 600°C. The alloy (Vicalloy I) is workable and machinable before the final temper.

If the V content is increased to 12 or 13%, the alloy is not only not a permanent magnet, but is non-magnetic after quenching. A considerable amount of cold drawing or cold rolling is necessary to transform the austenitic alloy into a ferritic magnetic one. Tempering at 600°C can then produce Vicalloy II with a $(BH)_{max}$ as high as 28 kJ $m^{-3}$ (3.5 MGOe) in wire, but with somewhat lower values in rolled strip in which the anisotropy is less favourable. The cold drawing operation is not very easy, as the alloy hardens to a state in which further drawing is only just possible.

If less than 9% V is used various semihard alloys with $B_r$ as high as 1.6 T and intermediate coercivities can be produced. These alloys are suitable for applications such as hysteresis motors. The genuine permanent magnet Vicalloy is used for some small magnets, for which strip or wire is required. If desired Cr may be used in place of some of the V.

There has been considerable speculation about the mechanism that produces the coercivity and anisotropy of Vicalloy. Work in Russia by Shur showed that considerable enhancement of permanent magnet properties could be secured by heat treating Vicalloy under tension. At first sight this experiment seemed to suggest that strains might contribute to the properties of Vically, but the result is also compatible with shape anisotropy. There is also a possibility that some of the phases become ordered.

## 4.8 Cu–Ni–Fe AND Cu–Ni–Co ALLOYS

Cu–Ni–Fe is chiefly notable for being slightly flexible in the form of tape or wire even in the permanent magnet state. Its normal composition is 60% Cu, 20% Ni, 20% Fe. Its preparation involved quenching from 1040°C, tempering at 650°C, cold drawing, and further tempering at 650°C. $(BH)_{max}$ values up to 12 kJ $m^{-3}$ (1.5 MGOe) have been obtained, but only after the alloy has been made anisotropic by considerable cold reduction. Evidence of a precipitation process or possibly a spinodal dissociation has been obtained. The material has been made to a limited extent in Germany some time ago, and more recently in the USA. Cu–Ni–Co may be regarded as Cu–Ni–Fe with Co replacing Fe, but it does not appear to be made commercially.

## 4.9 PtCo

PtCo was discovered in the late 1930s and was improved by various workers during the following 20 years. The composition is approximately equiatomic PtCo, and is consequently about 75% Pt by weight. After solution treatment at 1000°C it has a low coercivity and is mechanically soft. To obtain the best results it must be cooled at a controlled speed, and tempered at a temperature of about 600°C. $(BH)_{max}$ values exceeding 80 kJ $m^{-3}$ (10 MGOe) have been obtained, and these values are closely approached in production. In spite of its great cost there has been a considerable amount of business in PtCo, and at one time permanent magnets represented a considerable fraction of the usage of Pt. The main demand was for hearing aids, watches, and space applications. Many of these applications are now better performed by $RCo_5$, but there may possibly be a few applications, particularly in medicine, where the resistance to corrosion of PtCo may be an advantage.

PtCo is very interesting theoretically. The main ferromagnetic phase is fcc with a moderate anisotropy. The heat treatment produces an ordered tetragonal phase, with a very large uniaxial anisotropy. McCurrie and Gaunt (1964) proposed an explanation of how the interaction of these two phases produced barriers to domain wall movement. It would perhaps be enlightening to look again at the facts and interpretations in this paper in the light of modern theories of wall pinning in rare-earth-cobalt alloys.

The negative cubic anisotropy of the main phase means that there are four easy directions of magnetization, and in consequence the ratio $J_r/J_s$ is unusually high for an isotropic material.

PtFe has some permanent magnet properties, but it is inferior to PtCo and in view of the cost of Pt is completely uninteresting commercially.

## 4.10 MnBi

MnBi is an intermetallic compound, which is somewhat difficult to produce in a pure state. It can be formed into a powder, aligned in a field and pressed. It has a high coercivity based on crystal anisotropy. At one time its $H_{ci}$ value of 600 kAm$^{-1}$ ($\mu_0 H_{ci}$ = 0.77 T) was the highest known. The crystal anisotropy and therefore the coercivity falls to zero at low temperatures. The Curie point is low and the material is subject to corrosion. It is unsuitable for use as a practical permanent magnet, but there has recently been some interest in MnBi thin films in connection with computer memories.

## 4.11 SILMANAL

Silmanal is an alloy of 86.8% Ag, 8.8% Mn and 4.4% Al. It is one of a number of alloys of Mn that is magnetic and can be classed as a Heusler alloy. It is easily workable and requires a heat treatment at a temperature of 250°C or lower. It has an intrinsic coercivity of 477 kAm$^{-1}$ ($\mu_0 H_{ci}$ = 0.67 T and is about 10 times greater than $B_r$). The only application of silmanal is in a magnetometer type of instrument for which both its low $B_r$ and high $H_{ci}$ are an asset. It causes little disturbance to the flux pattern it is required to measure, and is not easily demagnetized. It does not appear to be offered commercially, and it is probable that the instrument manufacturers who have used it made their own material.

## 4.12 COMALLOY

Comalloy or Comol is a ternary alloy of about 12% Co, 17% Mo and balance iron. The carbon content is kept low, but small quantities of Mn or Ti may be beneficial. The magnetic properties are somewhat similar to those of 35% Co steel, and it has been used as a substitute or replacement for that steel at times when Co has been in short supply. It is workable but permanent magnet properties are produced in a completely different manner from that used for steels. The alloy is quenched from a temperature between 1200°C and 1300°C, after which it should be in a non-magnetic austenitic state. The quenching process cannot be performed sufficiently quickly on samples more than 2 or 3 cm in diameter. A magnetic phase and permanent magnet properties are generated by tempering for 2 to 4 hr at a temperature between 680 and 700°C. Recently any Comalloy magnets made in the UK have probably been sintered.

Hadfield (1962) mentions certain Fe–Mo–Co and Fe–W–Co alloys as being similar to Comalloy.

## 4.13 MnAl

In a narrow composition range approximating to $Mn_{1.11}Al_{0.89}$, there is a compound with a high uniaxial anisotropy. By a rather complicated process, involving cold work as well as heat treatment, a magnet with $(BH)_{max} = 28$ kJ m$^{-3}$ (3.5 MGOe) has been made, and theoretically somewhat higher values may be possible. The alloy continues to be the subject of academic research; Jakubovics et al (1975) used it for a study of the influence of anti-phase boundaries on the nucleation and pinning of domain walls. Although Kojima et al (1975) have obtained 48 kJ m$^{-3}$ (6 MGOe) on a Mn–Al–C alloy, the prospects of practical application of any modification of this alloy are poor. It is even more unlikely that an alloy of Mn–Al–Ge which has been the subject of some research, but which is 100 times more expensive than Mn–Al, will be used.

## 4.14 COBALT MAGNETS

Permanent magnets based on hexagonal cobalt are only an idea, not even an experimental reality. Metallic cobalt has a saturation polarization of 1.75 T, and in the hexagonal form an anisotropy field $H_a$ of about 480 kAm$^{-1}$ (6000 Oe). As a preliminary step in an investigation of possible ways of harnessing these properties carried out on behalf of the Cobalt Information Centre, I calculated the $(BH)_{max}$ values that might be obtained with various assumptions (McCaig 1966b). These estimates varied from 530 kJ m$^{-3}$ (66.5 MGOe) for the most favourable assumption of 100% density and alignment, and $H_{ci} = H_a$, to only 18.3 kJ m$^{-3}$ (2.3 MGOe) for an unaligned powder with two thirds packing density, and a less optimistic estimate of $H_{ci}$.

We commenced experiments, but met considerable difficulties. Most of the methods of producing fine powders of Co lead to the low coercivity fcc form. If the hexagonal type is produced it may be in the form of elongated particles with shape and crystal anisotropy opposed. It also appears that I was using a published estimate of the size of single domain particles in cobalt (Went et al, 1952) that is now generally regarded as too high.

One possible solution to the problem of obtaining a fine dispersion of hexagonal cobalt is to precipitate it in some other alloy. Masumoto and co-workers in Japan found permanent magnet properties in Co–Al alloys (Malcolloy). With about 15% Al and a water quench from 1380°C, followed by 30 hr at 500°C, they obtained $B_r = 0.42$ T, $(BH)_{max} =$ 13.6 kJ m$^{-3}$ (1.71 MGOe), and $H_c = 96$ kAm$^{-1}$ (1200 Oe). Johnson, Rayson and Wright (1972), who give references to the numerous papers by Matsumoto et al, deduced that the coercivity of this alloy can indeed be attributed to a hexagonal cobalt precipitate.

I find experts on the subject are divided on the subject of whether $RCo_5$ compounds are a modification of hexagonal cobalt, but their discovery made any further search for magnets based on hexagonal cobalt seem unnecessary.

Schüler and Brinkmann (1970) p.236 mention permanent magnets in the (Fe, Co)P system, but with a $(BH)_{max}$ of only 16 kJ m$^{-3}$ (2 MGOe).

## 4.15 STAINLESS STEEL

The 12% Ni, 12% Cr, and 18% Ni, 8% Cr stainless steels are austenitic and non-magnetic as supplied by the manufacturers. Cold work tends to precipitate a magnetic phase with considerable coercivity. These materials were investigated with a view to their use as recording materials, before the use of oxide particles in plastic ribbon became universal. With intense cold drawing to very fine wires appreciable permanent magnet properties appear. If this property had been discovered 20 or 30 years earlier, stainless steel might have been used as magnets. The source of the coercivity lies in elongated ferromagnetic percipitates produced by cold drawing.

As it is this permanent magnetism is rather a nuisance. These stainless steels would be very useful as tough non-magnetic metals, but after any form of machining by the customer they are liable to become slightly magnetic. If a strong completely non-magnetic metal is required, e.g. in the construction of minesweepers or in some scientific instruments, higher Cr stainless steels are more satisfactory.

# Chapter 5
# Stability of permanent magnets

The user of a permanent magnet is concerned to know under what conditions and for how long the magnet will continue to fulfil the purpose for which it was designed. Any permanent magnet will be demagnetized if it is exposed to certain conditions such as very high temperatures or very high magnetic fields. Provided it is properly made and treated, the life of a modern permanent magnet is to the best of our knowledge infinite. With rather elaborate precautions even the older and inferior permanent magnet steels could be embodied in high class scientific instruments, which maintained their accuracy for many years. With modern magnets the necessary precautions are much less exacting.

There is not, however, any one permanent magnet material that is the most stable in all possible circumstances. The material that is best for withstanding high temperatures is not necessarily the best for withstanding low temperatures or extraneous magnetic fields.

The degree of stability required depends on the purpose for which the magnet is used. A loss of say 10% in the magnetization of a holding magnet used as a door catch or to fix articles to a steel surface would scarcely be noticed, but a change of even 1% in the strength of a magnet in a high grade ammeter might be unacceptable. Many experimental measurements have been carried out to help the magnet designer to predict the behaviour of magnets in various specified circumstances. The results of these measurements depend not only on the material, but the shape of the magnet, minor variations in its heat treatment and various measures that can be taken to improve the stability. In order to make sense of a great mass of experimental data, different processes that can influence the strength of a magnet need to be distinguished. To facilitate the subsequent discussion some terms the meanings of which are not completely obvious must first be defined.

Any magnet after being fully magnetized is likely to lose strength slowly with time. Such after-effects measured in the first few seconds or minutes after the last field change are usually described as magnetic viscosity. One method adopted by some instrument manufacturers to avoid errors due to such losses is (or at least was) to keep the magnets for several months after magnetizing before calibrating the instrument. This process is known as ageing or sometimes natural ageing. Artificial means of speeding up this process such as deliberately demagnetizing the magnets slightly by means of a small demagnetizing field or exposing them to a temperature cycle are often employed. Such stabilization processes are also often referred to as ageing.

The most serious, although not the most common, cause of a change in the strength of a magnet is an alteration in its phase composition or structure. Such alterations are known as metallurgical changes. Hardened steel magnets are in a metastable state, and the metallurgical transformation of martensite to ferrite may proceed slowly even at room temperature, and considerably more quickly at temperatures of 100 to 200°C. Some powder materials may be degraded by oxidation at room temperature, but in most permanent magnet materials metallurgical changes occur only at considerably higher temperatures.

When the temperature of a magnet is changed, there is usually a reversible decrease of magnetization with increasing temperature. This change is called reversible, because the magnet returns to its original strength, when it returns to room temperature. Reversible changes in magnetization are caused by reversible temperature variations in the spontaneous magnetization. If a magnet is cooled below room temperature there is of course a reversible increase in magnetization.

In addition to the reversible temperature changes in magnetization, there are irreversible losses. These losses may be caused by an accelerated natural ageing caused by thermal activation of domain processes at high temperatures. As the Curie point is approached irreversible losses also occur, because the exchange forces are no longer able to maintain magnetic order. Another cause of irreversible losses in magnetization is a change in the shape of the hysteresis loop with temperature, and in particular a reduction in the coercivity. In some materials the coercivity is lower at high temperatures, and in others it is lower at low temperatures than at room temperature. Thus in some materials irreversible losses may result from cooling as well as heating. These losses are irreversible in the sense that they are not restored by returning the magnet to room temperature. Unlike the metallurgical losses they can be restored by remagnetizing the magnet at room temperature. This use of the terms reversible and irreversible is admittedly rather confusing, but it is difficult to see how without considerable circumlocution this confusion can be avoided.

The measurement of the various magnetization changes described above often requires precise measurements made in difficult conditions.

Some of the methods by which these experimental difficulties have been overcome are described in the next Section. Sections then follow in which the results obtained for the various processes outlined above are described and as far as possible explained for various materials. One section is devoted to each process rather than each material. The practical limitations which these results impose on the use of each material are pointed out when possible, together with hints for the design of satisfactory stable magnets. A final section summarizes how stability characteristics influence the choice of magnet material for different applications.

In the opinion of the author some claims that the results published about the stability of permanent magnets are sufficient to enable the designer to predict the behaviour of his product accurately are overstated. Slight changes in shape, size, composition or treatment may cause the commercial product today to differ significantly from the particular samples on which the published results were based. The figures given in this chapter should be treated as a rough guide, and should be checked on a prototype of the actual equipment.

## 5.1 STABILITY MEASUREMENTS

Experiments on the stability of permanent magnets require a number of different types of measurement. Firstly it may be necessary to measure the strength of a permanent magnet before and after it has been exposed to some adverse condition such as high or low temperature, mechanical vibration or proximity to other sources of magnetic field. Such experiments do not differ from any other measurements of the strength of a magnet, except that they must often be carried out with more than usual accuracy. It is not easy to read any instrument such as a ballistic galvanometer, fluxmeter or Hall device meter with an accuracy of better than about ½%, and this is rather unsatisfactory for measuring changes of only one or two percent.

The second type of experiment requires measuring the strength of a magnet, while it is actually in some high or low temperature environment. There are then other complications in addition to the need for high accuracy.

We used a number of methods to overcome these difficulties, some of the most successful being pioneered by A. G. Clegg. The method used to increase the sensitivity and measure changes in magnetization of 0.1% or less, was to make a differential measurement in which the magnet under test, and subjected to various influences, was compared with a similar magnet that had been carefully stabilized and was protected from any disturbance. More recently Dietrich (1968) has used the very accurate method of nuclear magnetic resonance to measure the fields in the gaps of assemblies.

Initially Clegg used an astatic magnetometer. In this instrument two similar magnets selected to have equal strengths are fixed vertically above one another in opposition by a light but rigid rod. The assembly of two magnets is suspended by a long thin thread. The object of this astatic arrangement is to eliminate the effect of changes in the Earth's field or caused by other apparatus in the building.

One of the suspended magnets is then placed under the influence of a control magnet and the magnet to be tested. The positions of these magnets are adjusted to give zero deflection. Small changes in the strength of the test magnet cause a relatively large deflection of the system, which is measured by a mirror lamp and scale. Clegg performed many useful measurements with this system, particularly experiments on the variation of the strength of a magnet at sub-zero temperatures. Even with an astatic magnetometer he encountered some difficulties in using the method in a laboratory in which large magnets were being moved about, and all manner of equipment was being switched on and off. It seems unnecessary therefore to enter into further details, such as the method of calibration, as more recently we have used a differential ballistic method.

Once again a standard magnet as well as the magnet under test is required. The latter may be in a bath of liquid or a furnace according to the nature of the test. Silicone liquids can be obtained for use at low or fairly high temperatures. An arrangement for use in a liquid bath between −80 and +100°C is sketched in Fig.5.1. For still higher temperatures an air furnace is necessary. Coils in series opposition, sometimes called bucking coils, are pulled off the two magnets simultaneously. When the method was first used, the only instrument of sufficient sensitivity was a long period ballistic galvanometer. The traditional type of moving coil fluxmeter was not sensitive enough for most purposes. With a ballistic galvanometer difficulty often arises, because the pulses from the two magnets do not have precisely the same profile. If double pulses in opposite directions occur instead of one single differential pulse, the ballistic galvanometer is unsatisfactory. Modern electronic integrating fluxmeters are now almost as sensitive as ballistic galvanometers, and handle double pulses much better. The need for a reference magnet may be eliminated by using a digital instrument.

A number of problems arise in connection with the insulation of the wire of coils that are exposed to high temperatures or bath liquids. The most satisfactory coverings for high temperatures are glass or asbestos.

If coils are exposed to temperatures of 500°C or more thermoelectric emf's often become troublesome. It is not easy to pin-point the origin of such emf's, but they certainly occur. They cause zero-

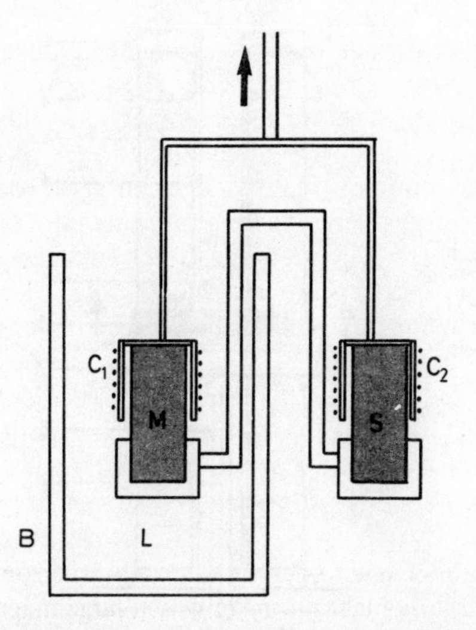


*Fig.5.1. Apparatus for differential ballistic measurement of changes of magnetization with temperature. M: Magnet under test in bath B of liquid L. S: Standard reference magnet.* $C_1$, $C_2$*: Similar coils connected in series opposition and removed together*

shift in a ballistic galvanometer and drift in any kind of fluxmeter.

For some measurements of fairly large changes in magnets at high temperatures over a long period of time, the measurements were made by means of a coil outside the furnace as shown in Fig.5.2. This coil was connected to an ordinary moving coil fluxmeter, and the readings obtained when it was pulled off the furnace were proportional to the strength of the magnet.

If it is necessary to investigate whether metallurgical changes have altered the hysteresis loop of a magnet, measurements can be made before and after the experiments by one of the methods described in Sections 8.4.1 to 8.4.3.

Measurements of the properties of a magnet at a temperature other than room temperature present more difficulties. $H_{ci}$ can be measured by the extraction method using a solenoid and coil, described in Section 8.4.5. For sub-zero temperatures the easiest method is to immerse the sample contained in a suitable holder in a low temperature bath, and quickly transfer it to the coil and solenoid for the extraction measurement. This method can also be used for temperatures a little above room temperature, but for really high temperatures, we used a solenoid containing a furnace, and separated from it by water cooling.

This method can also bc used to measure complete hysteresis loops or demagnetization curves at high temperatures. An open circuit method

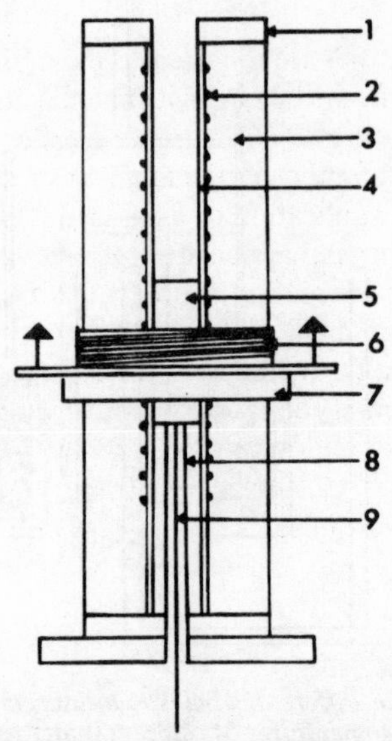


*Fig.5.2. Apparatus for measurements at high temperatures. (1) Brass case with Syndanyo end plates. (2) Non-inductive heating elements. (3) Vermiculite packing. (4) Silica tube 0.5 in ID. (5) Sample magnet. (6) Removable search coil (connected to fluxmeter). (7) Clamp to locate search coil. (8) Ceramic spacer. (9) Thermocouple*

with a calculated $H$, corrected for the self-demagnetizing field of the sample is usually necessary.

## 5.2 NATURAL AGEING AT ROOM TEMPERATURE

Figure 5.3 shows the flux loss of assemblies using mild steel poles as measured between 1 day and 1000 days after magnetizing. The measurements were made by Webb at the National Physical Laboratory and no artificial stabilization was used. The ratio $(B/\mu_0 H)$ is given in brackets after the name of each material, and is underlined for materials working at $(BH)_{max}$. These magnets were kept in carefully controlled conditions and some artificial ageing would be necessary to obtain this degree of stability under normal conditions of operation.

All the curves except those for the steels in which structural changes occur, show a linear change of magnetization when plotted on a logarithmic time scale. This behaviour is expected for a process activated by thermal fluctuations. The anisotropic materials appear much better than the isotropic ones. In fact the poor showing of isotropic Al–Ni–Co is rather surprising.

Usually magnets that are required to maintain their strength accurately, as for instance in measuring instruments, are knocked down by say 1 to 5% after magnetization. That is to say they are deliberately demagnetized by this amount by the application of a magnetic field. Sometimes a single application of a demagnetizing field is used for this knock down, but the first application of a demagnetizing field does not produce complete stabilization. A better method is to apply an alternating field that is gradually reduced to zero. Obviously the starting value of this alternating field determines the percentage knock down. If the amount of demagnetization is insufficient, it is necessary to start with a slightly higher alternating field, while if the demagnetization is too great it is necessary to remagnetize and try again.

Figure 5.4 illustrates the kind of changes that might occur during the first one and a half cycles of a reducing alternating field. After mag-

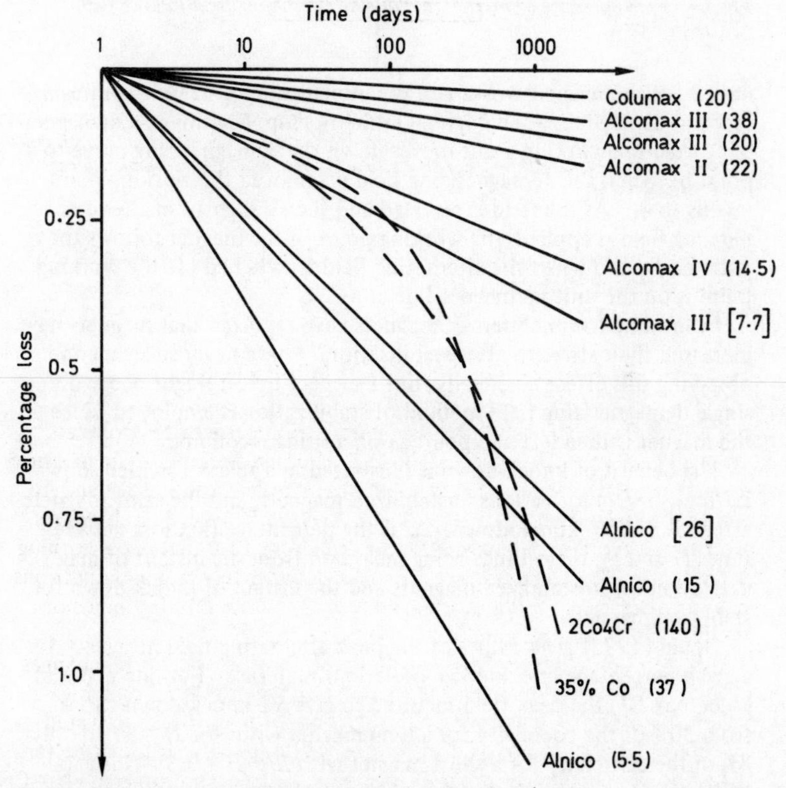


*Fig.5.3. Time change of gap flux according to Webb. ($B/\mu_0H$) in round brackets corresponds approximately to $(BH)_{max}$*

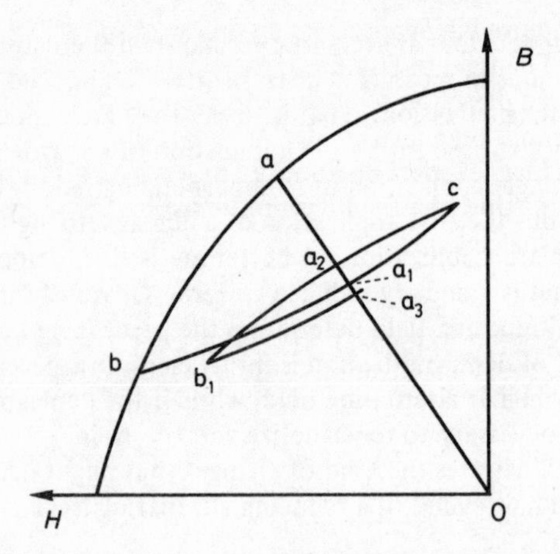


*Fig.5.4. Working point of magnet in reducing alternating field follows path* $aba_1ca_2b_1a_3$

netization the magnet works at a the intersection of its unit permeance line and demagnetization curve. On the first application of a demagnetizing field the working point moves down the demagnetizing curve to a point b. When the demagnetizing field is removed the working point recoils to $a_1$. As the field is reversed and then a slightly smaller demagnetizing field is applied, the working point of the magnet follows the path, $ca_2b_1$ and when the alternating field is switched off the working point is on the unit permeance line at $a_3$.

From time to time users of magnets have reported that magnets have increased their strength after stabilization. I have never succeeded in observing this effect personally, but I suspect that it might occur if a single demagnetizing field method of stabilization is employed, since the magnet is then left at a point $a_1$ on a rising recoil line.

The benefit of knock-down is illustrated in Table 5.1, which shows $\Delta J/\log_{10}(t_2/t_1)$ for various unstabilized magnets, and the same magnets after 1% and 5% knock-down. $\Delta J$ is the percentage flux loss between times $t_1$ and $t_2$, these times being measured from the instant of magnetization for unstabilized magnets and the instant of knock-down for stabilized magnets.

Gould (1973) gives values of the peak alternating field necessary for a given percentage knock-down of various materials. For one material (Alcomax III) the peak field required to give 5% knock-down varies from 30% of the coercivity for a long magnet with $B/\mu_0H = 37$ to only 8% of the coercivity for a short magnet with $B/\mu_0H = 9.5$. This variation is nearly as great as that found among the other materials for most of which figures are given for only one value of $B/\mu_0H$. Thus the knock-down field required depends as much on the shape as on the material of the magnet. The field required is greater if it is applied at right-angles to the preferred direction of an anisotropic magnet. Data of this kind can do little more than give an idea of the equipment required for stabilization. If a magnet must be knocked-down by a given percentage, or to a given flux value, the field required must be found by experiments on the actual magnets.

## 5.3 DEMAGNETIZATION OF MAGNETS BY DEMAGNETIZING FIELDS

The observations of the last section are also useful in assessing the resistance of magnets to accidental demagnetization by external fields. In no case did a demagnetizing force less than 1.5% of the coercivity produce a knock-down of 1%, and this result should also hold for fields accidentally encountered. If magnets are expected to encounter demagnetizing fields, they should be stabilized by knock-down in slightly greater fields. Temperature cycling described in the next Section gives some protection, but is mainly useful for stabilizing against temperature changes.

Variations in the self-demagnetizing field of a magnet, which can occur as a result of thoughtlessly removing a magnet from an assembly

**Table 5.1**

| | | $\Delta J/\log_{10}(t_2/t_1)$ | | |
|---|---|---|---|---|
| *Material* | *Working point,* $-B/\mu_0/H$ | *Unstabilized* | *1% knock-down* | *5% knock-down* |
| Columax | 20 | <0.01 | – | – |
| Alcomax II | 22 | 0.04 | <0.01 | – |
| Alcomax III | 38 | 0.025 | <0.01 | – |
| Alcomax III | 20 | 0.035 | <0.01 | <0.01 |
| Alcomax III | 7.7 | 0.125 | <0.01 | <0.01 |
| Alcomax IV | 14.5 | 0.10 | <0.01 | <0.01 |
| Hycomax III | 5.6 | 0.06 | – | – |
| Alnico | 26 | 0.23 | 0.02 | – |
| Alnico | 15 | 0.25 | 0.03 | 0.025 |
| Alnico | 5.5 | 0.38 | 0.07 | 0.04 |
| 35% Co steel | 37 | 0.28 (mean) | 0.07 | 0.02 |
| 2% Cr, 4% Cr steeel | 140 | 0.26 (mean) | 0.12 | 0.03 |

In this table, $\Delta J$ = % flux loss between times $t_1$ and $t_2$, both times being measured from the instant of magnetization in the case of unstabilized magnets, from the knock-down time for stabilized magnets.

are a special case similar to demagnetization by external fields. Magnets other than those of high coercivity can suffer serious demagnetization as a result of such removal from an assembly.

A serious cause of demagnetization of magnets, which can in a sense be attributed to self-induced fields is contact with ferromagnetic bodies. I performed a few experiments on this subject a considerable time ago (McCaig 1956). Rectangular bar magnets were tested before and after placing and moving in various ways on a large steel block. Demagnetization was never more than 1.2%, provided only the ends of the bar made contact with the block. The results were very different if the side surfaces of the bar touched the block. The worst effect was a demagnetization of 40% produced on an Alcomax III (material A2) magnet when each of its four sides in turn was placed on the block and slid off the edge in the direction of its length. This is actually the most natural and easiest way of removing a magnet. Most of the effect of any operation occurs in the first two or three times it is carried out. After ten operations no further change can be detected. About 3 mm of covering is sufficient to protect a magnet from this hazard.

I carried out this experiment on magnets of only two materials (A1 and A2, i.e. Alnico and Alcomax III). I regret now that I did not have the experiment repeated on more modern magnets. I would expect some demagnetization, although less than with Alcomax, to occur with high coercivity Al–Ni–Co alloys, but the effect to be small or absent in ferrites and $RCo_5$. The experiment is very simple and requires only a fluxmeter, a home made coil and the magnets. More difficult experiments are often undertaken as undergraduate or even as sixth form projects.

From time to time I have received complaints from magnet users of magnets becoming seriously demagnetized. For 28 years there was no such complaint that I was able to confirm, or was unable to explain away as the result of mishandling magnets such as placing them on a steel surface, dismantling an assembly, or touching with a screwdriver.

In 1974, some magnets were returned by a customer to a member of the PMA with the complaint that they had suffered substantial demagnetization. The magnets were referred to our Laboratory for examination. We magnetized them, tested them and placed them packed in cotton wool in separate boxes in a cupboard. A few months later they were taken out and retested. Some of them were found to have lost 20 to 30% of their strength. My feelings were similar to those I should have experienced, if I had discovered that a television performer had bent my best silver spoons in my sideboard drawer.

It is of course possible that one of my assistants made a mistake or even played a practical joke. I should have liked to have repeated the experiment myself, keeping the magnets locked in my own desk, but by this time such an experiment was impossible because the Laboratory was about to close.

If the changes were genuine, they need not upset faith in the stability of most magnets. The particular magnets were small C-shaped magnets of Alcomax. The field applied during the heat treatment process may have followed a different path from that used for final magnetization. It is conceivable that the magnets were magnetized in an unstable direction that differed from the preferred direction. There might then have been a spontaneous change to a lower energy state, involving magnetization closer to the preferred direction.

It has seemed necessary to recount the above incident, because if it had not occurred, I should have stated categorically that a loss of magnetization of the above amount could not occur in unmolested magnets.

## 5.4 STRUCTURAL CHANGES

As already mentioned structural changes occur slowly in the permanent magnet steels even at room temperature. Thus steels require ageing at 50 to 100°C to stabilize against these metallurgical changes, as well as against magnetic changes.

Although structural changes can be promoted in certain Ni–Fe alloys by exposure to sub-zero temperatures, they have not been noticed in any permanent magnet material. The Al–Ni–Co alloys are metallurgically stable up to about 500°C. Between about 500°C and 600°C changes are to be expected. The final tempering of the anisotropic Al–Ni–Co alloys is usually in this range. Most permanent magnets of class A2 would benefit very slightly from a much longer tempering than commercial magnets usually receive, the benefit being so slight that the extra cost is not justified. Consequently the coercivity of some such magnets may actually increase slightly if they are used for a time at 550°C.

The effect of holding at temperatures between 550 and 650°C is very complicated. At 550°C the Curie point of the second phase and hence the temperature at which the coercivity is a maximum are reduced to below room temperature. At 600°C the Curie point of the second phase and hence the temperature of the coercivity maximum increase above room temperature, as mentioned in Section 4.2.5. Put another way, if a magnet of Alcomax or Hycomax is used at 600°C the coercivity at the temperature may change little, while that at room temperature falls drastically. Such a magnet may function satisfactorily, while it is held at a high temperature, but become demagnetized if it is cooled to room temperature. Treatment at about 560°C at least partially repairs damage done at temperatures above 600°C.

Ferrite magnets suffer no structural changes below at least 1000°C. They are of course completely demagnetized if they are heated above their Curie point, which is about 450°C. They are the only type of permanent magnet that can be safely demagnetized by heating.

Cr–Fe–Co appears to be usable to about the same temperatures as Al–Ni–Co.

Although structural changes are not expected in CoPt below 500°C, it appears unusable above 200 to 300°C for other reasons.

Sintered $RCo_5$ magnets may be used up to 200 to 300°C, although oxidation may then cause trouble. There are certainly other structural changes at temperatures above 500°C, but the exact temperature at which these commence is not very certain and may depend on which rare-earth metal is used. The limit of use of $RCo_5$ is probably determined by changes in crystal anisotropy rather than structure.

Bonded magnets are usually limited by the temperature properties of the bond. For plastic bonds the maximum safe temperature is not at present usually greater than 100°C. Lodex is a fine particle material, bonded with low melting point metals, and the maximum recommended temperature is again about 100°C.

There seems to be little information available about the high temperature behaviour of Vicalloy, although there is no obvious reason for expecting it to be unstable.

## 5.5 REVERSIBLE AND IRREVERSIBLE CHANGES WITH TEMPERATURE

Reversible changes in the strength of a magnet with temperature are caused primarily by changes in the spontaneous magnetization with temperature. They can conveniently be represented by a temperature coefficient expressed as a percentage change in $B$ or $J$ per °C. For some materials and particularly the anisotropic Al–Ni–Co alloys, the process is somewhat more complicated, as the temperature coefficient is found to vary with the dimension ratio of the magnets, sometimes being smaller for shorter samples. This behaviour is rather puzzling; the facts will be stated and then some possible explanations will be discussed.

Irreversible changes in the strength of a magnet with temperature are caused by thermal fluctuations – in other words a great acceleration of magnetic viscosity after-effects at high temperatures, by changes in coercivity or more generally changes in the shape of the demagnetization curve with temperature. Irreversible changes take place during the first temperature excursion, and can be almost completely eliminated by cycling the material through a slightly wider temperature range than that through which they are to be used. Stabilization by an alternating field gives some protection against irreversible temperature changes, but is less efficient than temperature cycling for this purpose. In general it appears that the most efficient method of stabilizing against any environmental change, is to apply that change itself. Irreversible losses are normally less in long samples than in short ones, as they are less if $B/\mu_0 H$ is large.

Attempts to tabulate the temperature behaviour of various permanent magnet materials have been made by Hadfield (1962), Parker and Studders (1962), Schüler and Brinkmann (1970) and Gould (1971). The data given in these books is taken from a considerably larger number of individual references which are usually but not invariably given.

A perusal of all the tables and charts in these books suggests a bewildering diversity of results, while a perusal of only one book may suggest a misleading simplicity and degree of certainty.

Instead of attempting to reduce the temperature behaviour of permanent magnets to tabular form, the policy in this book is to devote a short section to each class of material. These sections attempt to explain to the reader the range of usefulness of each material, and the precautions that are recommended to secure satisfactory service. A few experimental results are quoted as illustrations or explanations of the probable behaviour, and an attempt is made to distinguish between what normally happens, and what may occasionally occur in a freak sample.

### 5.5.1 Magnet steels

Magnet steels being metallurgically unstable are not recommended for use above 50°C. They should be annealed at a slightly higher temperature to promote stability by allowing some of the metallurgical changes to occur. A typical reversible temperature coefficient is $-0.03\%\,°C^{-1}$ between 0 and 50°C. There are no anomalous effects at low temperatures.

### 5.5.2 Isotropic Al–Ni–Co

Materials in class A1 (isotropic Al–Ni–Co alloys with 0 to 12% Co) have coercivities that fall steadily with increasing temperature. No irreversible losses are to be expected as a result of cooling, but the reduction of coercivity on heating adds to the losses in magnetization produced by thermal fluctuations. Table 5.2 taken from Clegg and McCaig (1958) shows how these losses vary with temperature and working point for materials of this class.

Although the losses are rather large these materials once they have been stabilized could be used up to 500 or 550°C, although the anisotropic materials described in the following sections are preferable.

At first sight one might expect the reversible temperature coefficient to be equal to the change of spontaneous polarization with temperature. This would lead to temperature coefficients ranging from $-0.031\%°C^{-1}$ for the material with 12% Co to $-0.046\%°C^{-1}$ for the cobalt-free grade. In fact Clegg and McCaig found values between 0 and 200°C ranging from $-0.014$ to $0.024\%°C^{-1}$ for these materials. The value was higher for $B/\mu_0 H = 37.4$ than for short magnets with this ratio = 9.5.

They attempted to account for the differences by taking into account the recoil permeability of the materials. They supposed that first the polarization falls by the expected percentage. As a result the self-demagnetizing force also falls, and the actual polarization recovers somewhat because the relative recoil permeability is greater than unity. This explanation predicts the observed type of variation of the temperature coefficient with working point, but fails to account for the whole discrepancy between the temperature coefficient of magnetization of magnets, and the temperature coefficient of spontaneous polarization. This discrepancy remains something of a mystery and it is surprising that more effort has not been made to solve it.

Naturally the temperature coefficient must increase with increasing temperature. In spite of the above mentioned discrepancies, the temperature coefficients of isotropic Al–Ni–Co alloys are reasonably predictable compared with those of some materials.

### 5.5.3 Anisotropic Al–Ni–Co alloys

The anisotropic Al–Ni–Co alloys behave differently from the isotropic ones, probably because of their higher cobalt content, rather than

**Table 5.2** IRREVERSIBLE LOSS OF MAGNETIZATION AFTER HEATING FULLY MAGNETIZED ISOTROPIC Al–Ni–Co ALLOYS CLASS A1

| | | *Percentage loss after heating to stated temperature (°C)* | | | | | |
|---|---|---|---|---|---|---|---|
| *Material* | $-B/\mu_0 H$ | 200 | 300 | 400 | 450 | 500 | 550 |
| 0%Co | 37.4 | 8.7 | 14.1 | 19.1 | 22.4 | 27.2 | 32.9 |
| Alni | 21.6 | 10.0 | 16.2 | 22.8 | 27.2 | 32.0 | 26.8 |
| | 9.5 | 11.2 | 17.3 | 25.7 | 30.8 | 37.0 | 42.0 |
| 12%Co | 37.4 | 6.0 | 9.2 | 134. | 15.8 | 18.2 | 20.8 |
| Alnico | 21.6 | 7.1 | 11.0 | 15.8 | 18.7 | 21.5 | 24.2 |
| | 9.5 | 5.7 | 10.7 | 15.0 | 18.0 | 20.8 | 23.2 |

because they are anisotropic. The coercivity temperature curve of fully treated anisotropic Al–Ni–Co alloys usually exhibits a broad maximum. Depending upon the precise composition and heat treatment of the alloy, the maximum may be above or below room temperature. The temperature at which the maximum occurs greatly influences the temperature characteristics of the magnets.

Figure 5.5, which is extracted from Clegg (1955) illustrates the behavious for an A2 type material, Alcomax III. If the temperature is reduced still lower to the boiling point of liquid nitrogen, the irreversible losses in magnetization, on returning to room temperature are still greater; 5.3% for a dimension ratio of 4.5 and 9.9% for a dimension ratio of 2.7. On the other hand if the magnets are first stabilized by a knock-down in an alternating field, losses due to cooling to –60°C do not exceed 1%.

It has been reported that the losses do not continue to increase if the cooling is extended to 4 K.

It would be quite meaningless to give tables of the low temperature losses as if they were figures to be used as design data. Clegg pointed out that these losses can be almost eliminated by an extra treatment at 530°C. The actual behaviour of particular magnets is likely to depend

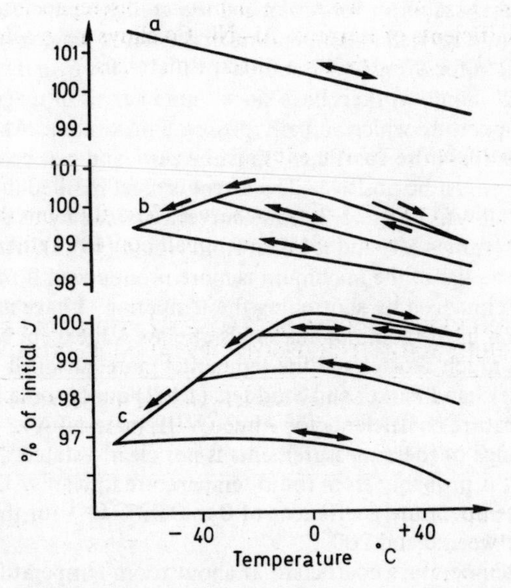


*Fig.5.5. Temperature variation of J for material Class A2 (Alcomax III). (a) L/D = 13.2 (b) L/D = 4.5 (c) L/D = 2.7*

very much on the particular heat treatment cycle adopted by the manufacturer. Clegg also measured demagnetization curves for the material at +15, −75 and −180°C as shown in Fig.5.6. On this figure the $B/\mu_0 H$ lines are superposed for various length to diameter ($L/D$) ratios. Irreversible losses after cooling are predicted if the demagnetization curve at the low temperature lies inside that at room temperature. This situation arises for working points below approximately the $(BH)_{max}$ point, but again it must be emphasized that these curves apply to a particular batch of material, and the curves for material from another source with nominally the same properties could differ appreciably from these. We did not make materials of the types containing Ti at the time Clegg carried out these experiments, but according to Parker and Studders (1962) the low temperature losses are rather large for material A3 (Alnico VI) but quite small for material A4 (Alnico VIII).

As the coercivity of the anisotropic Al–Ni–Co material falls only a little with increasing temperature, the irreversible losses after heating are much less than with the isotropic types, and on materials of type A2 Clegg found no irreversible losses greater than 1.2% after exposure to 300°C, and 2.7% after exposure to 550°C. There is, however, some evidence that the losses may be slightly greater in some high coercivity high Ti containing materials such as A4 or A5, being of the order 1.5 to 3.0% after 300°C. On the other hand there is some evidence that these materials are slightly superior for prolonged use at high temperatures.

Clegg found reversible temperature coefficients ranging between $-0.022\%°C^{-1}$ for $B/\mu_0 H = 37$ and $0.013\%°C^{-1}$ for $B/\mu_0 H = 9.5$ with material A2. Some workers have since found curves of magnetization against temperature which actually possess a maximum. At this maximum the temperature coefficient must be zero, and just below the maximum it must be positive. The effect is most marked in some Ti containing alloys. Figure 5.7 shows curves obtained by us on a material of class A3 (Alnico VI) and is taken from Gould (1971). As can be seen from the figure the maximum is more pronounced if $B/\mu_0 H$ is small and can be enhanced by shortening the tempering. I have not seen this phenomenon in the common materials of class A2, except when the sample was much shorter and the tempering more curtailed than normal. On the other hand Parker and Studders (1962) quote some large positive temperature coefficients for Alnico VIII, material A4. The temperature range of these measurements is not clearly stated, but from the context is probably from room temperature upwards. Gould (1971) suggests a temperature coefficient of 0 to $0.02\%°C^{-1}$ for this class of material between 0 and 200°C.

A zero temperature coefficient at about room temperature appears a very desirable property, but it seems very difficult to reproduce this behaviour consistently from batch to batch of material, and no manu-

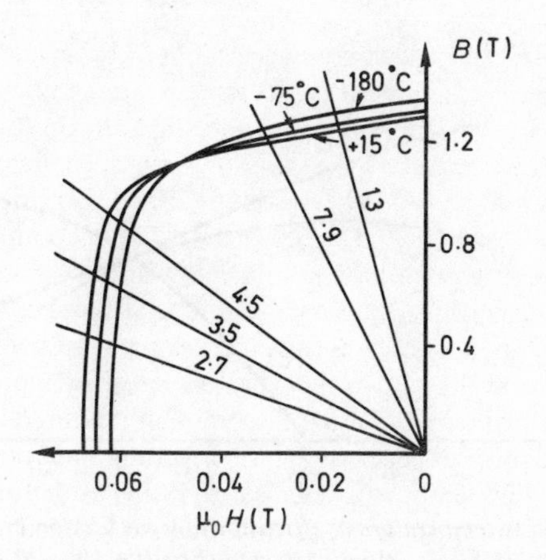


*Fig.5.6. Demagnetization curves at −180, −75 and +15°C for material of Class A2. L/D values are shown on the load lines*

facturer is as far as I know guaranteeing this attribute.

The most probable explanation of the maximum in the reversible magnetization curves of some materials is one suggested by Tenzer (1957), and supplemented in private discussions. He supposes that the magnetization of the low Curie point second phase is reversed by the self-demagnetizing field of the magnet. The shorter the *L/D* ratio of the magnet, the greater is this self demagnetizing field, and the more probable is this magnetization reversal. This explanation accounts for the positive temperature coefficient and maximum in the magnetization temperature curve being more pronounced in short magnets. It also accounts for the observation that the maximum in the magnetization temperature curve is more likely to be observed after a curtailed heat treatment, which is likely to leave a material with a higher Curie point. Tenzer also quotes an opposite example of an exceptionally large negative temperature coefficient of magnetization being found in a very long magnet. I have come across one example in which this large coefficient appeared in a commercial magnet. In a very long magnet with a small demagnetizing field, the magnetization of the second phase may actually increase the magnetization of the whole magnet below its Curie point.

### 5.5.4 Cr–Fe–Co

In the last few months of the existence of our laboratory, we investigated the temperature stability of some experimental Cr–Fe–Co alloys

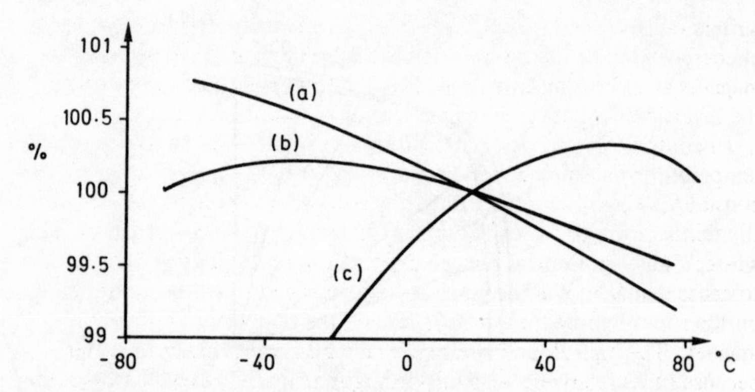


*Fig.5.7. Positive temperature coefficient of magnetization in material A3 (Alnico VI with 1% Ti). (a) Fully treated ($B/\mu_0H = 19$). (b) Fully treated ($B/\mu_0H = 8.7$). (c) Tempering shortened ($B/\mu_0H = 8.7$)*

(McCaig, 1975). The range of usefulness, the temperature coefficients, and the way in which they depend on dimension ratio, clearly parallel the behaviour of anisotropic Al–Ni–Co alloys. Even positive temperature coefficients and maxima in the magnetization temperature curves were found. There is probably a slightly greater irreversible loss in magnetization after short samples have been exposed to temperatures of 500°C for a short time. This loss is associated with a small decrease in coercivity with temperature, but stabilization in an alternating field gives complete immunity. Irreversible losses after cooling below room temperature may exist, but are very small.

Experiments of the above type will need to be repeated on commercial magnets if and when they are introduced. The above experiments do show that the alloy can be developed to replace Al–Ni–Co alloys for applications requiring good temperature stability.

## 5.5.5 Ferrites

The spontaneous polarization $J_s$ of ferrites falls almost linearly with temperature between –100 and +400°C. Expressed as a percentage of the value of room temperature, the change is about $-0.19\%\,^{\circ}\mathrm{C}^{-1}$, and this seems to be about the accepted value of the temperature coefficient of magnetization of ferrite magnets, irrespective of grade and dimensions. As explained in Section 4.6.3, the intrinsic coercivity of a material based on crystal anisotropy is a function of $2k_1/J_s$, and in barium and strontium ferrite this ratio increases with temperature. In consequence $H_{ci}$ increases with temperature; values quoted by different writers range from 0.2 to 0.5% per °C. As a consequence of this increase in coercivity there are no irreversible losses in magnetization of ferrite magnets after heating to at least 300°C, although at this temperature the reversible reduction in magnetization amounts to about 60%.

In contrast serious demagnetization can result from exposure to low temperatures, because at sub-zero temperatures the reduction in $H_{ci}$ is considerable. Unless appropriate precautions are taken in the design, the temperature to which equipment may be exposed in transit on a winter's night in Central Europe or North America are sufficiently low to cause damage. Whether such losses occur and to what extent depends on the room temperature properties and the dimension ratio of the magnet. The high $B_r$ and $(BH)_{max}$ grade F2 is most likely to suffer because of its relatively high self-demagnetizing field as well as its rather low room temperature coercivity. The most useful advice seems to be that quoted by Parker and Studders (1962) as the recommendation of an American manufacturer. To avoid the irreversible losses as the result of exposure to −60°C, the minimum recommended dimension ratio $(L/D)$ is 1.2 for the anisotropic high $B_r$ grade (F2), 0.5 for the anisotropic high coercivity grade (F3) and 0.4 for the isotropic grade (F1). There are probably now slightly improved versions of the high coercivity anisotropic material that permit slightly smaller values of $(L/D)$.

### 5.5.6 Rare-earth cobalt

A number of investigators have made brief references to the temperature behaviour of $RCo_5$ magnets, but by far the most systematic publications have come from the University of Dayton: Mildrum, Hartings and Strnat (1974) and Mildrum, Hartings, Wong and Strnat (1974). Their work has the advantage of being an independent survey of a considerable number of commercial samples from two different suppliers, although it is pointed out that the magnets were received in 1972, and since that year methods of production have been improved.

They found considerable variations in both the room temperature properties and stability performance of magnets from the same supplier, but little correlation between the stability and initial properties. There were considerable differences between the properties of the magnets from the two manufacturers, who used different methods of production, as discussed in Section 4.7.1. There was little difference, however, between their stability characteristics.

The anisotropy field and coercivity of $SmCo_5$ falls with increasing temperature, and no irreversible losses are expected or observed as a result of exposure to low temperatures.

The procedure used to investigate performance at high temperatures was as follows. Magnets with working point $B/\mu_0 H$ of approximately 1.0 were used. The flux density was measured in the as-magnetized state at room temperature by pulling off a coil connected to an integrating fluxmeter with digital read-out. The magnet was then heated to a given temperature at which it was held for ¼ hr; it was then cooled to room temperature and the measurement of B was repeated. This procedure was repeated for increasing times, sometimes exceeding 100 hr at the same temperature. The experiment was carried out for a number of different temperatures, several magnets being used for each temperature.

Figure 5.8 is a typical curve showing fairly schematically the variation of $B$ with time for unstabilized magnets. There is an initial drop in $B$ from a to b, most of which occurs in the first 15 min. Thereafter there is what the authors describe as a plateau from c to d, along which $B$ drops by only one or two percent. Finally at d a more rapid and catastrophic drop commences. The time interval from b to d is only about 10 hr at 300°C, but increases rapidly as the temperature is reduced. Table 5.3 summarizes the results obtained by Mildrum, Hartings, Wong and Strnat (1974) for unstabilized magnets. Table 5.4 gives results obtained by the same authors for magnets stabilized at a higher temperature. The two figures given for the losses indicate the range of values obtained on different magnets.

The most recent and comprehensive measurements of the stability of rare-earth magnets are given by Mildrum and Wong (1976). The information is too detailed to permit more than a brief summary. $SmCo_5$ is found to have a reversible temperature coefficient between −0.032 and −0.039%°C$^{-1}$ at room temperature, and an average value

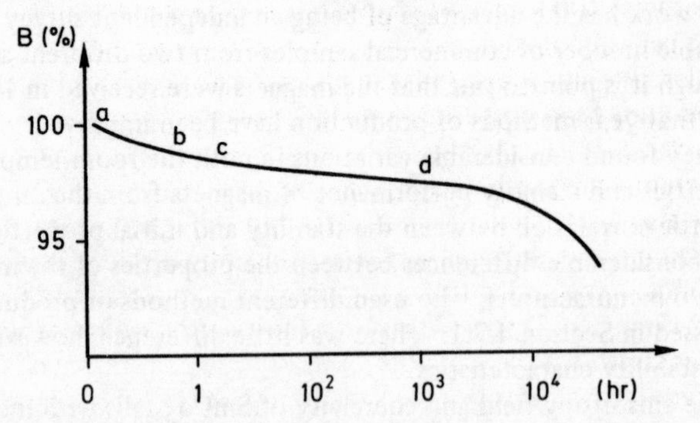


*Fig.5.8. Typical effect of time at temperature above 200° C on room temperature B of unstabilized $SmCo_5$ magnets*

**Table 5.3** PERCENTAGE FLUX LOSS ON NON-STABILIZED SINTERED $SmCo_5$ MAGNETS AFTER HEATING

| *Temp. °C* | *Initial loss at room temp. after ¼ hr at temp. (%)* | *Holding time at temp. (hr)* | *Final loss at room temp. (%)* |
|---|---|---|---|
| 25 | nil | 4200 | 0.18 to 0.4 |
| 150 | 1.6 to 3.8 | 3000 | 1.2 to 1.8 |
| 200 | 2.9 to 5.3 | 3000 | 0.9 to 2.3 |
| 250 | 4.3 to 12.1 | 3000 | 4.1 to 10.1 |

of $-0.046\%°C^{-1}$ between 25 and 250°C. If some misch-metal is used the coefficients are understandably higher, because of the lower Curie point, but high coefficients are also found with magnets containing Pr. $SmCo_5$ is also as good or better than the other materials for use at high temperatures, and provided it has been pre-stabilized can be recommended for use for 1000 hr at 250°C. There is a good correlation between the high temperature stability of a magnet and $H_k$, the demagnetizing force that reduces $J$ to $0.9\ J_s$. Similar measurements have recently been made by Clegg and Keyworth (1976).

Results of experiments on the stability of polymer bonded magnets at temperatures up to 100°C are given, but feed-back from these experiments may have already enabled manufacturers to improve on this performance.

New materials containing Gd or Ho, according to Jones and Tokunaga (1976), have a very low temperature coefficient. So far, however, it is not possible to obtain these materials with $(BH)_{max}$ greater than 100 $kJm^{-3}$ (12.5 MGOe).

We carried out some experiments on not very good experimental samples. We also found that the losses did not correlate with the room temperature loop properties, provided $\mu_0 H_{ci}$ was greater than 1 T. For lower values of $H_{ci}$ the irreversible losses after a short exposure to a high temperature are much greater.

We also found similar values of the reversible temperature coefficient on a sample containing only misch-metal as its rare-earth content.

It must be expected that quite different stability parameters will be found if radical changes are made in the rare-earth content. As evidence for this statement the structural instability of some magnets containing Pr referred to in Section 4.5.1 may be cited. It is of course equally possible that rare-earth magnets with improved stability characteristics may be discovered.

**Table 5.4** PERCENTAGE FLUX LOSS ON PRE-STABILIZED SINTERED $SmCo_5$ MAGNETS AFTER HEATING

| *Prestabilization* | | | *Ageing* | | |
|---|---|---|---|---|---|
| °*C* | *(hr)* | *Loss (%)* | *(hr)* | °*C* | *Loss at room temp. (%)* |
| 150 | 1 | 2.2 to 4.4 | 4200 | 25 | 0.05 to 0.12 |
| 200 | 1 | 3.1 to 4.5 | 4200 | 25 | 0.05 to 0.12 |
| 250 | 1 | 4.6 to 7.8 | 4200 | 25 | 0.00 to 0.21 |
| 200 | 2 | 2.7 to 7.9 | 3000 | 150 | 0.15 to 0.38 |
| 250 | 2 | 4.3 to 10.7 | 3000 | 200 | 0.64 to 0.77 |
| 300 | 2 | 5.5 to 13.4 | 3000 | 250 | 2.14 to 3.57 |

### 5.5.7 Other materials

The coercivity of Lodex ESD magnets increases at low temperatures, but they are not recommended for use at high temperatures. I have been unable to find a statement of their reversible temperature coefficient.

PtCo is stable between –200 and +100°C, but irreversible losses as great as 30% may occur if it is heated to 300°C, although there are no structural changes at this temperature. Parker and Studders (1962) show what appears to be a sudden change in the reversible temperature coefficient from $-0.015\%°C^{-1}$ between 20 and 100°C to a value many times greater at low temperatures.

I have found no data on Vicalloy, except that it is usable up to 400 to 500°C.

Cunife is said to be structurally stable up to 400°C, but to show considerable losses at this temperature.

## 5.6 MECHANICAL EFFECTS

As old textbooks recommend hammering a bar in the Earth's field as a method of magnetizing, and hammering while the bar is held in an east–west direction as a method of demagnetizing, the uninitiated may expect this section to be very important. These old experiments were performed on relatively low coercivity steels and mechanical shock and vibration are much more likely to break than seriously demagnetize modern permanent magnets.

Hadfield (1956) made a very thorough study of this subject, using ellipsoids designed to work at $(BH)_{max}$. By dropping each ellipsoid 200 times from a height of 3 ft on to hard wood, he produced a loss of about 5% in a 2% Co 4% Cr steel, but only about 0.5% in Alcomax III (material A2). It would probably be very difficult to measure any change at all in more modern materials with higher coercivity.

## 5.7 NUCLEAR RADIATION

There are a number of reasons why magnetic materials may be required to be used in nuclear reactors. Nuclear radiation may inflict some damage on soft magnetic materials, making them harder and increasing their coercivity. So far no changes in the properties of permanent magnets have been proved that cannot be attributed to incidental heating. Indeed in one experiment a control magnet, which was supposed to have been kept at the same temperature as one exposed to nuclear radiation suffered greater damage. Cobalt-containing magnets do of course become highly dangerous materials after exposure to neutron radiation.

## 5.8 CORROSION

Al–Ni–Co alloys are more resistant to corrosion than ordinary steel, although less resistant than stainless steel. They are satisfactory in most normal environments, and are only slightly discoloured by heating in air to 500°C. Ferrites are not subject to oxidation, but can be dissolved in certain acids. Rare-earth magnets are more subject to oxidation than the Al–Ni–Co alloys.

## 5.9 CONCLUSIONS ON STABILITY

The object of this final section on stability is to advise the reader how to choose the most suitable material, and how to secure stable operation in a particular environment. Thus it summarizes the conclusions to be drawn from the preceeding sections.

If the main hazards the magnets will meet are external fields from any source or random contact with ferromagnetic bodies, the most suitable materials in ascending order of cost and performance are isotropic ferrite, high coercivity anisotropic ferrite and rare-earth cobalt.

If the magnets are required to operate over a wide temperature range, anisotropic Al–Ni–Co magnets are indicated. Demagnetization by exposure to low temperatures has been mentioned in connection with these alloys, but will not occur provided the magnets are designed to work at a point with $B/\mu_0H$ a little above that corresponding to $(BH)_{max}$.

If the magnets are required to work at high temperatures they must be over-designed (made a little larger than for room temperature working) to allow for the expected reversible and irreversible losses. A similar over-design is necessary for most constructional materials required to work at high temperatures. If they are required to work at temperatures up to 500°C, they may be stabilized by first exposing them to a slightly higher temperature.

Magnets may be used for certain purposes at temperatures up to 600°C or even a little higher. At 600°C certain Al–Ni–Co magnets have a life comparable with that of an electric light bulb, and may be used for applications that do not require great constancy of flux density, such as manipulating objects in a furnace. They should be designed with a considerably greater $L/D$ ratio than would be required for $(BH)_{max}$ working at room temperature. Material of type A4, Hycomax III, is probably better than that of type A2, Alcomax II or III. The advantage is at least partly due to the fact that the optimum $(L/D)$ at room temperature is about 2 for the former and 4 for the latter material. Thus it is easier to obtain a high $L/D$ to protect against demagnetization without making unduly long magnets in the former material.

If magnets are to be used at temperatures of 600°C it is a good idea to keep them when not in use in a temperature of say 540 or 560°C. In this way they are protected against demagnetization which as explained in Section 5.4, may occur, when they are cooled to room temperature, as well as possibly recovering their properties to some extent. The demagnetization could be reversed by remagnetizing at the high temperature, but that is an inconvenient procedure.

If stability of properties is required at room temperature as in a scientific instrument, an anisotropic Al–Ni–Co alloy is again to be preferred. At first sight the range of temperature coefficients quoted: 0 to 0.02%°C$^{-1}$ appears alarming. This range only appears so great because it is in fact so small. If a magnet is designed for use at say 20°C, an uncertainty of ±0.01%°C$^{-1}$ in its temperature coefficient means an uncertainty of ±0.1% in its strength at 10 or 30°C. The change in the properties of many other materials that are used in instruments such as the resistance of copper wire is much greater than 0.1%, and in a normal laboratory the temperature is unlikely to stray outside the limits 10 to 30°C, unless there is a failure in a heating or air conditioning system.

The wide limits also allow for supplies from different manufacturers at different times. Magnets of a particular type produced in one batch from a given manufacturer should behave more consistently. The purchaser may have to test the magnets himself, possibly on a prototype of his own instrument.

An additional reason why the figures given in a book of this nature may not apply in a practical situation is that most of the measurements have been made on simple shapes such as bars or ellipsoids. It is not certain that say a C shaped magnet or an assembly with steel pole-pieces will behave in exactly the same manner as a bar of the same material, even if it is working at the same value of $B/\mu_0 H$.

The temperature coefficients of magnet assemblies are often neutralized by the incorporation of compensating alloys in the magnetic circuit. Compensating materials are alloys or ferrites with a low Curie point and consequently a permeability that changes rapidly with temperature. They may at the same time compensate for other temperature sensitive components in the equipment.

If alloy magnets are used precautions must be taken against their demagnetization by stray fields, and they should be protected from contact with steel objects and other magnets. It should be remembered that if untrained staff are given the opportunity to play with magnets, they probably will.

Care must be taken in servicing instruments containing magnets. Obviously assemblies containing Al–Ni–Co magnets that have been magnetized after assembly cannot be taken apart without a high probability of demagnetization. Even a skilled technician making adjustments to other parts of an instrument containing a magnet may accidentally allow a tool such as screwdriver or spanner to come in contact with the sides of a magnet. It may be worth making special non-magnetic tools for such work.

Ferrite magnets with the exception of the high $B_r$ anisotropic grade F2, require fewer precautions. F2 is also the material most likely to be demagnetized by exposure to low temperature. It is however a very popular material most likely to be met, because it is probably the cheapest permanent magnet material in terms of cost per unit of $(BH)_{max}$.

If rare-earth cobalt magnets are used it should be noted that their temperature coefficients are greater than those of Al–Ni–Co although still much less than those of ferrites. Accidental demagnetization except by heating is almost impossible. There has been considerable practical experience in using these materials in the temperature range 200 to 300°C, but as can be seen from Section 5.5.6 this is about the limit. The manufacture of rare-earth magnets is still developing, and the magnets available when this book is read may be quite different from those available when it was written.

# Chapter 6
# Design methods

For a few applications the designer is concerned directly with the distribution in magnitude and direction of the polarization in the magnet, in order to calculate the force or torque exerted on the magnet by some external field. More often the designer requires to calculate the field and possibly the field gradient produced by the magnet at some points outside the magnet or in a gap. This calculation also requires knowledge of the polarization within the magnet.

In the theoretical chapters magnets were regarded as made up of many tiny current loops. Practically, the mathematical description of the field produced by a dipole and a small current loop are the same. The atomic view of the origin of magnetism is very important theoretically, and for understanding the behaviour of magnetic materials, but for calculating external fields continuous models usually suffice.

The most realistic model is to suppose that each element of volume $\Delta V$ inside the magnet has a dipole moment $J\Delta V$, where $J$ is the average polarization within the element. The field at any external point is then the vector sum of the contributions of all the elements of volume. Sometimes this provides a practical method of calculating the field, and it is certainly the most direct method of dealing with non-uniform magnetization. The mathematics associated with this model tend to be rather complicated, and if it is used to calculate the self-demagnetizing field inside a magnet, difficulties arise because the contributions of the material close to the point at which the field is required rise to infinity.

The mathematics can often be simplified by using a pole model or a surface current model. The choice of one or other of these models should be made on purely utilitarian criteria of the ease with which they permit the correct result to be calculated, and not with any belief that either is more in harmony with our view of the nature of magnetism. Both actually involve very crude approximations.

The assumption of the pole model is that if dipoles are placed end to end in a chain (Fig.6.1) the N pole of one dipole neutralizes the S pole of the next, and only one N pole and one S pole are left at the ends of the chain. The same concept arises just as naturally from current loops. Each current loop is regarded as one turn in a long thin solenoid; a long thin solenoid is equivalent to a needle shaped magnet with a N pole at one end and a S pole at the other. The polarity on an element of area $\Delta A$ on the end of the magnet is then $J\Delta A$, or if the normal to the end surface is inclined at an angle $\theta$ to the direction of $J$, the polarity on $\Delta A$ is $J\Delta A \cos\theta$.

The advantage of replacing dipoles by poles is that the field due to a dipole obeys an inverse cube law, and depends on the angle between the direction of the dipole and the line joining it to the point at which the field is required, while the field due to a pole obeys a simple inverse square law and is directed along the line joining the pole to the point.

The third model replaces the individual current loops in a plane by a single turn. The individual tiny, widely separated current loops in one cross-sectional plane of a magnet (Fig.6.2(a)) are first greatly expanded so that they fill the whole plane as in Fig.6.2(b), and finally the currents in adjoining loops are neutralized to leave a single loop as in Fig.6.2(c). The step from (a) to (b) is of dubious validity, and makes it appear that there is no self-demagnetizing field, although the step from (b) to (c) is quite clear. The whole magnet is next replaced by a solenoid. There is no doubt that the solenoid model gives correct results for fields outside the magnet. There is little to choose between the pole and solenoid models for calculating external fields due to uniformly magnetized magnets. It happens that I am more familiar with the mathematics of the pole model, which has the additional advantage of being usable inside the magnet. These models are discussed by Craik and Harrison (1974) in connection with calculating the fields produced by uniformly magnetized cylinders.

Complications arise in either the pole or solenoid model, when problems with non-uniform magnetization are met, since the poles or currents can no longer be supposed to be situated solely on the surface of the magnet.

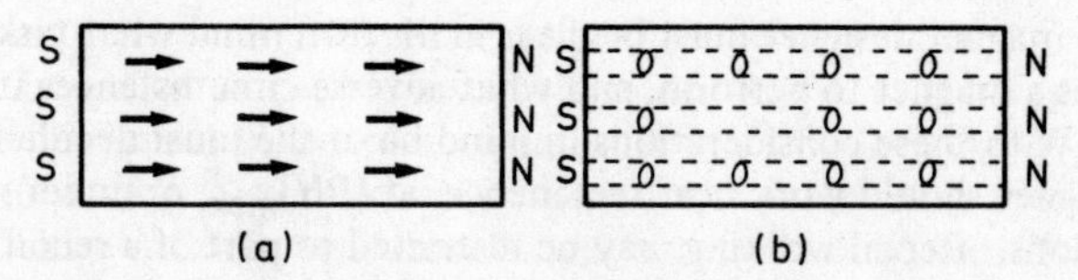


*Fig.6.1. A magnet can be represented by a distribution of N poles on one end and S poles on the other, whether the atomic magnets are regarded as forming chains of dipoles as in (a) or as turns of needle shaped solenoids as in (b)*

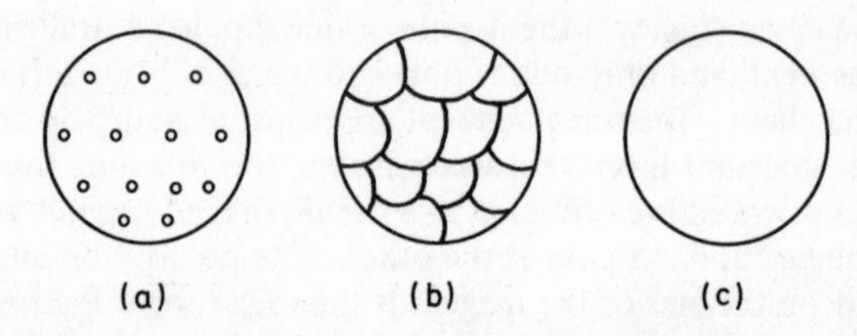


*Fig. 6.2. Surface current model. (a) Tiny, widely separated atomic circuits. (b) Grossly expanded loops filling whole section. (c) Single loop replacing expanded loops*

The use of the models can sometimes be simplified by introducing the scalar potential $\psi$. The integration of this scalar quantity for contributions from an extended source is simpler than adding the vector field quantities directly. If an expression for $\psi$ can be obtained $\boldsymbol{H} = -\text{grad}\,\psi$.

Alternatively a vector potential $\boldsymbol{A}$ such that $\boldsymbol{B} = \text{curl}\,\boldsymbol{A}$ is discussed in most advanced works on electromagnetism, and is used in some modern design calculations.

Sections 6.1, 6.2, and 6.3 deal with calculations using dipoles, poles and solenoids as models, respectively. Section 6.4 is concerned with calculations of the self-demagnetizing factors of ellipsoids and bars, and hence with estimating their working points.

Many magnets and magnet assemblies are used to produce flux in a more or less well-defined gap. These problems are discussed under the heading of the 'Magnetic Circuit' in Section 6.5. The magnetic circuit uses Ohm's law as an analogue of the magnetic problem, with reluctance and permeance used as analogous to resistance and conductance. In this section the concept of useful and leakage flux is discussed and some suggestions for tackling the very important but difficult problem of estimating leakage flux are made.

Dynamic working in which the reluctance of a circuit changes, as in a lifting magnet or varying external fields, are applied as in a motor or generator, are discussed in Section 6.8. Other sections of this chapter deal with a number of miscellaneous topics such as the use of computers in magnetic design, the possibilities of flux concentration and temperature compensation.

The magnet designer must be clear in his own mind what task he requires a magnet to perform, and what adverse circumstances it must resist. With these considerations in mind he or she must decide whether the magnet should work near remanence, at $(BH)_{max}$, or under recoil conditions. Recoil working may be restricted to part of a recoil line within the second quadrant or extend into the first or even the third quadrant. These considerations must be combined with a knowledge of whether the main aim is to maximize performance, minimize cost or minimize weight, in order that the most suitable material can be chosen. The geometry of the magnet may then dictate whether the magnet is best treated as a magnetic circuit or as a bar magnet with a self-demagnetizing coefficient. After answering these questions and making these decisions, the designer can choose the appropriate methods to calculate the required dimensions of the magnet and find the appropriate section in the following pages that describes the method to be followed. Sometimes it may be necessary to make calculations for different materials or different geometries to decide which is likely to give the best result. Finally before ordering a large consignment of magnets an experimental verification of the expected performance should be made, possibly on a scale model.

## 6.1 CALCULATION OF MAGNETIC FIELDS FROM DIPOLES

The magnetic dipole moment of a small volume of material $\Delta V$ m$^3$ with polarization $J$ tesla is $J\Delta V$. According to elementary magnetostatics the potential at a point P at a distance $r$ metres from this dipole, in a direction defined by an angle $\theta$, as shown in Fig.6.3 is

$$\psi = \frac{J\Delta V \cos\theta}{4\pi\mu_0 r^2} \quad \text{A} \tag{6.1}$$

The magnetizing force in the direction OP is

$$H_{OP} = -\frac{d\psi}{dr} = \frac{2J\Delta V \cos\theta}{4\pi\mu_0 r^3} \quad \text{Am}^{-1} \tag{6.2}$$

In the direction PQ at right-angles to OP, the component of magnetizing force is given by

$$H_{PQ} = -\frac{1 d\psi}{r d\theta} = \frac{J\Delta V \sin\theta}{4\pi\mu_0 r^3} \quad \text{Am}^{-1} \tag{6.3}$$

If flux densities in tesla are required $\mu_0$ is omitted from the denominators of Equations 6.2, and 6.3, while in CGS units $J$ becomes the 'intensity of magnetization' and $4\pi\mu_0$ is omitted from the denominator, whether $B$ or $H$ is required.

In theory it is possible to integrate Equation 6.1 over the volume of any magnet, if the distribution of $J$ is known, and then differentiate to find the components of $B$ or $H$. In practice the integration can only be carried out numerically, except for a few simple shapes and distributions.

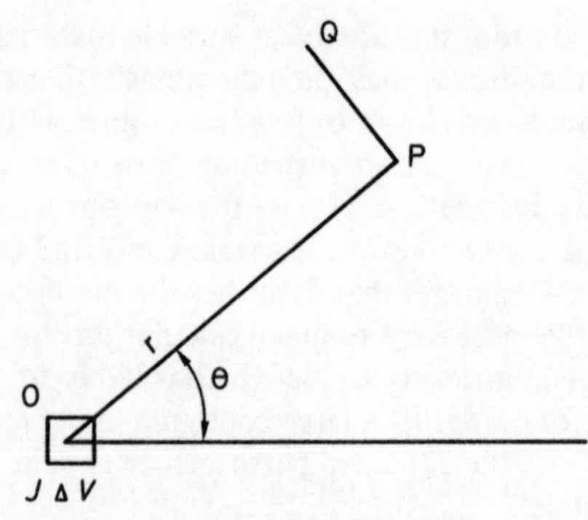


*Fig.6.3. Calculation of potential and magnetizing force at P due to magnetic moment JΔV.*

Modern computers have made numerical integration a reasonable proposition and the method is probably the most promising one for estimating the fields produced externally by magnets in which $J$ although not uniform has a distribution which is known or can be guessed.

## 6.2 CALCULATION OF MAGNETIC FIELDS FROM POLES

The old method of treating a magnet as having two point poles was not too bad an approximation for the long needle shaped magnets that were necessary with the old low coercivity steels, but it is quite unsuitable for use with the shapes in which magnets are made in modern materials. We can sometimes assume that there is a uniform distribution of poles on the ends of the magnet. As stated above the pole strength on area $\Delta A$ on a surface whose normal makes an angle $\theta$ to the direction of the polarization $J$, is $J\Delta A \cos\theta$. If $\theta = 0$ the potential at distance $r$ from such a small element of polarity is

$$\psi = J\Delta A/(4\pi\mu_0 r) \tag{6.4}$$

and the magnetizing force $H$ is given by

$$H = -\frac{\mathrm{d}\psi}{\mathrm{d}r} = \frac{J\Delta A}{4\pi\mu_0 r^2} \quad \mathrm{Am^{-1}} \tag{6.5}$$

To obtain $B$ in tesla multiply by $\mu_0$ and so eliminate $\mu_0$ from the denominator.

$$B = \frac{J\Delta A}{4\pi r^2} \quad \text{T} \tag{6.6}$$

In CGS units

$$B = H = \frac{J\Delta A}{r^2} \text{ gauss or oersted} \tag{6.7}$$

It is assumed that the reader can make similar adjustments to the ensuing equations. To obtain $B$ multiply $H$ by $\mu_0$, and to obtain the CGS value of $B$ or $H$ multiply $H$ by $4\pi\mu_0$.

The next step is to integrate Equation 6.4 to obtain the potential at a point P due to all the surface poles. This integration is a surface integration in contrast to the volume integration of dipoles described in the previous section, and it can be performed without recourse to numerical methods in a fair number of circumstances. The components of the magnetizing force are again obtained by differentiating with respect to the appropriate co-ordinate. The magnetizing force can sometimes be derived directly by integrating Equation 6.5, without calculating the potential.

The surface pole method will now be illustrated by calculating the field produced by a plane circular surface on a normal passing through its centre. This is an important practical case, because the field on the axis of a uniformly magnetized cylindrical magnet can be treated as produced by two such circles at its two ends.

The calculation is carried out with reference to Fig.6.4. The surface

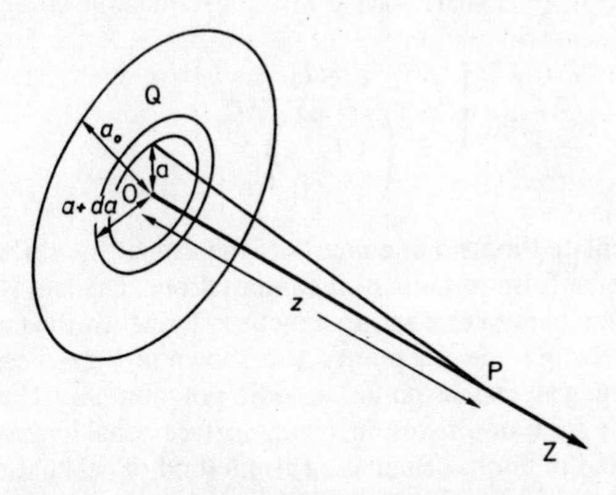


*Fig.6.4. Symbols used in calculation of field on axis of circular surface of uniform polarity. (Note OZ is perpendicular to circle)*

is a circle of radius $a_0$ and centre O. We require the potential $\psi$ and magnetizing force $H$ at a point P on OP normal to the plane of the circle. The co-ordinate of the direction OP is taken as $z$. Consider an element of area $\Delta A$ around a point Q on the surface at radial distance $a$ from O. The pole strength on $\Delta A$ is $J\Delta A$ and the distance OP is $r$. From Fig.6.4 it follows from the theorem of Pythagoras that $r = (z^2 + a^2)^{1/2}$. This relation holds for any point on an annular area between radii $a$ and $a + \mathrm{d}a$, so it is convenient to take such an area as our element of area $\mathrm{d}A$. This makes the value of $\mathrm{d}A = 2\pi a\mathrm{d}a$. The contribution of this annular element to the potential at P is

$$\mathrm{d}\psi = 2\pi aJ\mathrm{d}a/(a^2 + z^2)^{1/2}4\pi\mu_0$$

Integrating the potential contribution from the whole surface we have:

$$\psi = \frac{J}{2\mu_0}\int_0^{a_0} \frac{a\mathrm{d}a}{(a^2 + z^2)^{1/2}} \tag{6.8}$$

$$= \frac{J}{2\mu_0}\left[(a^2 + z^2)^{1/2}\right]_0^{a_0} = \frac{J}{2\mu_0}\left\{(a_0^2 + z^2)^{1/2} - z\right\} \tag{6.9}$$

By symmetry $H$ has no components perpendicular to $z$ and the value of $H$ in the direction of $z$ is

$$H_z = -\frac{\mathrm{d}\psi}{\mathrm{d}z} = -\frac{J}{2\mu_0}\left\{\frac{z}{(a_0^2 + z^2)^{1/2}} - 1\right\}$$

$$= \frac{J}{2\mu_0}\left\{1 - \frac{z/a_0}{\left(1 + \dfrac{z^2}{a_0^2}\right)^{1/2}}\right\} \tag{6.10}$$

To calculate the field at a point off the axis of the surface or cylinder requires some form of numerical computation. The late N. Davy drew my attention many years ago to a method by which the components of the magnetizing force at a point $z,y$ as shown in Fig.6.5 can be calculated by using an expansion in Legendre polynomials. The result is given in the form of a series, involving surface zonal harmonics, which can be found in books of tables. This method of calculation is convenient, except when $r$ is about equal to $a_0$, in which case the series does not converge sufficiently quickly.

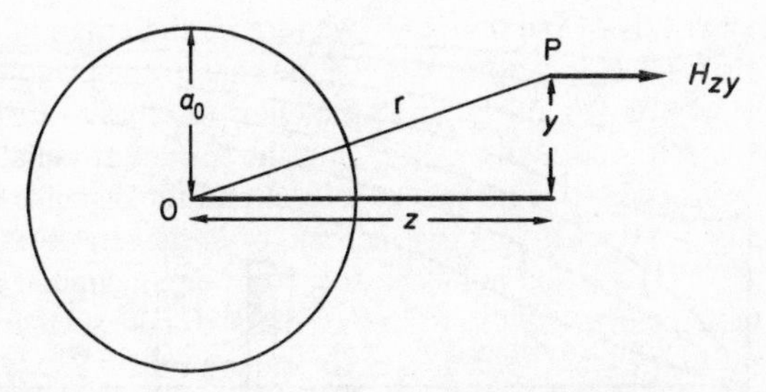


*Fig.6.5 Field at point off the axis of circle of uniform polarity (z is normal to plane of circle)*

An alternative method of calculating the field off the axis was communicated to me privately by J. C. Williamson in 1958. At that time Williamson was employed by one of the Member firms of the PMA. This method is in effect a numerical integration over the circular surface in Fig.6.5, with the complication that to obtain each term that enters this integration it is necessary to look up from tables the value of a complete elliptic integral of the second kind. This method is in general more tedious than the method using surface zonal harmonics, but is useful when the series in that method fails to converge.

Figure 6.6 gives ratios of the axial field off and on the axis, which I calculated (McCaig, 1961), using a combination of the two methods.

I do not think that either Davy or Williamson claimed that their method was original, but I am unable to give their references. Most of the formulae are contained in a recent publication by Craik and Harrison (1974). Radial components of the field can be calculated by similar methods if required.

Rather surprisingly the integrals for this potential and the magnetizing force produced by rectangular surfaces with uniform polarity can be evaluated precisely. Again the method was pointed out to me by Williamson many years ago. The derivation is rather involved and only the final formulae are quoted below; the symbols are explained in Fig.6.7.

$$H_x = \frac{J}{2\mu_0} \log_e \left[ \frac{y + b + \{(y+b)^2 + (x-a)^2 + z^2\}^{1/2}}{(y-b) + \{(y-b)^2 + (x-a)^2 + z^2\}^{1/2}} \times \frac{(y-b) + \{(y-b)^2 + (x+a)^2 + z^2\}^{1/2}}{(y+b) + \{(y+b)^2 + (x+a)^2 + z^2\}^{1/2}} \right] \qquad (6.11)$$

$$H_y = \frac{J}{2\mu_0} \log_e \left[ \frac{(x+a) + \{(y-b)^2 + (x+a)^2 + z^2\}^{1/2}}{(x-a) + \{(y-b)^2 + (x-a)^2 + z^2\}^{1/2}} \times \right.$$

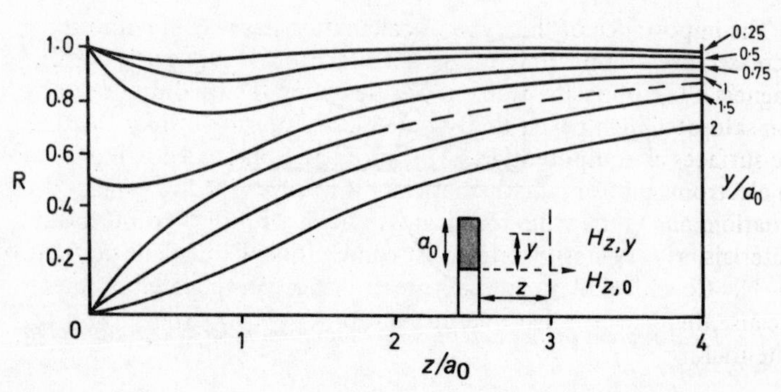


*Fig.6.6. Ratio of components of fields off and on the axis of circular surface: $H_{z,y}/H_{z,o}$*

$$\left.\frac{(x-a)+\{(y+b)^2+(x-a)^2+z^2\}^{1/2}}{(x+a)+\{(y+b)^2+(x+a)^2+z^2\}^{1/2}}\right] \tag{6.12}$$

$$H_z = \frac{J}{2\mu_0}\left[\tan^{-1}\left(\frac{(x+a)(y+b)}{z\{(x+a)^2+(y+b)^2+z^2\}^{1/2}}\right)\right.$$

$$+\tan^{-1}\left(\frac{(x-a)(y-b)}{z\{(x-a)^2+(y-b)^2+z^2\}^{1/2}}\right)$$

$$-\tan^{-1}\left(\frac{(x+a)(y-b)}{z\{(x+a)^2+(y-b)^2+z^2\}^{1/2}}\right)$$

$$\left.-\tan^{-1}\left(\frac{(x-a)(y+b)}{z\{(x-a)^2+(y+b)^2+z^2\}^{1/2}}\right)\right] \tag{6.13}$$

These calculations allow all three components of the field at any point due to a rectangular surface of uniform polarity to be calculated. The values for one point can be calculated in a few minutes with tables or a simple calculator. We once had a computer programmed to give us several thousand values. One of the better programmable desk calculators would probably now suffice for this purpose.

The calculation of fields by integrating the contribution from surface poles is not confined to cylinders or cuboids. I have used it for instance in problems involving conical pole-pieces. A ring can be treated as the difference between two cylinders, one with the outer and one with the inner diameter.

The importance of this type of calculation has waxed with the increased use of uniformly magnetized ferrite and rare-earth cobalt magnets. The opposite problem of calculating the field produced by non-saturated high permeability pole pieces is dealt with by treating the surfaces as equipotentials. This problem is discussed in many books on electromagnetism. Mathematically it involves solving Laplace's equation, and is the same for magnetic fields with high permeability materials, as for electric fields with conductors. Permanent magnets of Al–Ni–Co and similar materials present more complicated problems, because their surfaces have neither uniform polarity nor uniform potential.

## 6.3 CALCULATION OF MAGNETIC FIELDS FROM SURFACE CURRENT MODELS

A uniform solenoid of rectangular or circular cross-section with surface current density $I$ Am$^{-1}$ is equivalent for the purpose of calculating flux density to a bar with uniform magnetization

$$M = I = J/\mu_0 \tag{6.14}$$

Note that the surface currents give no indication of the magnetomotive force that a magnet can produce in a magnetic circuit, and that a uniform solenoid represents a magnet with a uniform $M$ or $J$, but a non-uniform $B$ or $H$. If a solenoid has a circular cross-section, $B$ on its axis is given by

$$B = \mu_0 I(\cos\theta_1 - \cos\theta_2)/2 \tag{6.15}$$

where $\theta_1$ and $\theta_2$ are defined by Fig.6.8. Although $H$ inside an empty solenoid is calculated from the ampere turns and dimensions of the

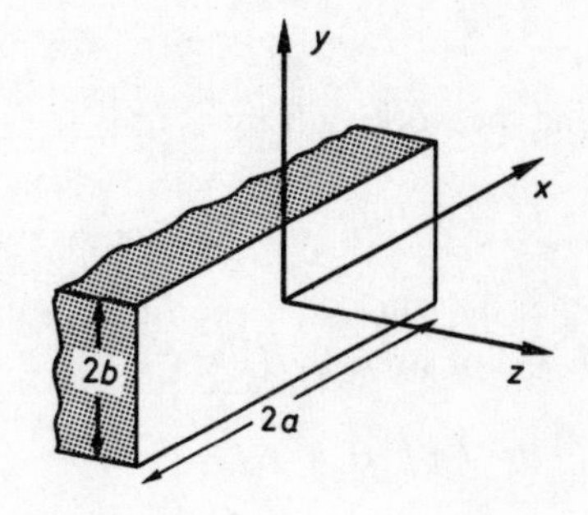


*Fig.6.7. Explanation of symbols used in Equations 6.11, 6.12 and 6.13*

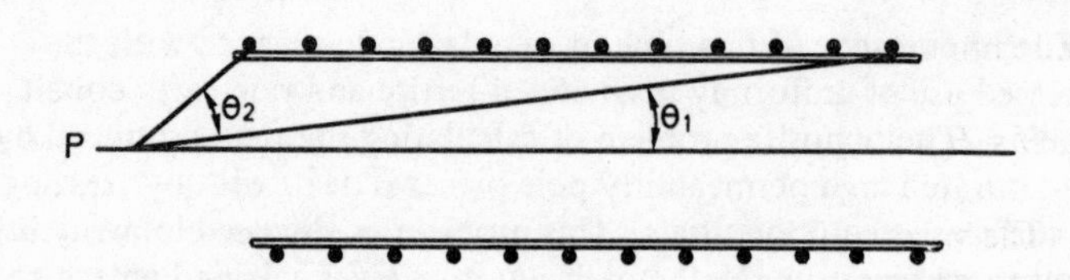


*Fig.6.8. Uniform solenoid model of magnet*

solenoid, the direction of $H$ inside a permanent magnet is in the opposite direction to that which would be produced by the equivalent solenoid.

Forces between magnets can in principle be calculated by integrating the interactions between the turns of their equivalent solenoids. The fields at points off the axis can similarly be found by summing contributions from all the conductors. Craik and Harrison (1974) use the vector potential and obtain solutions in terms of elliptic integrals or Legendre polynomials in much the same way as for pole models. They appear after using both methods to find the pole model slightly more adaptable. The most significant comment they make is possibly 'Discussion of the relative correctness of these two representations is considered sterile; all that is important is their relative usefulness'. I happen to be more accustomed to using the pole model; others may prefer the current models for similar reasons.

## 6.4 SELF-DEMAGNETIZING FACTORS

The concept of a self-demagnetizing factor applies strictly to ellipsoids. If an ellipsoid has a uniform polarization $J$, it has a uniform self-demagnetizing field. This field is derived from $J$ by using the self-demagnetizing factor $N$, which is a purely geometrical quantity determined by the ratio of the axes of the ellipsoid.

In the absence of any external fields the working $B$ of the ellipsoid is given by:

$$B = J - NJ \tag{6.16}$$

and the working $H$ by

$$H = -NJ/\mu_0 \tag{6.17}$$

If the ellipsoid is exposed to an external magnetizing force parallel to $J$, and of strength $H_e$ $\text{Am}^{-1}$

$$B = J + \mu_0 H_e - NJ \tag{6.18}$$

and

$$H = H_e - NJ/\mu_0 \tag{6.18a}$$

The value of $N$ for ellipsoids of revolution along the axis of symmetry, when $p$ is the ratio of the axes is given by Equations 6.19 and 6.20.

$$N_a = \frac{1}{p^2 - 1}\left[\frac{p}{\sqrt{p^2 - 1}} \log_e \{p + (p^2 - 1)^{1/2} - 1\}\right] \tag{6.19}$$

for prolate ellipsoids ($p > 1$), and

$$N_a = \frac{1}{1 - p^2}\left[1 - \frac{p}{(1 - p^2)^{1/2}} \cos^{-1} p\right] \tag{6.20}$$

for oblate ellipsoids ($p < 1$).

The formulae break down for a sphere for which $p = 1$, and for which $N$ is actually 1/3.

Values of $N$ for spheroids have been tabulated by Stoner (1945) and for the general ellipsoid by Osborn (1945).

In CGS units with $J$ the intensity of magnetization $H = -4\pi NJ$. Some writers give tables of $N$ and others tabulate $4\pi N$. In both cases the table may be described as a table of demagnetizing factors. Readers are warned to be careful to make sure which method is being used.

In a magnet with any shape other than an ellipsoid it is still possible for $J$ to be uniform if the magnet has a square loop or a high coercivity, or even if it is subjected to a saturating magnetizing force, but it is almost impossible for $B$ and $H$ to be uniform. If $J$ is uniform the non-uniform self-demagnetizing force can be calculated by means of the pole model using the appropriate equations of Section 6.2.

Figure 6.9 shows the directions of the self-demagnetizing force $H$ and flux density $B$ in a bar with uniform $J$. A uniform solenoid with $M$ ampere-turns per metre represents a magnet with uniform magnetization $M$ $\text{Am}^{-1}$, where $J = \mu_0 M$. Obviously it cannot represent a uniform flux density unless it is an infinite solenoid. It does have the same non-uniform distribution of $B$ as the equivalent magnet, and one could possibly regard the self-demagnetizing flux density as the end correction of the finite solenoid, but this is not really the easiest way of calculating or understanding the significance of this field.

Uniform magnetic polarization $J$ occurs very commonly in modern ferrite and rare-earth materials. Three particular cases are worth noting. Just inside the end of a long bar with polarization $J$, $B = \frac{1}{2}J$ and

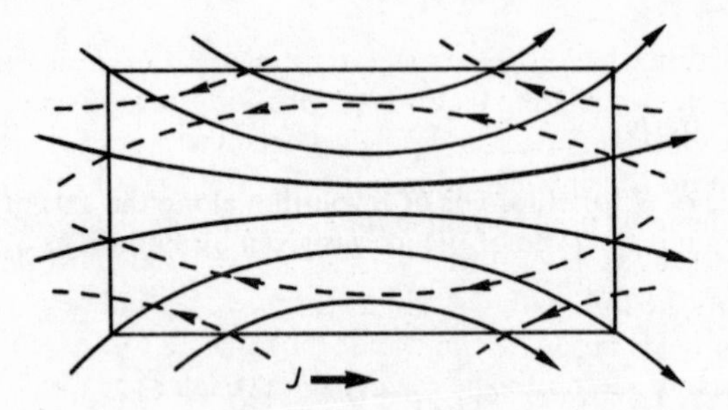


*Fig.6.9. Direction of H (dotted lines) and B (full lines) in magnet with uniform J*

$H = -\frac{1}{2}J/\mu_0$. In a thin film of large area magnetized through its thickness, $B = 0$ and $H = -J/\mu_0$; the value of $J$ can still be substantial. A sphere being an ellipsoid can be uniformly magnetized even if it has not a very high coercivity. Within a sphere $B = 2J/3$ and $H = -J/3\mu_0$.

In older permanent magnets such as those of Al–Ni–Co alloys the fact that $H$ is non-uniform, makes $J$ non-uniform as well. The situation is then very complicated.

In all magnets other than ellipsoids the concept of a self-demagnetizing factor is only an approximation. If we attempt to measure $B/\mu_0 H$, the result depends to some extent on the method of measurement we choose. So called magnetometer methods measure the magnetic moment of the magnet, which is the volume integral of the component $J_z$ over the whole magnet. Ballistic methods with a coil and fluxmeter may measure the mean value of $B$ over one cross-section if a short coil is used or over any fraction up to the whole length of the magnet if a longer coil is used. Some writers refer to the magnetometric and ballistic demagnetizing factors; the ballistic value probably refers to measurements over a short length in the middle. $H$ cannot normally be measured in the interior of a magnet. Devices used to measure $H$ include flat coils, magnetic potentiometers and solid state devices such as Hall effect elements. They all measure $H$ or more correctly $B$ just outside the magnet. Some measure it at what is almost a single point and some over a considerable area of the surface. Theory requires that the tangential component of $H$ is continuous on both sides of a surface, so that it can be legitimately claimed that $H$ has been measured just inside the surface. In a permanent magnet material with low permeability it cannot be claimed that the value measured is the same as that well below the surface.

In theory $B$ and $H$ can be calculated at any point inside the magnet, provided the properties of the material and its previous history are known. Except in a few simple cases this calculation requires a sophisticated computer programme.

The question which is the most correct method of measuring or calculating the demagnetizing factor is probably the wrong one to ask. It is better to consider which method gives the most accurate and useful information for the application in mind.

One of the most used formulae for estimating demagnetizing factors or $B/\mu_0 H$ ratios was derived by Evershed about 1920. He calculated the reluctance of the flux path outside the magnet, instead of directly calculating the fields inside. He imagined a bar magnet to have two spherical poles each of radius $r$. The reluctance of a spherical shell of radii $r$ and $r + \mathrm{d}r$ (Fig.6.10) is $\mathrm{d}r/4\pi\mu_0 r^2$. Hence the reluctance from a sphere of radius $r'$ to $\infty$ is

$$R = \int_{r'}^{\infty} \frac{\mathrm{d}r}{4\pi\mu_0 r^2} = \frac{1}{4\pi\mu_0 r'} \tag{6.21}$$

The reluctance from one spherical pole to the other is taken as twice the reluctance from one spherical pole to infinity, and is therefore $1/2\pi\mu_0 r'$. Alternatively the permeance P is $2\pi\mu_0 r'$.

The next step is to introduce the surface area $S$ of each pole. For a sphere of radius $r'$, $S = 4\pi r'^2$, so

$$P = 2\pi r' \mu_0 = 2\pi\mu_0 (S/4\pi)^{1/2} = (\pi S)^{1/2}\mu_0 \tag{6.22}$$

The next somewhat imaginative step taken by Evershed was to use this value of the permeance for magnets of various shapes, identifying $S$ the area of the spherical pole in the model with half the surface area of the actual magnet. If $A_m$ is the cross-sectional area of a bar and $B_m$ is the flux density in the magnet, the flux is $A_m B_m$. If $L_m$ is the length of the magnet, and $H_m$ is the working magnetizing force, and magnetomotive force is $H_m L_m f$ where $f$ is a number less than unity that depends on the

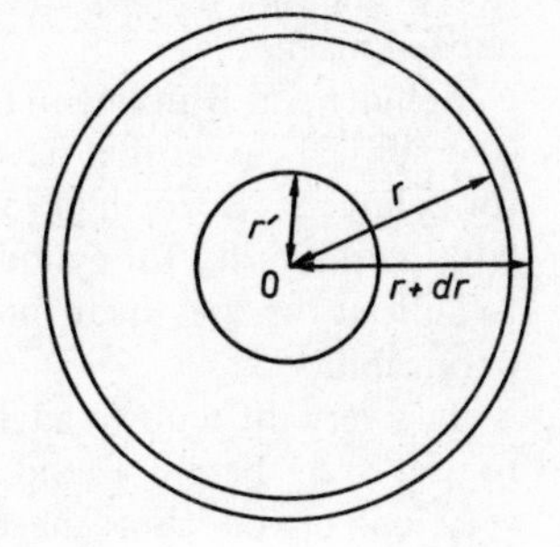


*Fig.6.10. Reluctance of spherical pole*

material. In the magnet steels used by Evershed $f$ was taken as 0.7, but the higher the coercivity the more nearly $f$ approaches unity. Equating the flux to the product of the magnetomotive force and the permeance we obtain:

$$\frac{B_m A_m}{\mu_0} = f L_m H_m (\pi S)^{1/2} \tag{6.23}$$

In the above work the transformation to CGS units only requires $\mu_0$ to be equated to unity.

In view of the apparent crudity of the assumptions, the success of Evershed's theory is remarkable. It is actually found to be more accurate and useful for magnets of the Al–Ni–Co type than for the steel magnets in use when it was derived. The assumption that the poles could be treated as spheres was further from the truth for the long needle shaped steel magnets of the 1920s than for magnets with the optimum length to diameter ratio for Al–Ni–Co alloys. Figure 6.11 shows curves of $B/\mu_0 H$, calculated according to Equations 6.19 and 6.20 for spheroids, and for cylinders according to Equation 6.23 with $f$ taken as 0.8.

Equation 6.23 can also be used for bars with square or rectangular sections. It predicts that for rectangular bars with a given length and area of cross-section, the ratio of $B/\mu_0 H$ increases slightly as the cross-section is changed from square to an elongated rectangle; i.e. from a square to a strip with the same area of cross-section.

Parker and Studders (1962) have also used Evershed's formula with suitable modifications to calculate the open circuit working points of quite complicated structures, particularly multipole rotors for motors and generators. This information is required if the rotors are to be operated on a recoil line starting from an open circuit working point. Certain machines are designed to be magnetized on open circuit, partly because the process is simpler, and partly because it is one way of stabilizing the magnet.

On the other hand the Evershed formula does not give a good prediction of the flux density and magnetizing force in a short uniformly magnetized magnet.

Schüler and Brinkmann (1970) quote a number of other formulae or empirical curves that have been suggested for estimating $N$ or $B/\mu_0 H$ for cylinders. Bozorth (1951) gives a figure showing a variation of $N$ with permeability for cylinders. For some dimension ratios $N$ is shown to be four times as great for infinite permeability as for relative permeability 5.

It is very difficult to advise the reader which of the various equations to use. We generally found Evershed's formula satisfactory, except for very long or very short magnets. If greater accuracy is required experiments with magnets of the material and dimension ratio to be used are desirable. Alternately a computer calculation may be tried.

## 6.5 MAGNETIC CIRCUITS FOR STATIC WORKING

There are many applications such as measuring instruments, loudspeakers and magnetrons in which the purpose of a magnet is to produce flux in a gap. If magnetization is performed after assembly, the gap is thereafter unchanged, and the magnet is not subjected to demagnetizing influences, arising either from external causes or its normal mode of operation, the application can be described as a static one.

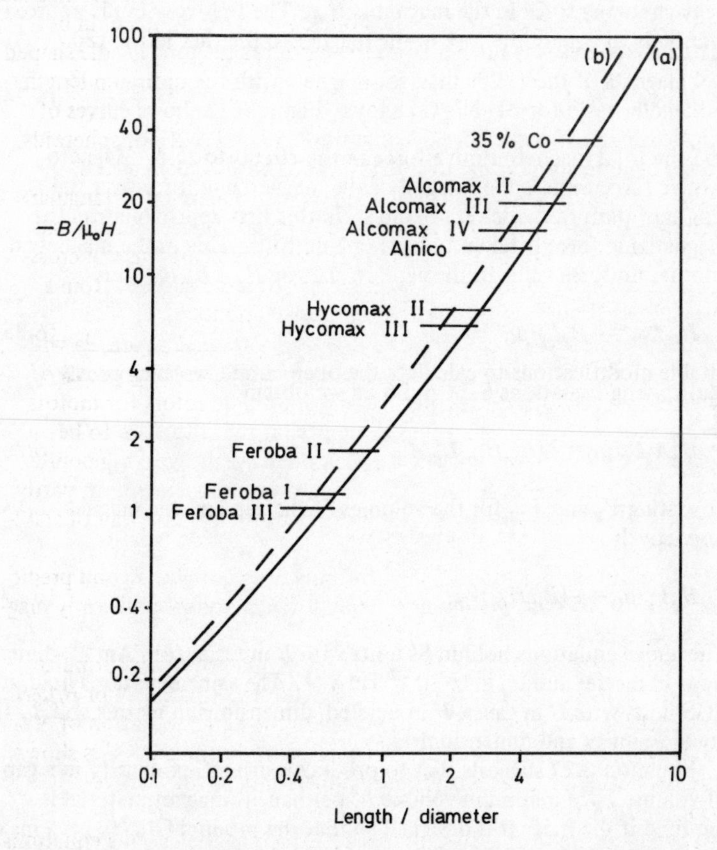


*Fig.6.11. Curve (a) $B/\mu_0H$ for spheroids. Curve (b) approx. $B/\mu_0H$ (with $f = 0.8$) for cylinders. Also showing $B/\mu_0H$ for $(BH)_{max}$ of some standard materials*

The designer of magnets for static applications should normally aim to produce the required flux in the gap for the minimum cost. Since permanent magnet material is much more expensive than the type of soft iron or mild steel used in permanent magnet assemblies, minimizing the cost is often equated with minimizing the volume of permanent magnet material used. This criterion is rather an over simplification, because it neglects the cost of machining and particularly the cost of fabricating the magnet in anything but the simplest shape.

The method of seeking the optimum design of a permanent magnet can be discussed with the aid of Fig.6.12. Here we consider a permanent magnet with cross-sectional area $A_m$ and length $L_m$, producing with the aid of mild steel polepieces flux in a gap of area $A_g$ and length $L_g$. The flux density is $B_m$ in the magnet and $B_g$ in the gap, and the working demagnetizing force in the magnet is $H_m$. The first very crude approximation is to equate the flux in the magnet to the flux in the gap.

$$B_m A_m = B_g A_g \tag{6.24}$$

and the total magnetomotive force in the circuit to zero. Magnetomotive force mmf is the integral of the magnetizing force over the length of path for which it operates. In this first approximation the magnetizing force is taken to have one uniform value in the magnet and another uniform value in the gap. In the gap $H_g = B_g/\mu_0$, thus

$$H_m L_m = -B_g L_g/\mu_0 \tag{6.25}$$

Multiplying Equations 6.24 and 6.25 we obtain

$$B_g^2 A_g L_g/\mu_0 = -(B_m H_m) L_m A_m \tag{6.26}$$

or writing $V_g$ and $V_m$ for the volumes of the gap and the magnet respectively:

$$B_g^2 V_g/\mu_0 = -(B_m H_m) V_m \tag{6.27}$$

The above equations hold in SI units with $B$ in tesla, $H$ in $\text{Am}^{-1}$, dimensions in metres and $\mu_0 = 4\pi\ 10^{-7}\ \text{TmA}^{-1}$. The same equations hold in CGS units with $B$ in gauss, $H$ in oersted, dimensions in metres and $\mu_0$ equal to unity and dimensionless.

Equation 6.27 suggests that to produce a given flux density in a gap of volume $V_g$, a minimum volume of permanent magnet material is required if the magnet is designed so that the product $(B_m H_m)$ is a maximum, that is to say if the magnet is designed to work at its $(BH)_{max}$ point.

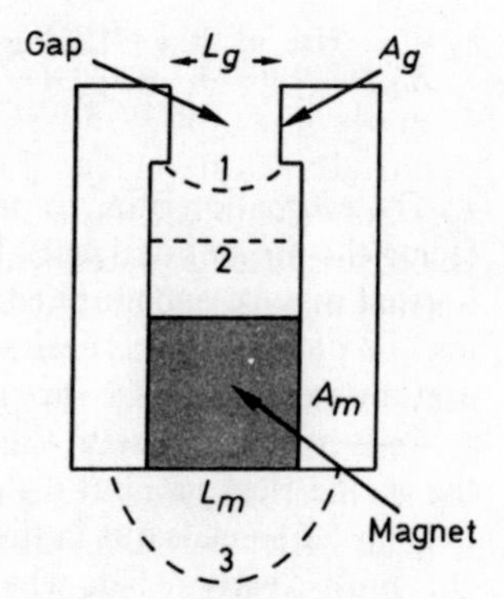


*Fig.6.12. Magnet assembly and leakage paths*

It is also useful to divide Equation 6.24 by 6.25 and rearrange to obtain

$$B_m A_m = \phi_m = H_m L_m / (L_g / A_g \mu_0) \tag{6.28}$$

In this equation $\phi_m$ is the flux in the magnet, $H_m L_m$ is the mmf which it exerts and $L_g/(A_g\mu_0) = R_g$ is the reluctance of the gap. This equation can be regarded as analogous to Ohm's law in electricity with flux equivalent to current, mmf equivalent to emf, and reluctance replacing resistance. If a magnetic circuit contains several reluctances $R_1$, $R_2$, etc in series:

$$\phi = \frac{\text{mmf}}{R_1 + R_2 + \ldots R_n} \tag{6.29}$$

The reluctance of a ferromagnetic material with a finite relative permeability differing from unity is $L/(A\mu_0\mu_1)$. Instead of reluctance it is sometimes useful to use its reciprocal permeance; $P = A\mu_0/L$. Like conductances in electricity permeances in parallel may be added so that

$$\phi = \text{mmf}(P_1 + P_2 + \ldots P_n) \tag{6.30}$$

The above theory neglects various correction factors. The most important is that much of the flux does not pass through the gap, but returns by leakage paths such as those shown by the dotted lines numbered 1, 2 and 3 in Fig.6.12. An attempt is made to distinguish between useful flux that passes through the gap, and leakage flux that passes outside the gap. Some writers make a further distinction and refer to that part of the leakage flux that originates close to the gap as 'fringing flux'. Mathematically it is usual to introduce a leakage constant $K_1$ that is defined as the ratio of the total flux to the useful flux, i.e.

$$K_1 = \frac{\text{Useful flux + Leakage flux}}{\text{Useful flux}}$$

The estimation of $K_1$ is perhaps the most troublesome problem facing the magnet designer. Before considering the range of values of $K_1$ that may be encountered, and the methods by which hopefully they may be predicted, it is desirable to consider how far the concepts of useful flux and leakage flux are justified.

Conventionally useful flux is identified with the flux passing through the geometrical gap, but if the gap is long, flux emerging from the pole-pieces spreads out in the centre of the gap, as shown in Fig.6.13. One must clearly decide whether the useful flux is that which emerges from the end face of one of the pole-pieces or that which crosses an equal area at the centre of the gap, indicated in the figure by PQ.

The useful flux may not actually be the same as the flux in the gap at all. It is possible to imagine an application such as a loudspeaker in which the coil extends beyond the geometrical limits of the gap, so that the useful flux is greater than the gap flux. Conversely in an application such as nuclear magnetic resonance, an extremely homogeneous magnetic field is required. To produce this homogeneous field, the area of the gap is made much greater than that of the specimen in which resonance is observed. The useful flux is only that which cuts the specimen and is a small fraction of the gap flux. The true effieicny of such a magnet can be increased by redesigning the pole-pieces to obtain the homogeneous field without the need to use such a large area. This redesign would not necessarily reduce the value of $K_1$.

### 6.5.1 Calculation of permeances and leakage factors

For many purposes the above concept of useful flux and leakage flux does suffice, and in any case to predict the working point of the magnet the total permeance must be calculated. Adding the gap and leakage permeances is one method of estimating the total permeance.

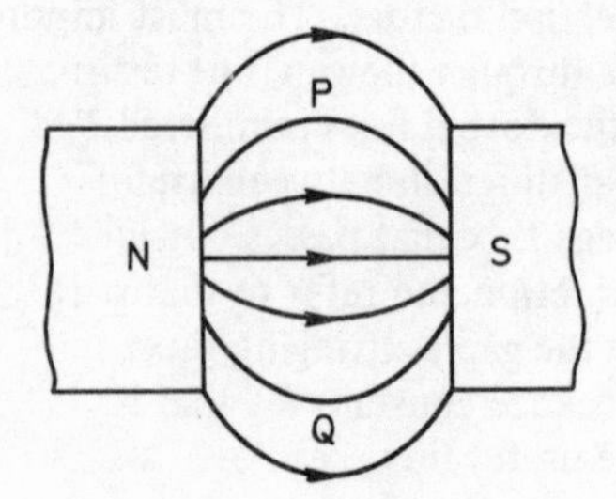


*Fig.6.13. Flux spreading at centre of gap*

In making such calculations, one regards mild steel surfaces as equipotentials, provided the steel is not saturated. Uniformly magnetized permanent magnets can sometimes be treated as having a uniform potential gradient, but in general the surfaces of permanent magnets have a potential gradient that is non-uniform. The optimum positioning of permanent magnet material in relation to the gap is an important design consideration.

A correction factor is also necessary to Equation 6.25, which should be written

$$H_m L_m + K_2 B_g L_g / \mu_0 = 0 \qquad (6.31)$$

The factor $K_2$ is a number usually lying between 1.1 and 1.4. It takes account of magneto-motive force lost in joints or in steel parts that cannot be regarded as having infinite permeability. Compared with $K_1$, $K_2$ is relatively unimportant.

Formulae for calculating the factor $K_1$ have been suggested by many writers, and collections of such formulae are given by Rotors (1941), Edwards in Hadfield (1962), Parker and Studders (1962) and Schüler and Brinkmann (1970).

Only in a minority of cases do these books suggest precisely the same formula, but sometimes the differences may be small and unimportant. Sometimes the form of the equations seems quite different. Experiments designed to discriminate between the different formulae are rare.

The simplest and most necessary procedures are those that give the permeance of the gap itself. For a gap between equal parallel surfaces opposite each other (Fig.6.14(a))

$$P = \mu_0 A_g / L_g \qquad (6.32)$$

This formula holds whether the gap is circular, square or any other shape, provided its area is calculated correctly.

The permeance of an annular gap between cylindrical surfaces of radii $r_2$ and $r_1$, subtending an angle $\alpha$ radians and with axial length $w$ (Fig.6.14(b)) is

$$P = \mu_0 w\alpha / \log_e(r_2/r_1) \qquad (6.33)$$

The permeance of a gap between inclined surfaces is shown in Fig.6.14(c) is

$$P = \frac{\mu_0 w}{\theta} \log_e \frac{r_2}{r_1} \qquad (6.34)$$

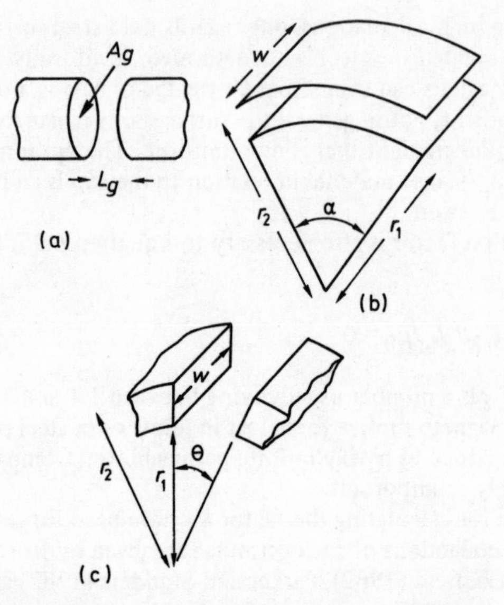


*Fig.6.14. Typical gaps for which permeance may be calculated*

One method of deriving leakage formulae is to divide the space around the magnet into sections, plausible assumptions about the flux paths being made. The permeance of each section is calculated from geometrical considerations, and each is treated as a contribution to the total permeance. Roters gives formulae on this basis for a large number of shapes, but to use the method is very laborious unless many of the contributions to the permeance are obviously small enough to be neglected.

Figure 6.15 illustrates a geometry for which it is useful to be able to calculate the leakage permeance. According to Edwards (Hadfield, 1962)

$$P = \frac{\mu_0 a}{\pi} \log_e \left( \frac{2b}{L_g} + 1 \right) \tag{6.35}$$

In a practical magnet there are usually a number of such pairs of surfaces for which the permeances must be added. Parker and Studders (1962) give for the same configuration the formula:

$$P = \frac{\mu_0 a}{\pi} \log_e \left[ 1 + \frac{2b}{L_g} + \frac{2(b^2 + bL_g)^{1/2}}{L_g} \right] \tag{6.36}$$

These two formulae have been quoted to illustrate the difficulties facing a designer who tries to make use of text-book formulae for this type of calculation. The additional term in Equation 6.36 compared with Equation 6.35 is likely to lead to values of $P$ about 20 to 30% higher. If the reader still wishes to try this method the most comprehensive collection of formulae is in the book by Roters (1941) as already mentioned.

In view of the uncertainty that remains even after the labour of adding all these permeance contributions, one may as well use empirical relations that attempt to predict the total permeance or leakage permeance for particular types of assembly. Such formulae may also take into account to some extent the non-uniformity of magnetization in some permanent magnet materials.

A very simple example of such a formula is quoted by Edwards (Hadfield, 1962) to give $K_1$ for a gap of length $L_g$ between circular pole-pieces of diameter $D_m$.

$$K_1 = 1 + \frac{7L_g}{D_m} \tag{6.37}$$

This equation is quoted with a note that some writers have found a steeper increase of $K_1$ for short gaps.

A comprehensive set of formula is quoted by Edwards (Hadfield, 1962) and attributed to Maynard and Tenzer. They are also quoted by Schüler and Brinkmann (1970) in a slightly simplified form with certain small terms omitted, and attributed to Tenzer. These formulae take account of the position of the magnets in relation to the gap, a very important factor in relation to leakage. The formulae are non-dimensional and valid in any self-consistent system of units. The meaning of the symbols is explained in Fig.6.16.

In Fig.6.16(a) the permanent magnet material is adjacent to the gap and leakage from the sides and base which are at the same potential can be ignored.

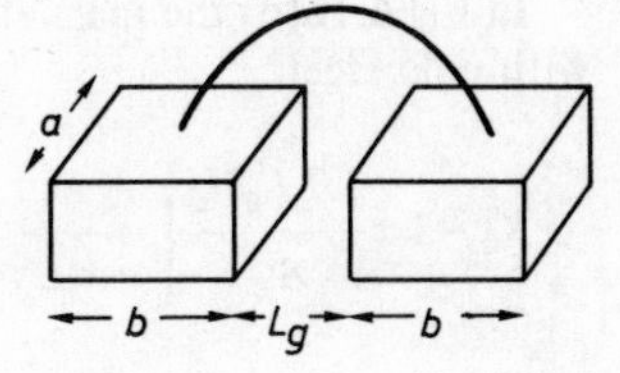


*Fig.6.15. Calculation of leakage between two surfaces*

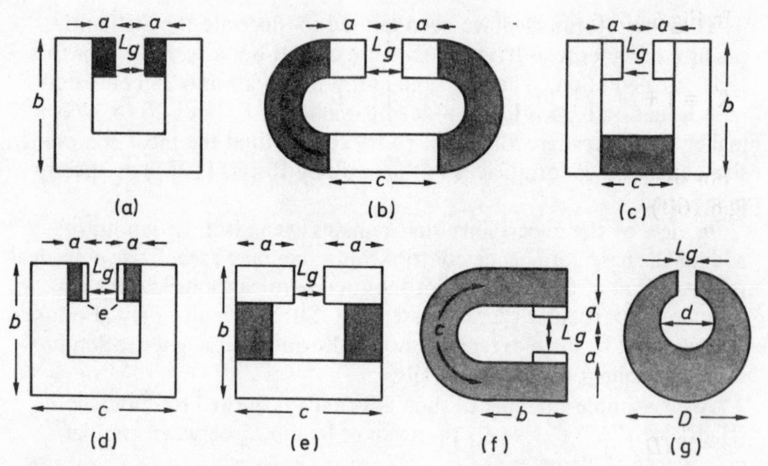


*Fig.6.16. Symbols used in Maynard and Tenzer's leakage formula (permanent magnet material shaded)*

$$K_1 = 1 + \frac{L_g}{A_g}\left(1.1\,U_a\,\frac{0.67a}{0.67a + L_g}\right)\left(1 + \frac{L_g}{a}\right) \tag{6.38}$$

In this and the following formulae $U$ with a suffix stands for the perimeter of the part whose length is indicated by the same symbol as the suffix. Hence in this case $U_a$ is the perimeter of the magnets.

In Fig.6.16(b) the magnets form the sides of the assembly and

$$K_1 = 1 + \frac{L_g}{A_g}\left(1.7\,U_a\,\frac{a}{a + L_g} + 0.64b\sqrt{\frac{U_b}{c} + 0.25} + 0.33\,U_b\right) \tag{6.39}$$

In Fig.6.16(c) a single permanent magnet is incorporated in the base of the assembly and:

$$K_1 = 1 + \frac{L_g}{A_g}\left(1.7\,U_a\,\frac{a}{a + L_g} + 1.4b\sqrt{\frac{U_b}{c} + 0.25} + 0.33\,U_c\right) \tag{6.40}$$

In Fig.6.16(d) the magnets are colinear with the gap but are faced with mild steel.

$$K_1 = 1 + \frac{1.7\,L_g U_a}{A_g}\left[\frac{e}{e + L_g} + 0.67\,\frac{0.67e}{0.67a + L_g + 2e}\left(1 + \frac{L_g + 2e}{a}\right)\right] \tag{6.41}$$

In Fig.6.16(e) the magnets are in the form of blocks on the sides:

$$K_1 = 1 + \frac{L_g}{A_g}\left(1.7\,U_a\frac{a}{a+L_g} + 1.4b + 0.94b\sqrt{\frac{U_b}{c} + 0.25}\right) \quad (6.42)$$

For the horseshoe magnet with soft iron pole-pieces shown in Fig.6.16(f)

$$K_1 = 1 + \frac{L_g}{A_g}\left(1.7\,U_a\frac{a}{a+L_q} + 0.94b\sqrt{\frac{U_b}{c} + 0.25}\right) \quad (6.43)$$

For the magnet of non-uniform section shown in Fig.6.16(g)

$$a = \frac{\pi}{6}(D + d - 2L_g)$$

and

$$K_1 = 1 + 1.1\frac{L_a}{L_g}U_a\frac{0.67a}{0.67a + L_g}\left(1 + \frac{L_g}{a}\right) \quad (6.44)$$

A mean value of $U_a$ is assumed for this magnet.

Edwards illustrates the use of Equation 6.42 (Fig.6.16(e)) with a numerical example, using Alcomax magnets. The predicted value of $K_1$ is 2.59, but the measured value is 2.8. He considers this degree of accuracy as good as could be expected. To illustrate the use of the above equations, let them be applied to the following circuit with the permanent magnet material in the different positions (a), (c) and (e) of Fig.6.16. The dimensions quoted would be plausible in cm but the units are immaterial. The circuit is to be made of permanent magnets and steel parts all of section 5 x 5, the length $L_g$ is to be unity, and the gap cross-section is obviously 25.

The total length of permanent magnet material is to be 10. When it forms the base as in (c), it is in one length, but when it is adjacent to the gap as in (a), or in the side arms as in (e) it is in two lengths of 5.

All the perimeters ($U$ terms) are 4 x 5 = 20. Most of the other dimensions necessarily follow, but in (c) the height $b$ is made 15, as is the length of the base in (e).

We find that in Fig.6.16(a) the length of each magnet adjacent to the gap is 5, and $K_1$ works out at 2.05. For Fig.6.16(c) with a single magnet in the base, $a = 4.5$, $b = 15$, $c = 10$ and $K_1 = 3.6$. In Fig.6.16(e) for the magnets in the side arms, the values to be inserted are: $a = 7$, $b = h = 5$, and the calculated value of $K_1$ is 2.7. This calculation confirms that the value of $K_1$ decreases, showing that the magnet design becomes more efficient, the nearer the magnets are placed to the gap.

A value of 2 or 3 is typical for $K_1$ in a reasonably efficient permanent magnet. Very high values of $K_1$, implying a very low efficiency are obtained in certain magnetron magnets, which require both a long gap and some flux concentration, produced by tapering the magnet. Figure 6.16(g) could be the silhouette of such a magnet. At the opposite extreme very high efficiencies with $K_1$ only a little greater than unity are obtained for the ferrite segment magnets incorporated in small DC motors, described in Section 7.4.3.1. These magnets have a large area adjacent to a small gap.

### 6.5.2 Analogue methods for determining permeance

The principal analogue method for estimating permeances and leakage factors makes use of a DC source, a potentiometer and conducting (Teledeltos) paper. The resistance between opposite sides of a square of any size of the paper is of the order 2000 to 3000 Ω. To investigate the flux path between mild steel poles, an outline of their surfaces is drawn on the paper with a silver conducting paint. A potential difference of a few volt is then set up between these, and a pencil connected to a potentiometer set to some known fraction, say 10%, of the potential difference is put in contact with the paper as shown in Fig.6.17. The point of the pencil is moved on the paper until a null deflection shows that a point has been found that is at the potential to which the potentiometer has been set. If a number of such points at the same potential are found and joined the result is an equipotential line.

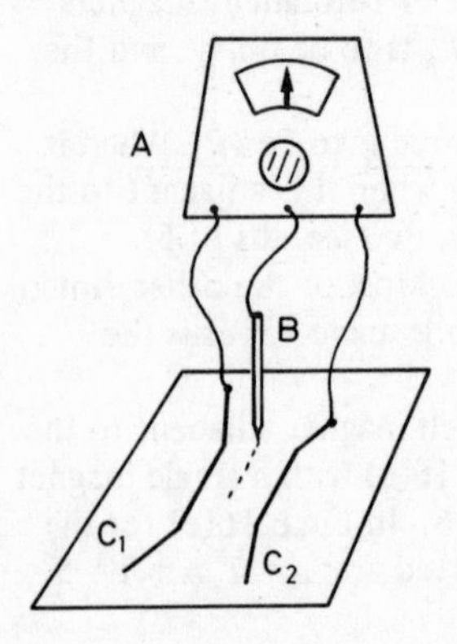


*Fig.6.17. Field plotting. A: Potentiometer. B: Pencil probe. $C_1$, $C_2$: outline of pole-pieces in silver paint*

It is usual to plot a number of isopotential lines at equal increments such as 10, 20, 30% of the total potential difference. The distance between these equipotential lines is a measure of the flux density. Two methods are suggested for estimating the flux density. The first is to draw in by hand 'curvilinear squares'. The meaning of this quaint term is explained with the aid of Fig.6.18. In a place where the field is uniform, lines drawn at right angles to the equipotentials indicate flux directions, or in the old language of magnetostatics are lines of force. These lines of force can be continued into areas, where the equipotentials are curved, by drawing them to cut the equipotentials at right angles. The resulting 'curvilinear squares' should be such that the mean lengths of opposite sides are equal. The density of the lines of force is a measure of the flux density. The leakage factor $K_1$ is the ratio of the total number of lines of force to the number in the gap.

A more precise method of drawing the lines of force and estimating the fluxes, is to cut away the shape of the magnet, and to paint curves corresponding roughly to the field direction at a large distance from the gap where the field is very small. Equipotentials drawn between these two curves can then be shown to represent the flux lines or lines of force as shown in Fig.6.19.

Field plotting is very satisfactory in a limited number of situations, possessing an approximation to a two-dimensional configuration. It is

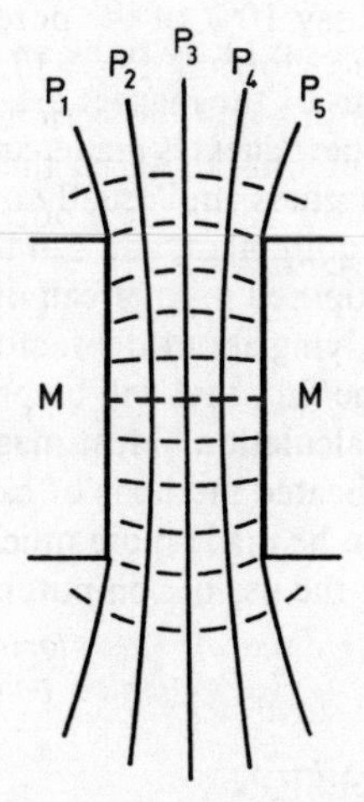


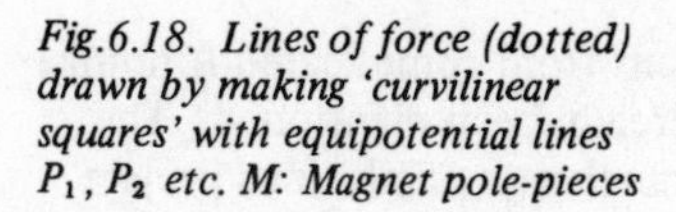
*Fig.6.18. Lines of force (dotted) drawn by making 'curvilinear squares' with equipotential lines $P_1$, $P_2$ etc. M: Magnet pole-pieces*

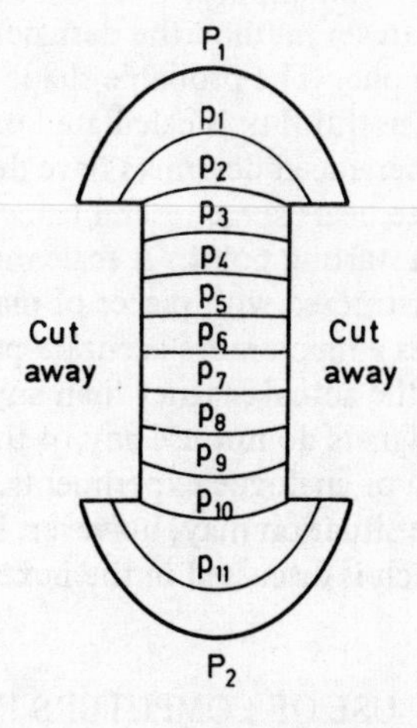


*Fig.6.19. Inverse plotting method. $P_1$, $P_2$ : Painted electrodes. $p_1$ , $p_2$ : Equipotentials that represent lines of force on original diagram*

very difficult to interpret the results for the more general three-dimensional case. The instruction booklet that came with an instrument, which we purchased, suggested that a problem, in which there was an axis of rotation could be solved by the use of conformal co-ordinates, but the makers were unable to supply us with any reference concerning this method. To find a satisfactory solution to this problem might be a suitable subject for post-graduate research in applied mathematics.

An ideal method of solving problems with an axis of rotation would be provided by paper the resistance of which varied inversely as the distance from a line representing this axis. I recollect a suggestion emanating from Russia trying to approximate to this ideal by reducing the resistance in steps by sticking layers of paper on top of one another.

Of course all the problems can be solved by the use of an electrolytic tank although apart from the general messiness, difficulties from contact potentials must be anticipated.

Another problem that can arise is that the surface of the permanent magnet part of a magnetic circuit is not an equipotential. We tried a suggestion that the paint line representing the magnet could be split and fed with a supply at different potentials, without much success.

## 6.6 PRACTICAL MAGNET DESIGN

Whatever method the designer uses the process is likely to be an iterative one. The probable shape and dimensions of the magnet are guessed; the useful flux is calculated or measured; a new guess is made, and so on. Experienced designers have details of old magnets, and usually one of these bears some resemblance to the new requirement, and can be used as a starting point. A scale model of the proposed magnet can often be constructed with pieces of magnet material lying about the factory, and gives a much more accurate prediction of the flux that will be produced by the actual magnet than any method of calculation. Most magnet designers do not use any of the more complicated methods of calculation or analogue experiments, if a model can be made more quickly. The situation may, however, be changed by the use of computers, which is discussed in the next section.

## 6.7 USE OF COMPUTERS IN MAGNET DESIGN

The use of a computer to make calculations from complicated formulae like Equations 6.11, 6.12 and 6.13 has already been mentioned. The use of computers for magnet design is a much more ambitious project.

The most promising method of designing a non-uniformly magnetized magnet seems to require dividing the volume into a large number of small elements to each of which a plausible value of $J$ is attributed. The magnetizing force in each element depends on the rest of the magnet and its surroundings. The value of $J$ and $H$ in any element must lie on an appropriate hysteresis loop or recoil line. A self-consistent solution must be sought by an iterative process. Crude attempts were sometimes made to find a probable distribution of polarization in a magnet before computers were available. The number of elements of volume that could be handled was limited and the results were not very encouraging.

A number of papers now appear each year in which the use of a computer to design a permanent magnet is described. A computer can deal with a far greater amount of data, and makes the problem appear more practical. The method appears in principle to be as described above, but the calculations are often carried out by the vector potential method. Two recent examples of such papers from which earlier references can be found are Binns, Jabbar and Barnard (1975) and Ledeboer and Schophuizen (1974).

Few of the older magnet designers are likely to have been trained in computer techniques, but it is probable that a firm that employs a computer programmer can make a worthwhile economy by the improved design of magnets required in large numbers.

Some computers or computer programmers require the properties of the magnet to be described by an equation rather than a curve. Magnets working on a recoil line can be described by a linear equation

$$B = B_p + \mu_0 \mu_r H \tag{6.45}$$

$\mu_r$ is the relative recoil permeability, $B_p$ is the flux density when the magnet is on closed circuit and $H = 0$, and of course in the second quadrant $H$ is negative. Ferrite and $RCo_5$ magnets often have an almost linear demagnetizing curve, and Equation 6.45 then describes the demagnetizing curve with $B_p = B_r$.

An old equation often used to approximate the hysteresis loops of alloy magnets is:

$$B = \frac{H + H_c}{H_c/B_r + H/B_s} \tag{6.46}$$

As this equation is homogeneous in $B$ and $H$ it should apply in SI or CGS units. In this equation $B_r$ is the remanence, $B_s$ is the saturation value and $H_c$ is the coercivity. This equation was used many years ago for justifying the Watson construction for finding the $(BH)_{max}$ point. (Draw lines through the $B_r$ and $H_c$ points parallel to the axes and intersecting at P. Join P to the origin O. OP cuts the demagnetizing curve close to the $(BH)_{max}$ point.)

Equation 6.46 is more satisfactory if it is confined to the second quadrant, and $B_s$ is taken as an arbitrary constant not necessarily equal to $J_s$. For many years the standard curves published by the Permanent Magnet Association were drawn with the aid of this equation, the constant being chosen so that the curves passed through the $(BH)_{max}$ point, $B_r$ and $H_c$. If this method is used the equation should be suitable for computer use. A curve constructed in this way is as likely to represent the demagnetizing curve of a particular sample of a material as accurately as any other typical curve for the material. Naturally a hysteresis loop measured on the actual sample will be more accurate.

## 6.8 DYNAMIC WORKING

A permanent magnet is subject to dynamic working if its working point changes as a result of its normal mode of operation. Such a change may occur as the result of a change in the external reluctance, as happens when a lifting magnet attracts a piece of steel, or as a result of fields produced by other parts of the apparatus, as in an electric motor or generator. Sometimes several processes occur together.

The working point of a magnet operating under dynamic conditions is not to be found on the outer hysteresis loop, but on an inner loop such as one of those shown in Fig.6.20. Such inner hysteresis loops may

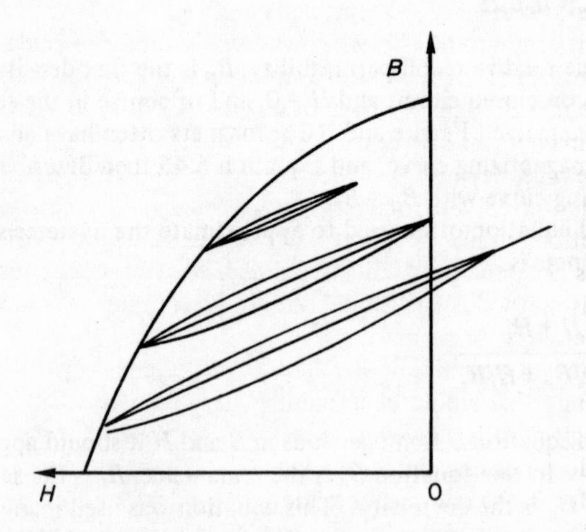


*Fig.6.20. Recoil loops and lines*

lie wholly in the second quadrant or extend into the first according to the operating circumstances. After a few cycles the loops are followed reversibly, provided the maximum demagnetizing force is not exceeded. The loops are very thin and for most purposes can be treated as straight lines, the slope of such lines being given by $\mu_0\mu_r$, where $\mu_r$ is the relative recoil permeability. It is usual to quote a single value of $\mu_r$ for a given material, although in fact $\mu_r$ depends somewhat on the point of the demagnetizing curve from which the recoil line starts; $\mu_r$ increases as the starting point moves down the curve towards the $(BH)_{max}$ point. Thereafter it may remain constant, although for some materials it falls slightly again as the coercivity point is approached.

In order to predict changes in the flux density of a magnet working on a recoil line, it is usually convenient to determine first the change in polarization $J$. Suppose a magnet is working with a gap at a point P on the $B$ recoil line, corresponding to Q on the $J$ recoil line (Fig.6.21). Suppose an additional demagnetizing force H′ is applied by means of a current flowing in a coil. The construction to find the new B working point P′ is to draw H′Q′ parallel to OQ, cutting the intrinsic recoil line at Q′. Draw a vertical line P′Q′ through Q′, cutting the $B$ recoil line at P′. P′ is the new value of $B$ and is not the same as would be obtained by drawing a line through H′ parallel to OP.

This complicated procedure is necessary because permanent magnet technologists have almost universally adopted a rather peculiar analogy with electricity to explain the behaviour of a permanent magnet. In effect a permanent magnet is compared with a black box containing a battery and an internal resistance. Such a black box could be described as a battery with a variable emf depending on the resistance of the external circuit. In current electricity we usually prefer to calculate the internal resistance of the battery and treat the emf as constant, unless of course it really falls with age or use.

A permanent magnet is usually treated as if it had no reluctance but a variable mmf depending on the external reluctance or permeance. It is possible to treat a permanent magnet as having reluctance; it then has something in place of mmf that remains constant so long as the magnet is not demagnetized and remains working on the same recoil line. This method is legitimate, but confusion may arise if it is not clear which method is being used.

I do not propose to rewrite the theory of permanent magnet design on new lines that would be unfamiliar to most permanent magnet technologists. It is, however, as well to remember that a permanent magnet does possess a high reluctance, but that this reluctance is allowed for in the conventional theory by assuming a change of mmf. In recent years a number of devices have been tried in which the flux of a high coercivity magnet is temporarily suppressed by a current carrying

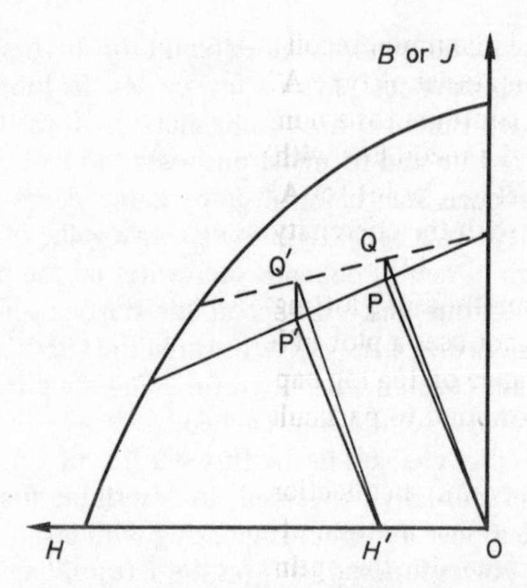


*Fig.6.21. Construction to find working point when additional magnetizing force is applied*

coil. Because of the high reluctance of a permanent magnet more ampere turns may be needed to control a permanent magnet in this way than to actuate an equivalent electromagnet.

It is important to know how to design most efficiently a magnet that must work under recoil conditions. The simplest circumstance to consider is a magnet that has to be magnetized on open circuit, and then assembled to produce flux in a gap. Referring to Fig.6.22, we can represent the open circuit working point of the magnet by P with flux density $B'$ and demagnetizing force $H'$. In this situation all the flux is leakage flux. When the magnet is assembled the working point moves up the recoil line to Q. The flux density of the magnet is now represented by the distance QW, but of this flux only QR is useful and RW is leakage. This statement is true provided the leakage permeance is unchanged by the act of assembling the magnet, which may not be true. The working magnetizing force is now represented by WO. The product QR.WO, which we may call $B_2H_2$ now replaces the product $BH$ in static working determining the flux in the gap. Hence the designer seeks to maximize the product QR.WO = QR.RS. For any given open circuit working point P, the best recoil working point can be seen from simple geometrical considerations to be obtained with Q bisecting the complete recoil line PV. There is also one particular point P on the demagnetizing curve, which gives a maximum value of the product $B_2H_2$ for the material, and this maximum recoil product is regarded as an important design parameter often listed in the properties of permanent magnet materials. The term recoil product rather than recoil energy is used here, because as with $(BH)_{max}$ the product is actually twice the energy.

Contours of the maximum recoil energy product are sometimes drawn as in Fig.6.22 for material of type A2. Parker and Studders (1962) give diagrams of these contours for a number of common materials. The point P that gives the recoil line with the maximum recoil product is a little below the $(BH)_{max}$ point for Al–Ni–Co alloys, and even lower, almost coinciding with the coercivity points, for some ferrites and $RCo_5$ compounds.

An alternative method of plotting recoil contours that may be more useful to the designer uses a plot of leakage unit permeance against useful unit permeance on log log paper, as shown in Fig.6.24. The diagonal lines correspond to particular values of the leakage constant $K_1$.

Space does not permit a collection of recoil contours for all the different materials by either method of plotting. The tables in Appendix 2 list the $B$ and $H$ point for the optimum recoil line, the relative recoil permeability and the maximum value of the recoil product when this information is available. The information given in these tables with the aid of Equations 6.45 and 6.46 should be sufficient to reconstruct the demagnetization curves and recoil contours. It will often be simpler and more accurate to seek information from the manufacturer about his own particular products.

To ensure that the leakage and useful permeances are both just right so that the magnet finally works at the optimum point on the optimum recoil line may be difficult to reconcile with the restraints imposed by the application on the dimensions of the various parts. There are also

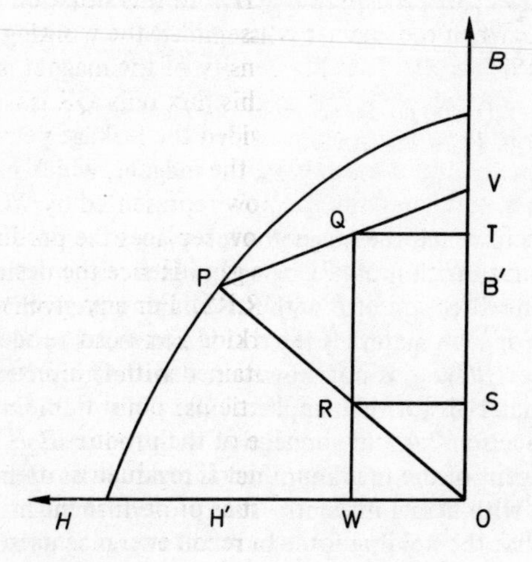


*Fig.6.22. Recoil product*

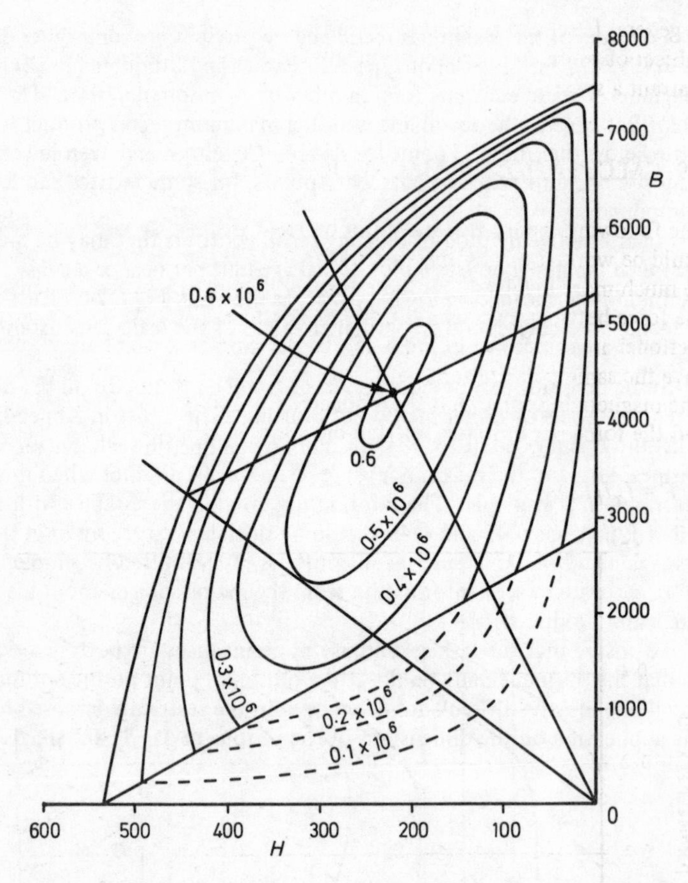


*Fig.6.23. Useful recoil product contours for material of type A2, in CGS units. (After Edwards, 1944)*

applications in which the magnet is taken by external fields into the third quadrant. With most $RCo_5$ and even ferrite magnets it is possible to reverse the direction of $B$ without reducing the ability of the magnet to recoil. For such materials the maximum recoil product becomes equal to $(BH)_{max}$. $(BH)_{max}$ is not, however, a sufficient criterion for the suitability of materials for such applications. This problem is discussed further in Section 7.4.3 in connection with motors.

The concept of the maximum recoil product is often introduced in connection with lifting magnets. This procedure might be correct if it were true that the holding force between two magnetized surfaces was proportional to $B^2$ in the gap between them. This widely quoted result is in general false, and holds only in certain special cases. This matter is the subject of so much misconception and is sufficiently important to warrant a special section.

## 6.9 CALCULATION OF MAGNETIC FORCES

The following theory is worked out on the basis of the pole model. It could be worked out by the current model, but the mathematics would be much more involved. Suppose for simplicity that we are calculating the force between two semi-infinite magnetic bars with the same cross-sectional area as shown in Fig.6.25. In the simplest case of all the bars have the same polarization, i.e. $J_a = J_b$ and they make good contact. The magnetizing force that is produced by A and acts on B is $J_a/2\mu_0$, and the force per unit area exerted on B is

$$F = J_a J_b / 2\mu_0 \tag{6.47}$$

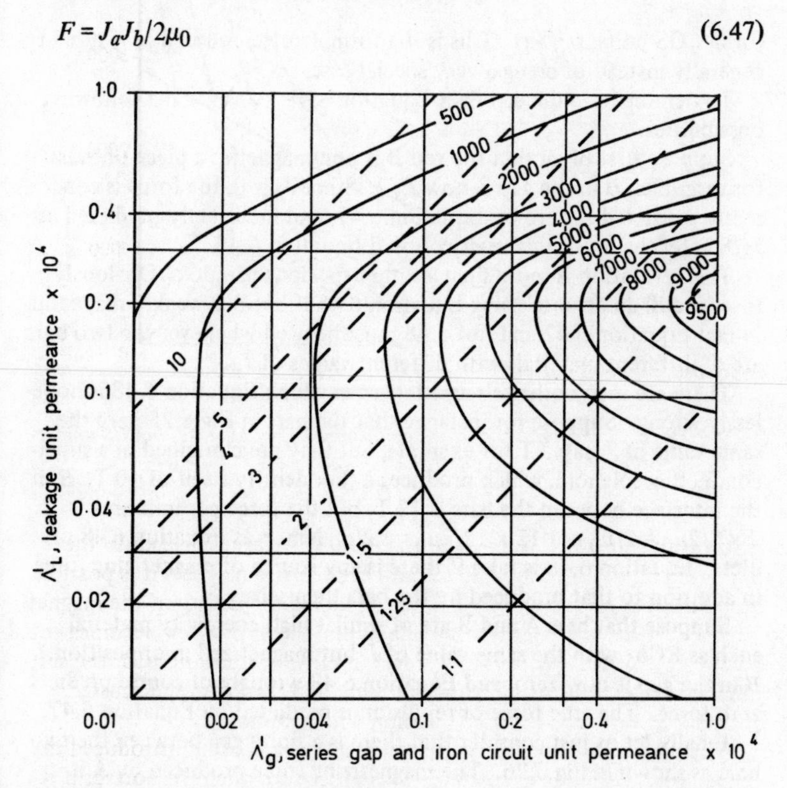


*Fig.6.24. Maximum values of recoil product in a generator using Alcomax III (Class A2) magnets*

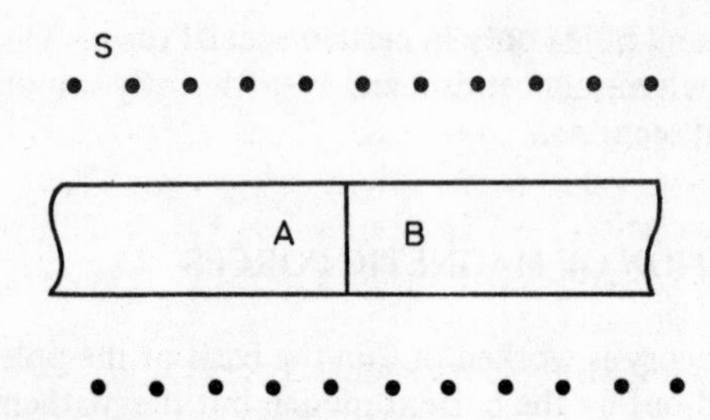


*Fig.6.25. Magnetic force between two bars*

In this simple case the components of $H$ cancel, and $J_a = J_b = B$. Hence Equation 6.47 can be rewritten:

$$F = B^2/2\mu_0 \tag{6.48}$$

(or in CGS units, $B^2/8\pi$). This is the formula often quoted as if it held generally instead of being a very special case.

To demonstrate the errors in Equation 6.48, consider the following possibilities.

Suppose first of all that the rod B is non-magnetic: a piece of brass for example. $B$ in the gap is now $J_a/2$. Since $J_b = 0$, the force is correctly predicted as zero by Equation 6.47, and incorrectly predicted as $J_a^2/8\mu_0$ by the most universally quoted Equation 6.48.

This refutation is more than a rather specious paradox. Obviously there is still an error in using Equation 6.48 if bar B is weakly magnetic, in fact Equation 6.47 and not 6.48 must be used whenever the two bars are of different materials with different values of $J$.

There are many other circumstances in which Equation 6.48 is hopelessly wrong. Suppose for instance that the bars in Fig.6.25 have the same value of $J$, say 2 T for example, but they are contained in a superconducting solenoid, which produces a flux density itself of 10 T. $B$ in the interface between the bars is 12 T, but the force per unit area is $2 \times 2/2\mu_0 = 2/\mu_0$ not $12 \times 12/2\mu_0 = 72/\mu_0$ Nm$^{-2}$ as Equation 6.48 predicts. Equation 6.48 is false if there is any source of magnetizing force in addition to that produced by the bars themselves.

Suppose that bars A and B are of similar high coercivity material such as $RCo_5$ with the same value of $J$, but magnetized in opposition. $B$ in the gap is now zero, and Equation 6.48 wrongly of course predicts zero force. The true force of repulsion is predicted by Equation 6.47.

Finally let us just consider that there is a finite gap between the two bars as shown in Fig.6.26. The magnetizing force produced by A is $J_a/2\mu_0$ at its surface but non-uniform and less than $J_a/2\mu_0$ at the centre of the gap and still less and more non-uniform at the surface of B. The magnetizing force that exerts a force on the poles on the surface of B, is that due to A integrated over the surface of B. There is no part of the gap at which B can be measured, for which $B^2/2\mu_0$ does not over-estimate the force per unit area, even if $J_a = J_b$. The larger the gap the greater the error in using Equation 6.48. This error is not quite so enormous as some of the others, but it does throw doubts on many attempts to calculate the force distance curve.

If a surface current model is used the force is calculated from the interactions of surface currents in two solenoids that represent the two bars. It leads to Equation 6.48, only if the solenoids are similar, carry the same current in the same direction, and are not separated by a gap.

The attainment of the maximum recoil product requires an appreciable gap, and many calculations relating the performance of lifting and attracting magnets to the maximum recoil product, assume Equation 6.48 to hold with such a gap and are therefore of dubious validity. I believe that the use of the maximum recoil product in connection with the design of attracting magnets is probably qualitatively but not quantitatively justified.

The quantitative theory predicts that to exert an attractive force at a given distance with a minimum volume of permanent magnet material, the magnet should operate at its point of maximum recoil product. If the gap is reduced to zero the force is predicted as four times greater. I have found considerable departures from this ratio in practice.

Although I doubt the numerical accuracy of maximum recoil product theory in predicting the performance of lifting magnets, I do not suggest that its principles should be completely ignored. It is probably good practice to arrange the open circuit leakage of a lifting magnet so that it works on the optimum recoil line. A valuable lesson from recoil theory is that if a magnet is likely to be demagnetized on open circuit so that it works on too low a recoil line, it is beneficial to redesign it in such a way as to increase the leakage, until the optimum recoil line is reached.

It is probably less important to try to make the working point lie in the centre of the recoil line. Frequently a magnet has to work at a variety of distances, and such magnets can take advantage of the larger holding forces pertaining to points on the recoil line close to the $B$ axis when the gap is small.

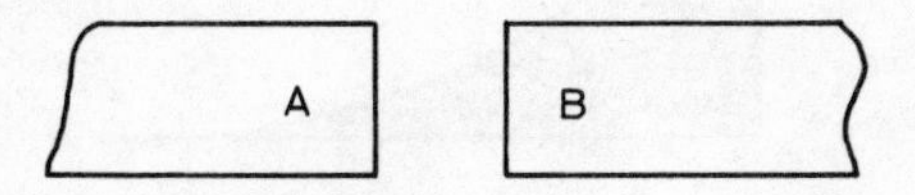


*Fig.6.26. Bars as in Fig.6.25, separated by gap*

## 6.10 ENERGY RELATIONS

Energy seems such a well-established concept that it may seem surprising that differing views about the energy relations of permanent magnets should have been expressed. First of all let us recall the essential facts about which no controversy exists.

When there is a magnetic field in free space there is energy $B^2/2\mu_0 = BH/2 = \mu_0 H^2/2$ $\text{Jm}^{-3}$ or in CGS units, $B^2/8\pi$ erg $\text{cm}^{-3}$. The integral of any of these quantities, say $\mu_0 H^2/2$ over the volume outside the magnet, is equal to the integral of $BH/2$ over the volume of the magnet.

If a keeper is removed from a magnet the work done is equal to the integral of the pull, gap curve, the area as shown in Fig.6.27.

What is not always made clear is whether the energy $BH/2$ inside the magnet exists in addition to the energy $H^2/2\mu_0$ outside or whether it is another method of calculating the same energy. There have been many attempts to calculate the force gap curve in terms of the increase in energy of the magnet or the space outside it. Some of these have been marred by unjustified mathematical approximations designed to lead to the formula $F = B^2/2\mu_0$, which was criticised in Section 6.9. It seems certain, however, that the work done is insufficient to produce the energies both inside and outside the magnet.

The German writers, Schüler and Brinkmann (1970) equate the work done in removing a keeper to $BH/2$ inside the magnet, but regard this energy as negative and the energy $\mu_0 H^2/2$ outside as positive. It follows that the total energy inside and outside the magnet is always zero, and is unchanged by work done on or by the keeper when it is attracted by or pulled off the magnet. Although this is not explicitly stated it is presumably postulated that the work done on pulling a keeper off a magnet serves to pump energy out of the magnet into space rather than increase the total energy.

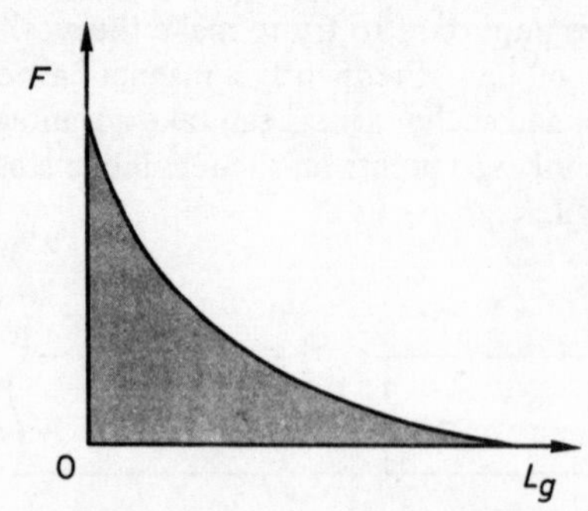


*Fig.6.27. Area under force, gap curve represents work done in removing keeper*

Koch (1974) has written provocatively and illuminatingly on this subject. His arguments can best be appreciated by considering first a simple case of a completely hard magnetic material with relative recoil permeability equal to unity. Rare-earth cobalt and some ferrite magnets fulfil this condition well. For such a material the open circuit working point can move very close to $H_c$, and it is clear from Fig.6.28 that the product $BH/2$ in the magnet and therefore the value of $\mu_0 H^2/2$ outside both pass through a maximum and fall again as the keeper is removed from the magnet. The areas of the rectangles $P_1B_1OH_1$, $P_2B_2OH_2$, $P_3B_3OH_3$ increase and then decrease in that order. Thus the work done in removing the keeper, which always increases, cannot be regarded as the increase in internal energy of the magnet $BH/2$, or of the external field energy $\mu_0 H^2/2$ or of their sum.

Koch overcomes this paradox by setting the sum of the internal and external energies to zero in the same way as Schüler and Brinkmann, but introduces a new term $NJ^2/2\mu_0$, which he calls the demagnetizing energy of the magnet.

The following comments are from a paper which I presented at the joint MMM and Intermag Conference in Pittsburgh in June 1976.

Consider that the energy term $NJ^2/2\mu_0$ applies to the volume of the magnet. Within this volume

$$NJ/\mu_0 = -H \tag{6.49}$$

$$J = (B - \mu_0 H) \tag{6.50}$$

Hence

$$\begin{aligned} NJ^2/2\mu_0 &= -H(B - \mu_0 H)/2 \\ &= -BH/2 + \mu_0 H^2/2 \end{aligned} \tag{6.51}$$

Both terms on the righthand side of Equation 6.51 apply to the volume of the magnet, but according to Koch

$$-\frac{BH}{2} \text{ inside} = \frac{\mu_0 H^2}{2} \text{ outside}$$

Thus $NJ^2/2\mu_0$ = the sum of $\frac{1}{2}\mu_0 H^2$ outside and inside the magnet. Thus another way of expressing Koch's solution to the paradox is to say that the work done in removing the keeper increases the field energy in all space inside and outside the magnet. The field energy inside the magnet is the energy that would exist in that volume if it consisted of free space, and is a result of the self-demagnetizing field only. In a thin plate of

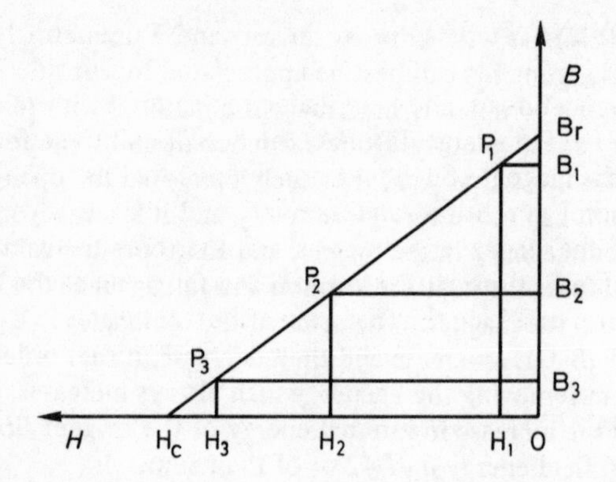


*Fig.6.28. Area of rectangles representing product (BH) at points on recoil line pass through maximum*

ferrite or $RCo_5$, which is the only kind of magnet we are at the moment considering, most of the energy may be inside the magnet. In older types of magnet with low coercivity very little of the total field energy was inside the magnet, and consequently no error was noticed in equating the work done to either the external field energy or ½*BH*. Koch also points out that in magnets with $\mu_1$ greater than unity, there is an additional energy term, which he calls the magnetizing energy. This additional energy is due to the increase in *J* as the magnet recoils, and is included in the calculation or measurement of ½*BH*.

Finally in the paper which was presented at Pittsburgh (McCaig, 1976), I have argued that the energy of a magnetic material is not zero at the remanence point. I have suggested that in a soft magnetic material this energy can be used if the flux is made to collapse, and probably is so used in computer memory stores, and DC pulse transformers, sometimes used to magnetize permanent magnets. If my view is confirmed the design of such transformers could be improved. The core of such a transformer could be made in two halves, the energy could be stored by magnetizing the core and released at a later time when required by separating the two halves. If the core were made of a permanent magnet material, the energy stored at remanence might be much the same, but more work would be required to separate the halves, and some of the energy would be used to create the external field. In short my view is that a magnet holding a keeper possesses positive energy. When the keeper is removed work is done and part of the energy ½*BH* appears as field energy in outer space. This view explains why Koch and Schüler and Brinkmann regard ½*BH* as negative.

## 6.11 REMANENT WORKING

For a few applications such as compass needles and magnets that have to produce a field at a large distance, the need is to maximize the moment of the magnet. Such magnets should work as close to the remanence point as possible, and if a material such as an Al–Ni–Co alloy is used, the magnet should be as long as is reasonably convenient. The same volume of magnetic material has a larger magnetic moment if it is in the form of a long thin bar than a short fat one.

## 6.12 MECHANICAL CONSIDERATIONS

The magnet designer must consider how a magnet can be fabricated and assembled as well as its desirable shape from the point of view of magnetic performance. Al–Ni–Co alloy, ferrite, and rare-earth cobalt magnets with any one dimension much smaller than the rest are difficult to make and fragile in use. Processes such as internal grinding are possible but expensive. It is almost impossible to make magnets with small diameter holes except by slow processes such as spark erosion. If magnets are required with such holes for fixing purposes, they can be cast or sintered with relatively large holes that can be filled with a low melting point metal and drilled.

If possible magnets should be assembled by clamping or the use of adhesives. Epoxy resins are very suitable. Soft solder can be used on some magnets although it is not ideal. Brazing is possible on Al–Ni–Co materials, but requires great care to avoid spoiling the magnet by overheating.

Plating may be used to improve the appearance of magnets. It is usually unnecessary for protection, and it is somewhat detrimental to magnetic performance to plate internal surfaces in an assembly through which flux must pass.

# Chapter 7

# Applications

Although this chapter on applications is the longest in the book, it has been necessary to keep it shorter than some other accounts. Schüler and Brinkmann (1970) devote as much space to applications as the whole length of this book.

To keep the account short I have assumed that the designer interested in a particular application is an expert on his own product. He knows the function it is required to perform, and the environment in which it will be required to work. He is familiar with all the difficulties and faults other than magnet ones likely to arise in the use of his equipment.

He is also assumed to have a rough idea of the task a permanent magnet is required to perform, but may have difficulty in specifying this quantitatively in the correct technical terms. He is expected to require help in choosing the best or most economical material to perform this task, and the purpose of this book is to give that help.

The earlier chapters have described the properties of materials and some general principles of design. Wherever possible repetition will be avoided, and the reader will be referred to appropriate sections of these chapters.

There is more than one logical order in which permanent magnet applications can be described. Some users might find it convenient to find a section dealing with their own industry, but one industry may use magnets for a great variety of purposes, while magnets of similar design may be used in quite different industries. It seems more sensible to classify magnets by the function that they perform. This classification is outlined in Table 7.1. The first column names general functions that magnets may perform. The second column subdivides these broad functions, describing different types of action, method or situation in which the function arises. The third column lists particular applications that make use of the functions described in the first two columns.

As far as possible the order of Table 7.1 is used in the following sections, but it must be understood that there is some overlap in these headings. Eddy current devices involve both the induction of a current and the exertion of a force on that current. Most genuine medical applications fall under one of the mechanical headings. A synchronous motor can be considered as a torque application or from the point of view of conversion of electrical into mechanical energy. Occasional departures from the order outlined in the table may be desirable, but it is hoped that they are adequately covered by the index.

In Table 7.1 the word iron has been used for brevity, when it would be more accurate to write 'any soft ferromagnetic or ferrimagnetic material'.

## 7.1 THE COMPASS AND OTHER TORQUE DEVICES

The compass, which indicates the direction of the horizontal component of the Earth's magnetic field, is the oldest practical application of magnetism. A rudimentary compass can be made by floating a bar magnet on a cork in a vessel of water. Since the Earth's field is practically uniform, the magnet experiences no translational force, but only a torque which tends to set it in the field direction.

Since torque is proportional to the product of the magnitudes of the magnetic moment and the field, the moment must be made as large as possible. Now the magnetic moment is equal to the integral, $\int J \mathrm{d}V$ over the volume of the magnet. With the older steels the only way of ensuring a large value of $J$ is to use long needle shaped magnets, and in the simplest type of compass such a magnet can also serve as a pointer. In such simple designs the suspension is generally a simple unipivot, the needle being prevented from coming adrift by the top plate.

Ship's compasses are often made with a number of needles attached to a card (sometimes called a rose) on which the points of the compass are marked. The rose may be partly floating on a liquid, which provides damping as well as taking most of the weight off the pivot. For needle shaped magnets the only modern material that may be an improvement on the steels is some high remanence grade of Vicalloy.

Some modern compasses use a diametrically magnetized ring of the anisotropic Al–Ni–Co alloy (material A2), and this design produces a much larger magnetic moment within limited dimensions.

Schüler and Brinkmann (1970) discuss some of the errors that can arise in a compass. If the rose is not perfectly balanced the vertical component of the Earth's field can cause an error, and even the pitching and rolling of a ship can cause a semi-permanent false reading. The iron parts of a ship may naturally become magnetized, and their influence on the compass must be neutralized by carefully placed permanent magnets.

**Table 7.1** APPLICATIONS OF PERMANENT MAGNETS

| *General property of magnets used* | *More specific method of application* | *Particular devices* |
|---|---|---|
| Mechanical forces exerted by or on magnets | Torque on magnet in magnetic field | Compass. magnetometer |
| | Attraction between magnet and iron or other magnets | Holding devices, notice boards, games, door catches. Magnetic filtration, retrieval devices, drives and couplings, window cleaning, switches, thermostats, thickness gauges, magnetic tools |
| | Attraction induced between other iron parts | Reed switches, railway warning systems, magnetic clutches |
| | Repulsion: magnet and magnet | Bearings (some bearings now use attraction) levitated transport, toys |
| | Electromagnet and magnet | Brake for coil winder etc. |
| | Induced repulsion | Sheet floater |
| Electromagnetic forces | Moving coil devices | Loudspeakers and telephone receivers. Instruments. DC motors |
| | Moving magnet devices | Synchronous and brushless motors, clocks, hysteresis motors |
| | Moving iron devices | Some telephone receivers, polarized relays |
| Electromagnetic induction | Relative motion coil and magnet | Flux measurement. Generators, sensing devices. Microphones, pick-ups |

| | | |
|---|---|---|
| | Eddy currents | Damping devices, brakes, speedometers |
| Forces on moving electrons and ions | Focusing | TV (now little used) electron microscope, klystron, travelling-wave tube |
| | Crossed field action | Blow-out on switches, magneto-hydrodynamics generator (see generators). Magnetron, mass spectrometer, ion pump, omegatron, TV picture shift |
| Magneto-optical effects | Kerr and Faraday effects | Domain studies, Memory stores, Other possible devices |
| Magnetic resonance | NMR | Chemical analysis. Accurate field measurements. Magneto-crystalline anisotropy |
| Action of magnetic fields on solid and liquid media | Magnets in magnetic circuits | Saturistors. Magnetizers and magnetic heat treatment |
| | Electronic and semi-conductor | Wave guides, circulators. Hall and magnetoresistance devices |
| Biological and chemical effects | | Research on influence of magnetic fields on life processes. Water conditioners |

In a transmitting compass (Hadfield 1962, p.370) a mumetal core inside a solenoid fed with 200 Hz current is placed under a master compass. A servomotor brings this core into line with the needle of the master compass, and relays a signal to a number of repeaters at different parts of the ship. Compass movements of as little as 0.5° can be detected without interfering with the accuracy of the master compass.

In the mid 1940s, P. M. S. Blackett became interested in the problems of geomagnetism, and can probably be credited with starting an upsurge of interest in the subject. He required sensitive magnetometers working on the torque principle to measure small variations in the strength and direction of the Earth's field. He concluded that to obtain sensitivity and a short oscillation period a large ratio of the magnetic moment to the moment of inertia was required. This condition is met better by a shape such as a sphere or a cube than a long needle. At that time it was difficult to suggest a permanent magnet that would remain strongly magnetized with such dimensions. Modern materials present no difficulties in this respect. The Al–Ni–Co rings used in some modern compasses do not fulfil Professor Blackett's specification precisely but come much closer to it than the old fashioned steel needles.

Although the name 'magnetometer' is given to instruments similar to compasses and capable of measuring very small fields, robust magnetometers suitable for measuring moderately high fields can be made, e.g. a tiny cylinder of Silmanal about 1 mm in diameter fixed to a pivoted axle, restrained by springs. Silmanal has a saturation polarization of only about 0.05 tesla, but can be exposed to a flux density of about 10 times this value in any direction without changing its magnetization. Instruments of this type are made with full scale deflection ranging from 0.01 to 0.5 tesla. The small magnetic moment of the Silmanal magnets is sufficient in these fairly large fields. A more strongly magnetic material would sometimes distort the field to be measured. The upper limit of these instruments can, however, be increased by using rare-earth-cobalt.

These instruments have been made in both Britain and America. The British type is held with the correct orientation to the magnetic field which is then read off directly on a dial. The American type is rotated in the field to be measured, until the reading on the dial is a maximum, which is the value of the flux density required. Synchronous motors are also a torque application, but are discussed under the heading of motors in Section 7.4.3.2.

## 7.2 MAGNETS FOR HOLDING AND ATTRACTING

The distinction between a holding and an attracting magnet is that the first named has to exert a magnetic force on a material with which it makes a good contact, while the second has to exert a force on a magnetic body at a considerable distance. The distinction is important, because the optimum design for the two purposes differs considerably, and it is necessary that the customer should be able to explain to the manufacturer exactly what the magnet is required to do.

Unfortunately this is a field of application in which difficulties of communication between user and magnet maker frequently arise. Industries such as the manufacture of scientific instruments or electronic equipment usually employ scientists or engineers trained to give precise technical specifications of their requirements. Holding and attracting magnets are often incorporated into articles such as toys, furniture and domestic equipment, whose makers may not always employ such technologists.

Holding magnets are used for keeping a great variety of objects tidily in place. Iron objects such as screwdrivers and spanners may be held on a tool rack such as that shown in Fig.7.1. A non-magnetic article may have a magnet incorporated in it, to hold it to an iron surface. An article that is not too thick may be trapped between a magnet and a steel surface. Papers may be held on an iron notice board by this method more conveniently than on a conventional notice board using drawing pins. Magnets painted with appropriate letters or symbols can also be used very conveniently with an iron notice board. This method is used by the BBC television weather men.

More powerful holding magnets are used, for example, in chucks in workshops and factories. The forces are often so great that problems arise in separating the objects held from the magnets. A number of methods have been devised for switching off a permanent magnet, and these methods are described in Section 7.2.5.

A permanent magnet holding with a good contact on a sheet of iron is a case for which the formula for the force

$$F = B^2 A / 2\mu_0 \text{ newton} \tag{7.1}$$

holds. If we substitute the flux $\phi$ for the product of flux density $B$ and area $A$, Equation 7.1 becomes

$$F = \phi^2 / 2A\mu_0 \tag{7.2}$$

(In CGS units with $F$ in dynes, $\phi$ in Maxwell, $B$ in Gauss and areas in $cm^2$ these equations become $F = B^2A/8\pi = \phi^2/8\pi A$.)

Equation 7.2 suggests that it is advantageous up to a point to concentrate the flux by tapering the pole-pieces and reducing the area of contact A. This step is useful only so long as the material of the pole-

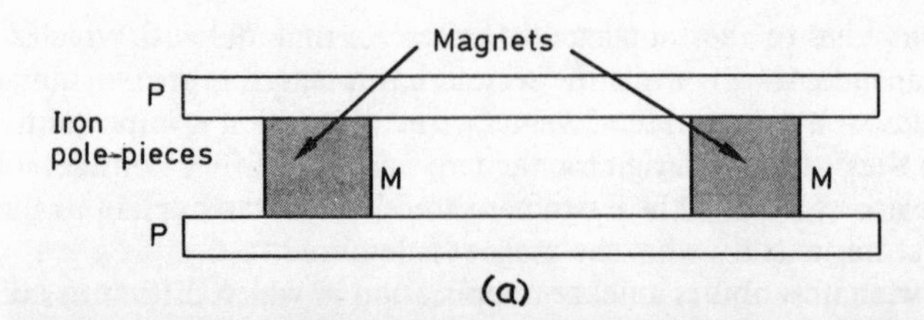


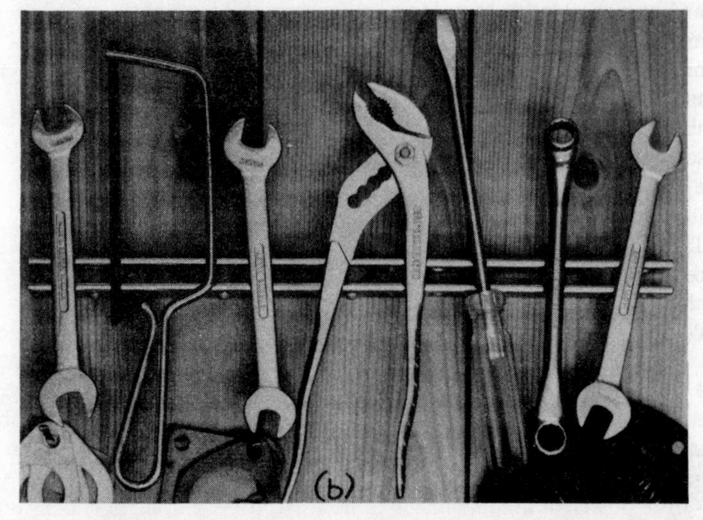


*Fig. 7.1. Tool rack. (a) sketch showing construction. M: magnets, P: iron pole-pieces. (b) Photograph of rack in use. (Courtesy of James Neill Ltd., Sheffield)*

pieces and the material held is capable of transmitting the flux. There is a difficulty in exerting a large force on a thin sheet of material, because it is incapable of passing a high flux. This difficulty can be overcome by using a magnet with many small poles close together.

Some designs of holding magnet are shown in Fig.7.2. The large horseshoe (a) is suitable for use with old-fashioned steel, (b) is suitable for use in an isotropic Al–Ni–Co alloy. Magnets of this shape are sometimes made in the anisotropic Al–Ni–Co alloys, but it is difficult to be sure that the field cooling is carried out efficiently. Nevertheless such a magnet is smaller and more powerful than a steel horseshoe. Design (c) is very suitable for use with an anisotropic alloy. If the area of the magnet is made larger this sandwich design is a very good way of using sintered ferrites, as it enables the flux density of the ferrite to be concentrated to any desired degree compatible with the properties of the pole-pieces. In (d) the magnet forms a centre pole and the outer pot is made of iron. To obtain sufficient flux the centre pole must be of a high remanence Al–Ni–Co alloy, which in this design is protected from demagnetization by contact with iron parts on its sides, the hazard mentioned in Section 5.3. In (e) the magnet forms the outer ring, while the centre-pole is of iron. This design is suitable for use with ferrite magnets, since these magnets are not susceptible to demagnetization by chance contacts with iron bodies and the magnets can be made as large as necessary to compensate for their low flux density. Finally (f) shows a common way of magnetizing rubber sheet. If it is magnetized in this manner, very little flux can be detected on the back surface. The bonded material can alternatively be magnetized with an iron backing in place. This method is preferable with the more expensive type of anisotropic rubber magnets, in which the flux does not easily curve round as it does in the isotropic material. The thickness of the material can be reduced in this method of application.

There has been a steady tendency for designs using ferrites to displace those using Al–Ni–Co for many holding operations during the last few years.

When it is necessary to attract bodies from a considerable distance,

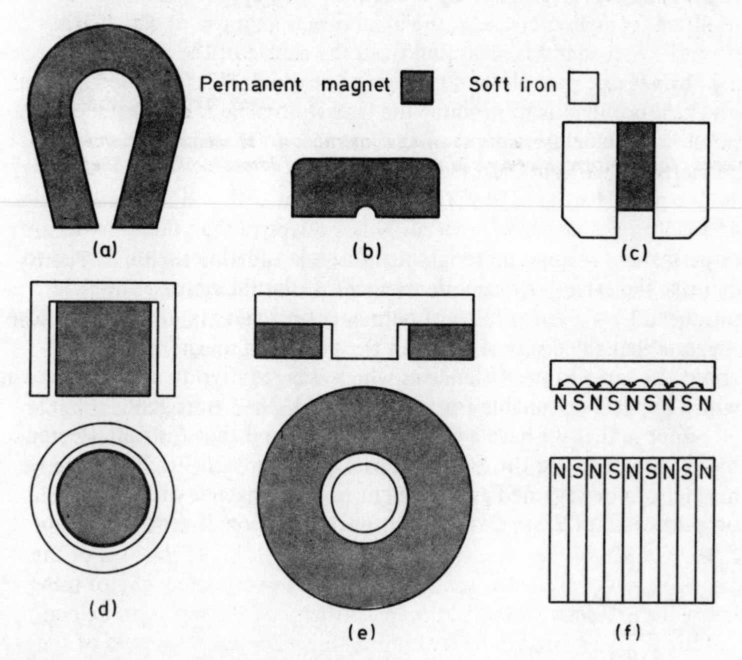


*Fig. 7.2. Different designs of holding magnets*

the design must obviously aim to maximize the field at a distance. If a simple bar magnet is used, the moment of the magnet is the primary consideration. In the older materials including to some extent the Al–Ni–Co alloys a large moment required a sufficient length to ensure the necessary value of $J$. With ferrites and $RCo_5$ this consideration does not arise. To show that for a bar magnet there is no intrinsic virtue in increasing the length, it is only necessary to quote the expressions for the field due to bar magnets of finite length that used to occupy a prominent place in school text-books.

'Gauss A position' $$H = \frac{2Jd}{4\pi\mu_0(d^2 - l^2)^2} \tag{7.3}$$

'Gauss B position' $$H = \frac{J}{4\pi\mu_0(d^2 + l^2)^{3/2}} \tag{7.4}$$

(In CGS units delete the term $4\pi\mu_0$ in the denominator.)

In these equations $J$ = magnetic moment, $d$ is the distance from the centre of the magnet and $l$ is half its effective length. In the equatorial case an increase in $l$ obviously reduces $H$. The apparent increase in $H$ resulting from an increase in the axial case is an illusion, which arises from the fact that $d$ is measured from the centre of the magnet. Increasing $l$ brings one pole closer to the point at which $H$ is measured, but the practical problem is to produce the largest possible $H$ at a distance measured from the closest point of the magnet. In the end-on case $l$ cannot be increased without increasing $d$ also.

Tyack in Hadfield (1962) states that the use of iron in tractive magnets is inadvisable. Iron is certainly less effective than the same volume of permanent magnet material, but if iron is added in such a way as to increase the effective magnetic moment, it should increase the field produced by a given volume of permanent magnet material. Tyack also suggests that the distance between the poles of a magnet should be about the same as the distance at which it is required to act, a statement which appears reasonable for a C shaped Al–Ni–Co magnet.

Suppose that we have a C shaped magnet and that $J$ is not affected by its curvature and the distance apart of its ends $2l$ (Fig.7.3). Such a magnet can be regarded as equivalent to a bar magnet with moment proportional to $2l$, say $2kl$. The magnetizing force it produces at point P is

$$H = \frac{2lk}{4\pi\mu_0(d^2 + l^2)^{3/2}} \tag{7.5}$$

To find the optimum value of $H$ differentiate with respect to $l$.

$$\frac{\mathrm{d}H}{\mathrm{d}l} = \frac{2k}{4\pi\mu_0(d^2 + l^2)^{3/2}} - \frac{6kl^2}{4\pi\mu_0(d^2 + l^2)^{5/2}}$$

$$= \frac{k}{2\pi\mu_0} \cdot \frac{d^2 + l^2 - 3l^2}{(d^2 + l^2)^{5/2}} \tag{7.6}$$

If $d^2 = 2l^2$ this derivative is zero and the second derivative is negative so that $H$ is a maximum. Thus the maximum $H$ occurs when $d = 1.41l$, i.e. $d$ is 0.7 times the distance between the poles.

In an Al–Ni–Co alloy $J$ is slightly increased by bringing the poles closer together, so it is not unreasonable to say that $l$ should be reduced to make $d = 2l$. In a ferrite or rare-earth cobalt alloy, one has the option of using a bar magnet. The moment does not depend on the ratio of the length to the area, and the only limitation to reducing the length and increasing the area is that it must not move the centre of the magnet away from P and so increase the value of $d$.

Some uses of holding magnets have already been mentioned. Other applications include holding type blocks or stencils in printing machines, fishing for iron articles lost in vats of liquid or in oil wells and a variety of toys.

For holding in contact there is no point in using two permanent magnets rather than one magnet adhering to iron, but a permanent magnet can exert a greater force on another permanent magnet at a distance than on a similarly shaped piece of iron.

A number of particularly important applications that present special problems will now be discussed in separate sections.

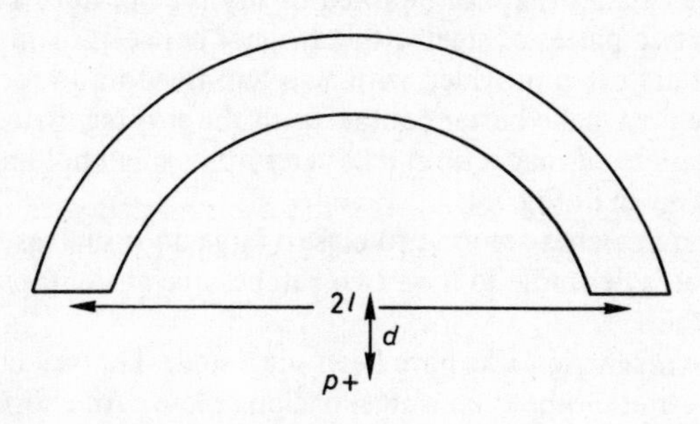


*Fig. 7.3. Deduction of optimum value of l*

### 7.2.1 Board games

As I am interested in a number of board games, I have a double reason for being interested in this subject, and I have been annoyed by the bad design of the magnets used in some of the commercially available products. I have a pocket chess set in which each piece contains a small sintered ferrite magnet adhering to the base by one pole. When the set is put away the pieces run together in chains. Other sets use thin magnetic rubber magnetized with both poles on the same side. This is a cheap and satisfactory solution, but some players may prefer to pay for rather thicker pieces easier to handle. There are numerous ways in which such pieces could be made with poles on one side only, so that the board could be put away without the pieces moving together. Magnetic pieces could be used in many other board games besides chess and draughts.

### 7.2.2 Door catches

Magnetic door catches are intended to keep a door shut when it is closed without slamming and without turning any knob, and equally to allow it to be opened by means of a reasonable pull. They are not intended to be used as locks. They are generally more satisfactory than mechanical devices intended to achieve the same purpose. The right size of catch must of course be chosen, and satisfactory functioning depends on proper fixing. A magnetic catch can cease to function if the hinges distort or the door warps, but it is usually less sensitive to such changes than most mechanical devices.

Refrigerator doors are usually closed with a strip of bonded ferrite, which acts as an air seal as well as a catch. The strip is magnetized with the flux following a U-shaped path, giving one long N and one long S pole along the length of the strip.

Separate catches that can be fixed to any type of door are often made with two pieces of steel with a magnet between them. The steel pole pieces are often provided with a certain freedom to rock that enables them to make better contact with the striking plate. An idea of the great variety of magnetic catches and other small holding magnets available is given in Fig.7.4.

If separate catches are used to close a large door such as that of a wardrobe, it is desirable to have two catches one at the top and one at the bottom.

Genuine magnetic locks have been suggested. The key could be a sheet of magnetic rubber on a steel backing plate. An enormous number of different patterns of magnetization could be imprinted in such a sheet, and the lock could be designed so that it was operated only by the correct pattern. I am unable yet to name any general commercial source for such locks and keys, but they are reported to be in use on miners' lamps.

### 7.2.3 Magnetic filtration and separation

Magnetic separators may be used to remove magnetic impurities from non-magnetic products, or to concentrate magnetic materials such as iron ore. A field gradient as well as a high magnetic field is required to exert a magnetic force, and so it is not a good idea to pass the material between the poles of a powerful magnet.

The force on a particle of volume $V$ and polarization $J$ is $JV(\mathrm{d}H/\mathrm{d}z)$

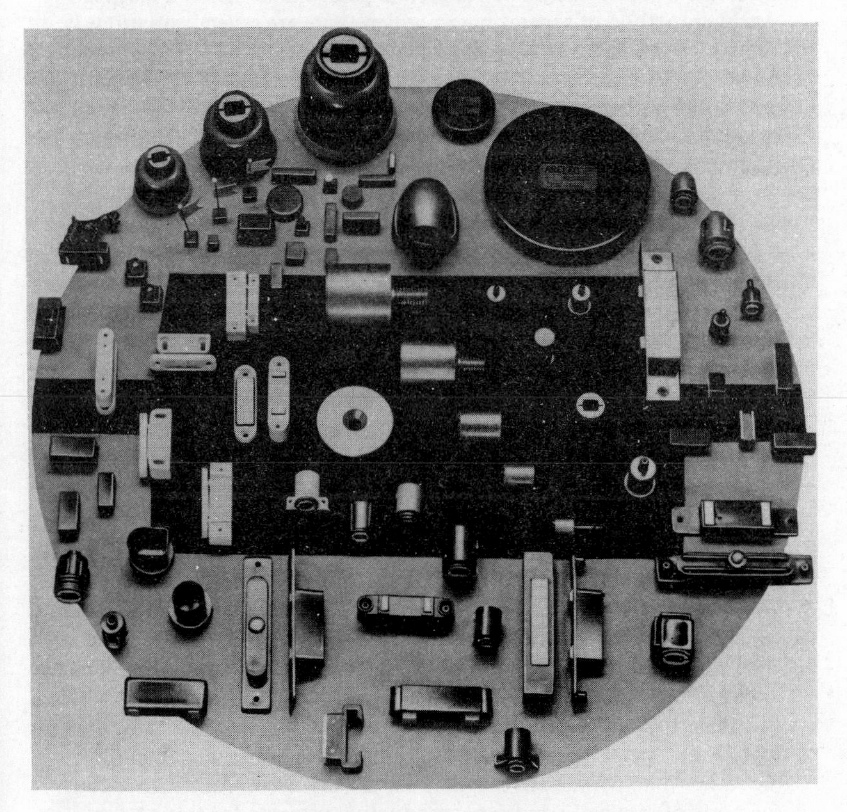

*Fig. 7.4. Array of small holding magnets, including range of door catches. (Courtesy of Magnetic Applications Ltd., London, and Aralec, France)*

and is in the direction of the field gradient, not of the field itself. If the relative permeability of the material is $\mu_r$, $J = (\mu_r - 1)\mu_0 H$, but $H$ is the effective field in the particle, that is the external magnetizing force reduced by the self-demagnetizing force of the particle. If the relative permeability is large, the effective relative permeability is equal to the reciprocal of the self-demagnetizing factor. Thus if the relative permeability is greater than about 20, its precise value is unimportant compared with the influence of the shape of the particle. On the other hand the shape of weakly magnetic particles is unimportant, and the actual value of their relative permeability controls the force, together with the produce $H(\mathrm{d}H/\mathrm{d}z)$. A number of formulae and graphs for calculating this product are given by Schüler and Brinkmann (1970) for magnets with a number of alternating poles of different shapes.

There are several designs of separator. In one type the material to be cleaned of ferromagnetic particles is passed underneath a magnet. Another method of separation, of which there are several variations is illustrated in Fig.7.5. The material is carried on the belt A over the pulley B. The non-magnetic material falls off at D. A fixed magnet C within the pulley makes ferromagnetic particles adhere to the belt as far as the point E to which they are carried by the roughness of the belt. At this point they fall off too.

A novel design of separator magnet is sketched in Fig.7.6. A and B are ferrite blocks magnetized through their thickness, and C is an iron backing plate. The unusual feature is the ferrite block D. One explanation of this block is that it acts as a magnetic insulator, and reduces the leakage flux between A and B. This is an idea that is mentioned from time to time in connection with a variety of magnetic devices. It is true that D reduces or even reverses the flux direction in the volume it occupies. Possibly this fact justifies calling it a magnetic insulator. Personally I regard this concept as unnecessary. In this and similar applications I think the extra flux obtained can be attributed to adding a piece of strategically placed permanent magnet material; in effect to using a bigger permanent magnet.

Large quantities of ferrite magnets are used to concentrate magnetic iron oxide from natural deposits and so make the exploitation of low grade ore economic.

The efficiency of filters can be increased by inserting an iron grid or steel wool into the flow path. Large fields and field gradients are produced in the vicinity of these rods or wires, and cause magnetic particles to adhere to them. The idea has recently been extended to permit the filtration of paramagnetic particles, but for this purpose superconducting rather than permanent magnets are usually used. A comparison of the costs of rare-earth and superconducting magnets for this kind of separation now appears justified.

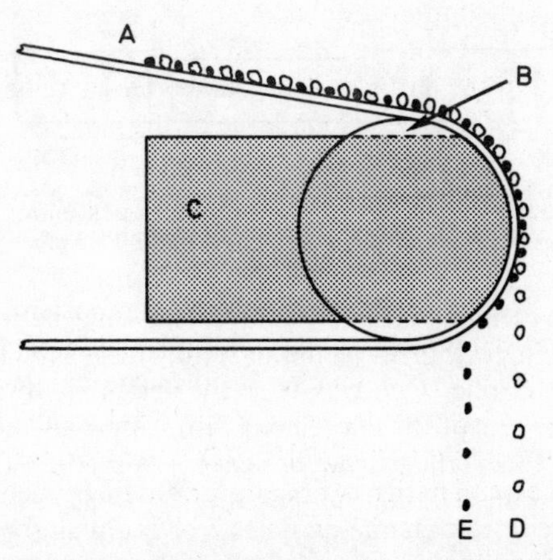


*Fig. 7.5. Separator using belt and non-magnetic pulley with fixed U-shaped magnet inside. A: Belt. B: Pulley. C: Fixed magnets with poles inside pulley. D: Purified material. E: Ferromagnetic contamination removed*

Separation and filtration problems involve many factors in addition to the purely magnetic ones. Such factors are the viscosity and rate of flow of the fluid, and the concentration of the material to be removed. These factors determine not only the magnetic fields that must be employed, but whether it is necessary to provide for frequent or continuous removal of the magnetic material.

### 7.2.4 Drives and couplings

A magnetic drive or coupling is a device in which a magnet promotes the motion of another body without physical contact. The two words 'drive' and 'coupling' are used synonymously by different authors in approximately equal numbers, but the British Patent Office prefers the word 'coupling'. Magnetic couplings can be made to work on the eddy current effect or the hysteresis effect, but these types of coupling can be more conveniently discussed in conjunction with other eddy current devices and hysteresis motors respectively.

This section is concerned with devices in which a magnet drives another magnet or suitably shaped piece of iron. In such devices the driven member must travel at the same speed as the driving member or stop altogether. They are therefore known as follower drives or synchronous couplings. There are a few applications in which linear motion is transmitted, but rotary machines will be considered first.

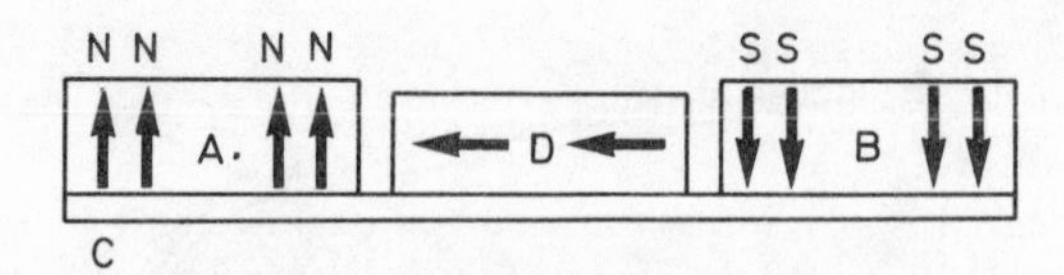


*Fig. 7.6. Separator magnet with device to reduce leakage flux. A, B and D: Ferrite magnets magnetized as indicated. C: Mild steel backing plate*

The most common reason for employing a magnetic coupling is to drive something within a vacuum or fluid without the need to provide a leak-proof bearing. Constraints on the design are often imposed by the shape and size of the container and thickness of the partitions, and the problem facing the magnet designer is to produce the required torque within these limitations and specifications.

Concentric drives are required to make something rotate inside a tube, while butt ended drives operate end-on through a plane partition. A number of possible designs for magnets for drives are shown in Fig.7.7.

A concentric drive such as (a) uses isotropic ferrite or alloy for the outer magnet and iron or hard ferrite for the inner member. Another concentric drive such as (b) uses isotropic ferrite rings with iron backing for both inner and outer magnets. The rings in (c) can be of anisotropic ferrite with iron backing to make an end-on coupling. The concentric drive drawn in (d) uses Alcomax (material A2) horseshoes; it can transmit a considerable torque but is rather bulky. The butt-ended design shown in (e) was developed by us. The magnets are trapezia of Hycomax III (material A4) on iron backing plates. A description of its performance is given by McCaig (1975); the design is protected by a patent assigned to the Cobalt Information Centre.

The optimum number of poles on any of these designs depends upon the radius. Details of the performance of concentric and butt-ended couplings using ferrite rings are given by Fahlenbrach and Baran (1965), Table 7.2, which compares some of these ferrite rings with our results with Hycomax trapezia, taken from McCaig (1975).

Some general observations about follower drives are:

(1) The end-on type is capable of transmitting greater torque within a given radius than the concentric type. The reason for this difference is that it is difficult to use radially magnetized magnets efficiently, because of the varying cross-section, and it is also difficult to produce anisotropic magnets with radial magnetization. The end-on type suffers from the drawback that it causes axial forces on the bearings.

(2) If there are no size limitations ferrite magnets almost always provide the cheapest solution, but Table 7.2 shows that, if as must sometimes happen, the couplings need to be accommodated within

a restricted space high coercivity alloys have an advantage. There must clearly be an opportunity for the use of the even better rare-earth-cobalt.

(3) Although it is possible to make one of the members of a drive of iron, the use of permanent magnets for both members gives a greater torque, partly because it usually increases the attractive force, but also because with proper design it introduces an extra repulsive force.

If the members of a drive of say type (e) are allowed to come to rest freely, they do so with unlike poles in opposition. The only force is that of axial attraction. If one member is turned while the other is held still an increasing circumferential attraction arises between each pole and the pole which it is leaving, and also a repulsion between each pole and the one it is approaching on the opposite member. These forces increase until the torque reaches a maximum and the driving member jumps to the next rest position.

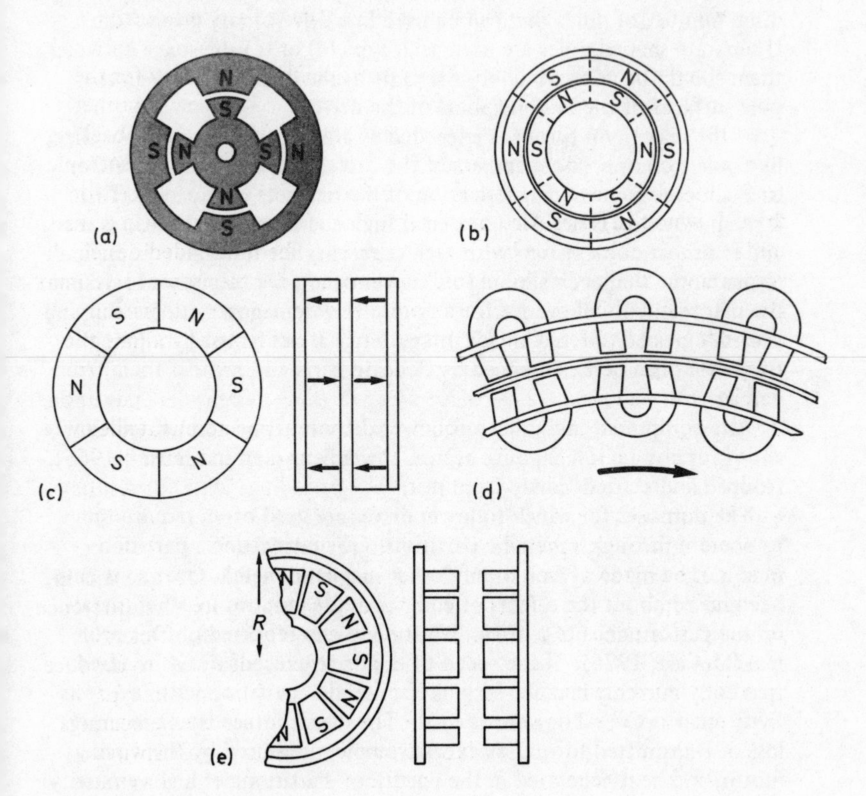


*Fig. 7.7. Magnetic couplings*

**Table 7.2** COMPARISON OF VARIOUS BUTT-ENDED FOLLOWER DRIVES $\delta = 10$ mm

| *Magnet arrangement* | *Outer radius* (mm) | *Mean radius* (mm) | *Length* (mm) | *Torque* (Nm) |
|---|---|---|---|---|
| 12 Hycomax rects | 70 | 54.5 | 18 | 3.43 |
| 6 Hycomax III & 6 MS rects. | 70 | 54.5 | 18 | 1.06 |
| 12 Hycomax III trap. | 70 | 54.5 | 19.4 | 6.35 |
| 12 Alcomax rect. | 70 | 55 | 38 | 3.73 |
| Ferrite ring | 70 | 52.5 | 20 | 2.90 |
| 12 Hycomax III trap. | 57 | 41 | 19.4 | 2.68 |
| Ferrite ring | 57 | 43 | 18 | 1.26 |

A high torque depends on having a large number of poles rather than poles of large area, but of course there is a practical limit to the maximum number of poles that can be used in a drive of any given radius. If separate shaped poles are used as in type (a) or (e) the space between them should be approximately the same as the distance between the pole surfaces of the two members of the drive.

If the maximum torque is exceeded so that the driving member slips, like poles, on opposite members of the drive, pass one another. After such slipping some demagnetization of the magnets may occur. This demagnetization is not very serious if high coercivity Al–Ni–Co is used, and is almost non-existent with high coercivity ferrites. As it is difficult to guarantee that such slipping will never occur, it is desirable to design the magnets with allowance for any resulting demagnetization. Slipping is sometimes considered an advantage, since it can be used to limit the forces that can be experienced by delicate parts and protect them from damage.

Any equipment including a follower drive must be accelerated slowly and if for any reason slipping occurs, the driving member must be stopped and started slowly from rest.

The purposes for which follower drives are used often require them to operate through a metallic partition. Obviously such a partition must not be made of iron or any other magnetic metal. Questions have been raised about the effects of eddy currents induced in other metals on the performance of a drive. We studied and reported on this problem (McCaig, 1975). If the speed of the drive exceeds a few hundred rpm eddy currents become serious if a metallic partition with low resistivity such as Cu, Al or brass is used. The disadvantage is not so much loss of transmitted torque, as excessive power required by the driving motor, and heat generated in the partition. Partitions of high resistivity metals such as austenitic stainless steel and Nimonic alloys can be safely used.

A very old and simple example of a magnet used to promote linear motion is provided by a maximum and minimum thermometer in which the small steel indicators are reset by means of a permanent magnet. There are numerous level indicators in which an external magnet follows the level of a float contained in an opaque vessel. Sometimes the magnet is required to perform some switching operation, and depending on the nature of this operation it may be desirable to use repulsion forces rather than attraction.

Jaffe and Herr (1976) described a magnetic drive that converts rotary to linear motion. The rotating member consists of two ceramic permanent magnet rings, magnetized as shown in Fig.7.8. The part that is driven backwards and forwards in a linear manner consists of two blocks of rare-earth cobalt also magnetized as shown, and held in a U-shaped aluminium structure. Sewing machines are one of the possible applications for such a device.

The French National Centre for Scientific Research showed couplings that also acted as step-up or step-down gears at the Physics Exhibition in London in 1966. Each tooth on each gear wheel is a small magnet, and as with an ordinary gear, the ratio of the number of teeth gives the inverse ratio of the number of revolutions made by the wheels. Since there is no physical contact, there is no friction, noise or wear.

An interesting application of permanent magnets is the ARCT Klinger Mono-Roller Magnetic Spindle illustrated in Fig.7.9. A textile thread is passed through the tubular spindle, which contains a diabolo shaped sapphire piece around which the thread is wound; the thread is twisted first in one direction and then in the other. If the thread is a thermoplastic and is heated, the twisting sets up strains, and gives the yarn the properties required for stretch fabrics.

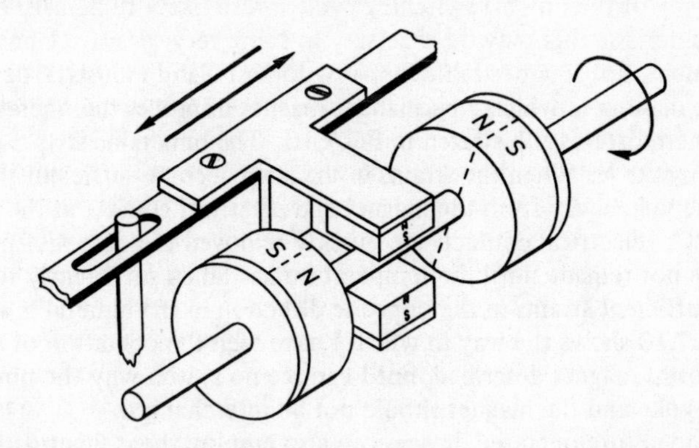


*Fig.7.8. Rotary to linear translator (Jaffe and Herr, 1976). (Courtesy of the Singer Company, Elizabeth NJ)*

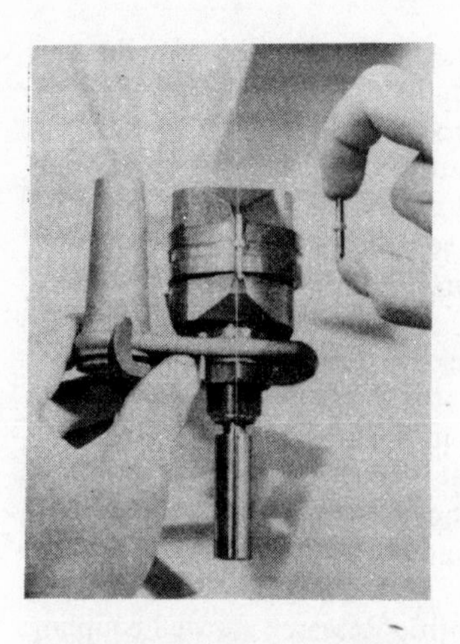

*Fig. 7.9. The ARCT – Klinger Mono-roller magnetic spindle. (Courtesy of Klinger Manufacturing Co., London)*

Exactly how the stretch fabric properties are produced will be more clear to the textile scientists. From the point of view of the magnetician the interesting fact is that the tubular steel spindle is held in contact with the large driving wheel by a magnet. There are no bearings at all and the spindle does not actually touch the magnet; spindle speeds up to several hundred thousand rpm can be attained.

Many small magnets are used in conjunction with thermostats. The typical thermostat for domestic equipment operates by expansion processes in a bimetallic strip designed to break an electric contact at some preset temperature. The difficulty is that the movement of the bimetallic strip is slow and gradual, so that acting by itself it produces only a slow movement of the contact and arcing is likely to occur. The frequency of the on–off switching cycle is also likely to be very rapid, a characteristic that may be necessary in some very sensitive laboratory equipment, but is undesirable in many domestic and industrial devices.

The manner in which a permanent magnet improves the operation of such thermostats is illustrated in Fig. 7.10. The bimetallic strip S carries an iron yoke Y. When the strains in the strip become sufficient they pull the yoke away from the magnet and electrical contact at the same time. The electrical contacts are quickly removed to a safe distance, and are not remade until the temperature has fallen sufficiently to produce sufficient strains in the opposite direction in the bimetallic strip. Figure 7.10 shows the way in which I have seen the operation of a thermostat magnet described, but I can see no reason why the positions of the yoke and the magnet should not be interchanged.

Temperature operated devices can also employ the magnetic interaction between a permanent magnet and a temperature sensitive magnetic material near its Curie point.

It has been suggested that permanent magnets could assist do-it-yourself cleaning of the outside of windows, or the cleaning of any window difficult to reach from the outside. The idea is that the leather or sponge on the outside would be held to the window and moved about by a magnet on the inside. The idea sounds promising but does not appear to have progressed far, probably because it is difficult to design a system easy to control. Certainly I tried to make a device for my own windows, and became well aware of the difficulties.

Thickness gauges have been used to measure the thickness of paint or electroplating on a steel surface. Their operation depends on measuring the force necessary to pull a magnet away from the surface.

The automatic warning system used by British Rail is a spectacular use of permanent magnets. The arrangement is sketched in Fig.7.11. The locomotive passes first over a large permanent magnet 1. The magnet is placed between the rails, with the top plate approximately level with their upper surfaces. The vertical field produced by this magnet tilts a pivoted magnet 3, in the receiver, usually housed in one of the bogies of the locomotive. To increase the torque on the pivoted magnet, iron parts 4 and 5 are provided to concentrate the flux. The pivoted magnet is about 17 cm above the track magnet, and without the receiver in position the flux density at this height is specified to exceed 0.0035 T. The large pole plates on the track magnet spread the flux, so that the detector is within the field for sufficient time to be operated, even when the train is travelling at more than 100 mph.

The permanent magnet on the track is quickly followed by an electromagnet 2. If the line is clear a signal is given by the electro-magnet,

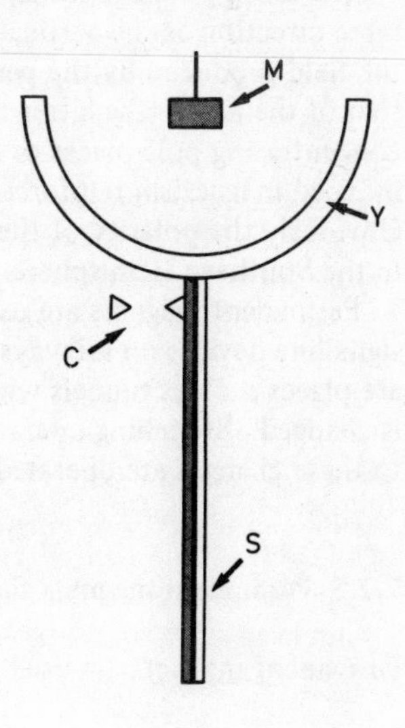


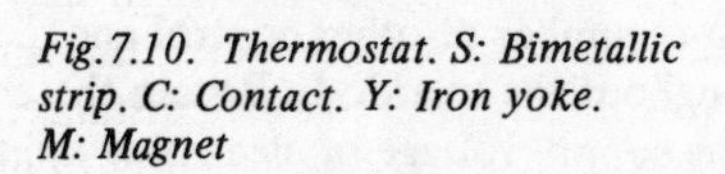

*Fig.7.10. Thermostat. S: Bimetallic strip. C: Contact. Y: Iron yoke. M: Magnet*

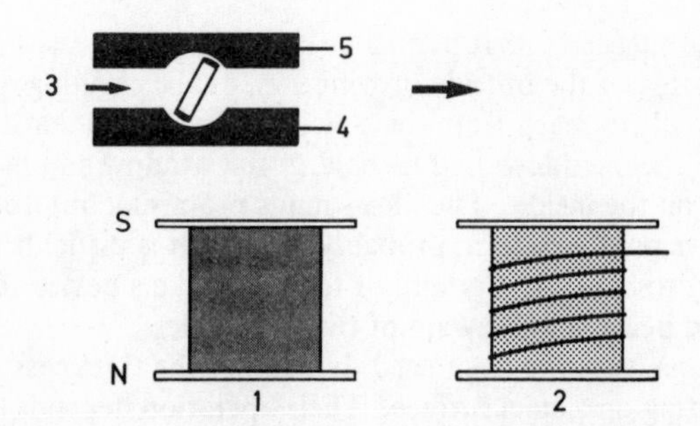


*Fig. 7.11. Automatic warning system, British Rail. (1) Permanent magnet at track level. (2) Electromagnet. (3) Pivoted permanent magnet with mild steel ends. (4), (5) Mild steel pole-pieces between which pivoted magnet has two stable positions*

which cancels the signal given by the permanent magnet. If the electromagnet is not energized, audible and visible danger signals are given in the cab and the brakes are actually applied. Thus a failure in the energy supply has the same effect as a deliberate danger signal, and the system fails safe.

The permanent magnets were originally made of a block of Alcomax with iron plates about 25 x 25 $cm^2$ in area. These magnets were replaced by ferrite magnets on electrified lines because Alcomax was occasionally demagnetized by the large magnetic fields produced by the heavy currents flowing in the rails. The ferrite magnets have now replaced the alloy magnets for all new installations.

The track permanent magnets are arranged to produce a field in the same direction as the vertical component of the Earth's field. Although the field produced by the permanent magnet is large compared with that of the Earth, the latter may induce substantial magnetism in the concentrating pole-pieces or core of the electromagnet. Any such induced magnetism reinforces the danger signal with this arrangement. Obviously the polarity of the permanent magnets should be reversed in the Southern Hemisphere.

Permanent magnets are used for a number of other control and signalling devices on railways throughout the world. In Britain there are places such as tunnels where the supply voltage of electrified lines is changed. Switching operations in the locomotive to accommodate to these changes are operated automatically by magnets.

### 7.2.5 Permanent magnets that switch off

Permanent magnets are used for vices, chucks, conveyers and cranes in workshops, factories, shipyards and harbours; for many of these applications they must hold steel objects with great force, and a problem arises when these objects must be released.

A simple method of separating a steel object from a magnet is a screw that acts as a jack to separate the two components. This method is quite extensively used, and works well when the forces involved are not too large.

A permanent magnet that can be switched off sounds a contradiction in terms. For more than a hundred years the ease with which an electro-magnet can be switched off was thought to be its own particular virtue, while the special advantage of a permanent magnet was just that it was permanent. The need to switch off permanent magnets scarcely arose until modern powerful magnets were invented.

Since a permanent magnet can be demagnetized, it is obvious in a sense that it must be possible to switch it off electrically. A large number of electrical methods of releasing the load from a permanent magnet have been suggested, and there is some doubt about which is the best. Purely mechanical methods which are much more widely used, will be described first. These methods do not demagnetize the permanent magnet, but direct flux away from the object held.

### *7.2.5.1 Mechanical methods of flux diversion*

One simple method of flux diversion is illustrated in Fig.7.12 (a) and (b). In (a) the flux produced by the magnet A passes through one of the pole-pieces B, then through C the iron workpiece to be held and back through the other pole-piece to the magnet. In this position the workpiece is firmly held.

In position (b) the magnet has been turned through a right angle. Half of the flux produced by the magnet passes into each pole-piece, and back to the opposite pole of the magnet without entering the work-piece, which is therefore released.

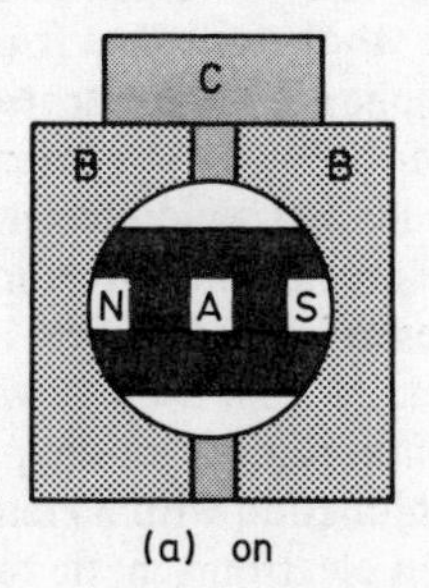


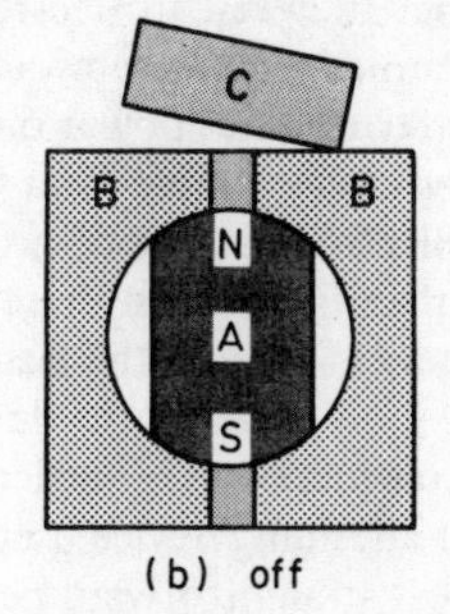


*Fig. 7.12. Simple device using flux diversion. A: Permanent magnet. B: Mild steel pole-pieces. C: Object to be held.*

The operation of switching on and off a magnet constructed on this principle is performed by merely turning a handle. Lifting magnets and chucks that are switched on and off by this methods are manufactured in a great variety of shapes and sizes, although the particular method shown in Fig.7.12 using only one magnet is confined to the smaller devices.

Figure 7.13 (a) and (b) shows a larger chuck, containing a number of anisotropic Al–Ni–Co magnets. The flux diversion principle operates as in the simpler Fig.7.12. This particular design of chuck would be suitable for use on a grinding or milling machine.

Alcomax or similar A2 type materials are suitable for chucks and other large switchable holding magnets, because they have a high flux density and enable a powerful magnet to be assembled within a small volume. The permanent magnet is well protected against demagnetization in the interior of the device.

Recently more use of ferrite magnets has been made in magnetic tools. The area of this low flux density material must be great so that the high flux density can be produced by flux concentration. Figures 7.14 (a) and (b) shows a typical design of chuck using ferrite magnets. Compared with a device using alloy magnets the direction of the permanent magnets has been rotated through 90°. There are now also two sets of magnets that slide relatively to each other as shown.

Permanent magnet tools that are switched on and off by mechanical means have the advantage of requiring no power supply. It is particularly advantageous not to require electrical connections for rotating machinery such as lathes. The limit of the size and strength of these tools is the force required to slide the magnets and iron parts for the switching action, and for larger tools permanent electromagnetic devices may be used.

### *7.2.5.2 Permanent electromagnetic tools*

The difference between an electromagnet and permanent electromagnet is that the former requires power for the whole time that it is operating, while the latter requires power only while it is being switched on and off. The advantage is more than just saving the cost of power. In the permanent electromagnetic device the leads and windings carry current only intermittently and can therefore be much lighter than in the simple electromagnet. If desired the leads can even be disconnected while the tool is being used. Finally the device does not fail if the power supply fails. This consideration is particularly important from the point of view of safety for a holding device used in conjunction with a crane.

In the most straightforward permanent electromagnetic tool the permanent magnet is magnetized when it is required to grip the load,

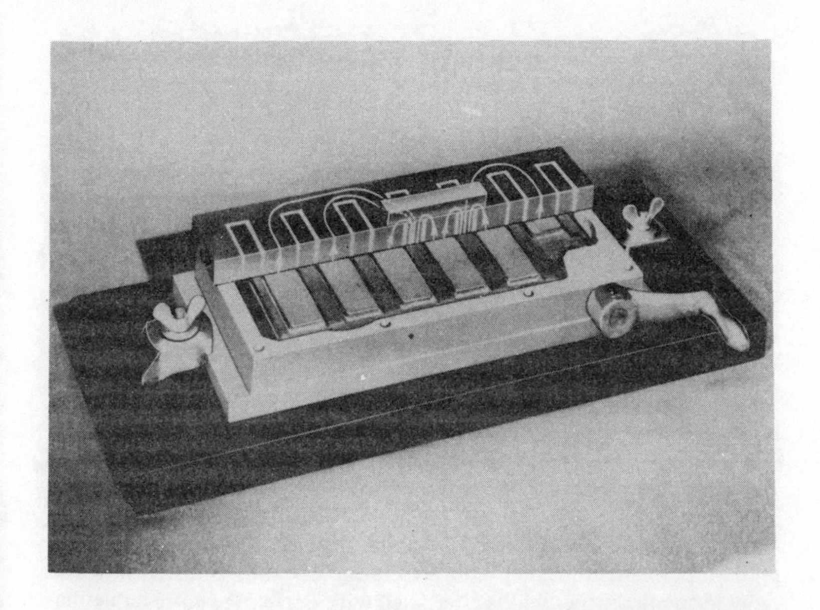

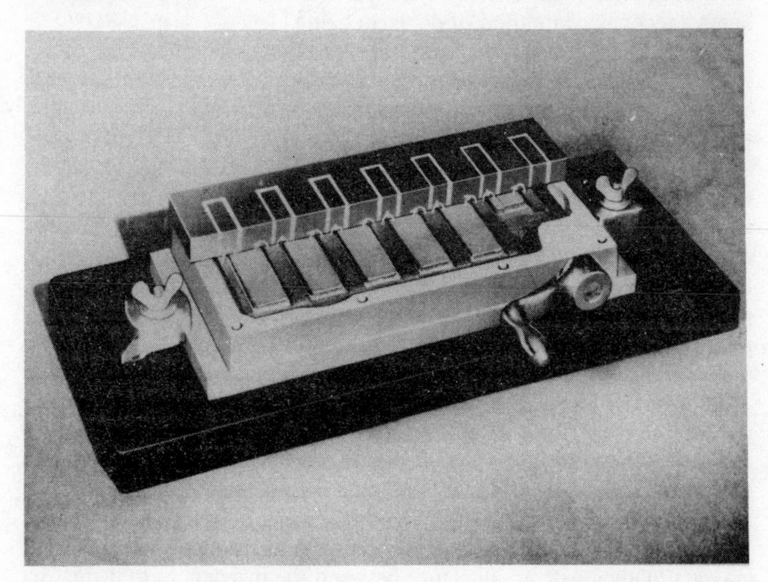

*Fig. 7.13. Principle of magnetic chuck; (a) on, (b) off. (Courtesy of James Neill Ltd., Sheffield)*

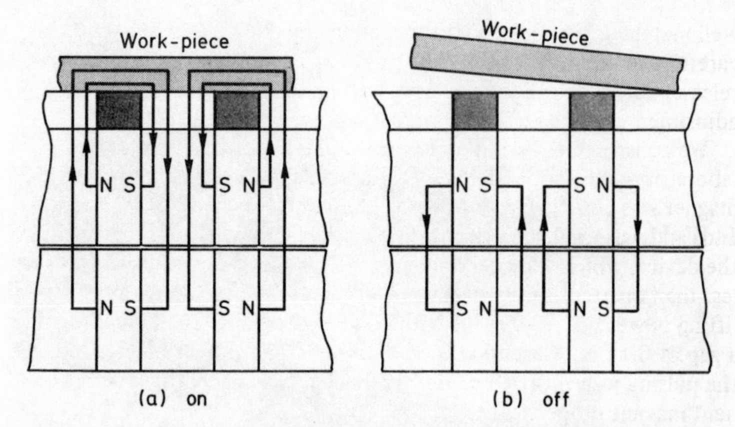


*Fig. 7.14. Part of chuck with ferrite magnets. The flux paths are shown (a) with the work piece held, and (b) after the magnets in the lower row have been displaced by one position, and the work piece is released.*

and demagnetized when the load is to be released. Demagnetizing a modern permanent magnet requires a rather carefully adjusted reverse field. Sometimes the field is alternated with decreasing pulses, this being a recognized method of demagnetizing.

An idea, which originated in France, is explained with the aid of Fig.7.15. In this arrangement, although only two are shown, there may be any even number of permanent magnets. To switch the magnet off, the direction of magnetization of half the magnets is reversed by passing a pulse through suitable coils. When all the magnets are magnetized in the same direction the flux is directed through the work-piece and the whole device is switched on. When the magnetization is reversed in half the magnets, the flux returns through the pole-pieces without entering the work-piece, and the tool as a whole is switched off. Although the pulse required to reverse the magnetization is a little greater than that required to cause demagnetization this method has the advantage that it does not require the pulse to be accurately adjusted. So long as the pulse reverses the magnetization its precise value is immaterial; it cannot be too great. Coils should actually be placed round all the magnets, so that those that are not reversed are remagnetized in the original direction. If this precaution is not taken the magnets that are not reversed may gradually become slightly demagnetized. This demagnetization does not seriously affect the lifting power of the device, but it may gradually create an inbalance between the magnets so that the flux through the work-piece is not precisely zero, when the tool is supposed to be switched off, so that the work-piece is not released. To ensure that the work-piece is always released and small articles cannot be picked up when the tool is supposed to be switched off, the magnets should be well matched, particularly at the remanence point. If the magnets are carefully chosen there should be no difficulty in securing complete release, but if the magnets are not well-matched, it is unlikely that any adjustments to the electrical system will cure the defect.

We constructed a small experimental model of this type in our laboratory, and it operated very satisfactorily. A larger version of this magnet was constructed by one of our Members (James Neill and Co Ltd) and was used for a time in their own factory. The total weight of the device, which was used on a crane, was 400 lbf of which rather less than one tenth was permanent magnet material (Columax). The lifting power was 10 000 lbf with good contact and 2000 lbf through a gap of 0.15 in. Certain refinements were said to have been made to the pulsing procedure to secure the release of loads with slight permanent magnet properties.

I was very impressed with a demonstration of this equipment and have never been quite clear why it has not been preferred to that in which the magnet is completely demagnetized. The difficulty is said to have been that of preventing it from attracting small objects when it was supposed to be switched off. I suspect the need to match the magnets may not have been fully realized.

Another idea that has received considerable publicity is based on the fact that certain grades of barium and strontium ferrite can be reduced to zero flux density by a demagnetizing force, but when this is switched off the magnet recovers almost completely. The idea is that a winding is placed on the magnet or pole-pieces, but is energized only while the load is released. There is no doubt that this mode of operation is possible, and would work even better with rare-earth cobalt.

There are however practical snags. The ferrite magnet must exert the same mmf as an electromagnet would to drive flux through the

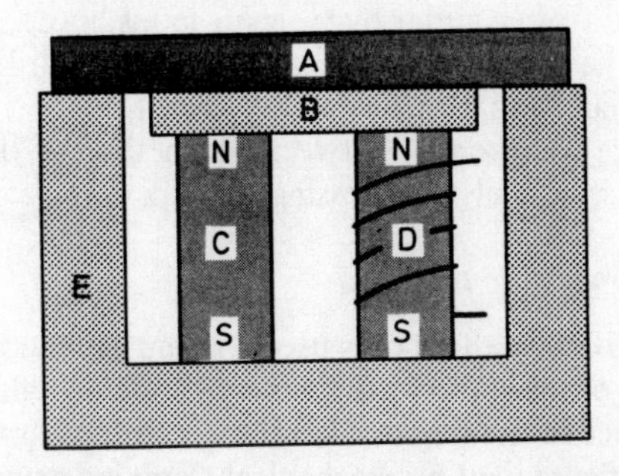


*Fig. 7.15. Flux reversal in half the magnets switches this chuck off. A: Workpiece. B: Mild steel top plate. C and D: Permanent magnets. (D has coil to reverse direction of magnetization) E: Outer case of mild steel*

same work-piece. When one tries to reduce the flux to zero it is the characteristic of a permanent magnet that it exerts a greater mmf. If the magnet was working efficiently in its holding mode, the mmf for release is actually twice as great. Thus the number of ampere-turns required to release the load with this system is double the number required to hold the load with a conventional electromagnet. The power may not be required for so long as in the case of an electromagnet, but it is required for more than a short pulse. If the coils are placed on the pole-pieces rather than the magnet, it may be unnecessary completely to demagnetize the latter, and the idea becomes much more promising.

There are also possibilities of electrically promoted flux diversion. One such method, suggested by the Westinghouse Corporation of Pittsburgh is depicted in Fig.7.16. If neither of the keepers A or B is in position, the first to be brought near the magnet is attracted. Once the flux is directed through A say, there is none left for B, which is consequently not attracted. It is however, possible by passing a suitable pulse through the coils C, D, E and F to direct the flux from A to B, so that A is released and B is attracted. In testing the idea I found it desirable to let the limbs G and H have slight permanent magnet properties, such as might be possessed by a tool steel rather than soft iron. For maximum efficiency I also found good contact between G and H and the keepers to be necessary. If these conditions were observed the pulses can be quite small; much less in fact than is required in any of the other devices. If the contact between G and H and the keepers was less good, I found it necessary to increase both the coercivity of the arms and the size of the pulses. Several variations in the arrangements of the coils have been suggested.

Westinghouse may have solved these problems by different methods. To demonstrate the method they made a film of a man walking upside down suspended from a girder by magnets in his boots. In order to take a step forward, he released one of his feet from the girder, by means of a switch in his hand and a battery in his pocket.

Possibly some compromise between the method of flux diversion and the method of completely suppressing the flux may prove fruitful.

### *7.2.5.3 Use of magnetic tools*

Whatever type of magnetic tool is used, certain general principles apply. The advantages of magnetic tools are speed and adaptability in setting up the work-piece in the required position, and the elimination of damage and disfigurement by mechanical clamping devices.

The limitation of magnetic tools is that they can only satisfactorily hold strongly magnetic materials. There are ways of circumventing this difficulty such as holding non-magnetic parts between magnetic ones or embedding them in a low melting point alloy or sulphur in an iron trough. These methods are useful in a workshop that deals mainly with magnetic materials, and only occasionally with non-magnetic ones.

Permanent magnet materials, including some tool steels, present a special problem. They can be held satisfactorily, but when the tool is switched off may continue to adhere by their own permanent magnetism. To release such objects some chucks and other magnetic tools are designed to produce a small demagnetizing field.

The design of a chuck and the force which it can exert depends on the nature of the work-piece. A chuck with large area poles is suitable for holding large thick objects. It is unsatisfactory for holding small objects that do not span at least two poles, or thin sheets that cannot carry all the flux entering from these broad areas. Adapter plates can be used to convert a tool with large surface areas of the same polarity into a much finer pattern, for holding small or thin objects. The reverse process cannot be done efficiently, mainly because a magnet suitable for a fine pole device lacks the mmf necessary for a coarse pole one. Sometimes objects that do not have a plane surface require a chuck with a specially designed contoured surface. Such specialized tools are most economically provided by adapter plates.

In most machining operations the force exerted by the chuck must be parallel to its surface. The magnetic force is normal to the surface, but friction converts it into what is usually a smaller tangential force.

The surface of a chuck should be kept clean, and any moving parts should be well lubricated. The magnets cannot normally be demagnetized in use, unless possibly a large fully magnetized permanent magnet is placed on its surface. Magnetic tools incorporating alloy magnets may not be completely dismantled without the possibility of demagnetization.

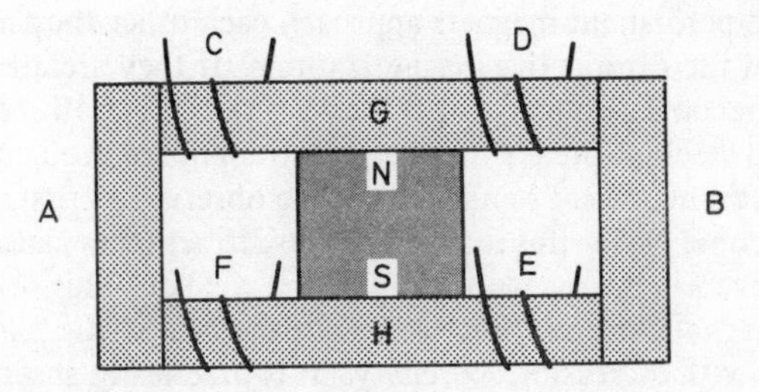


*Fig. 7.16. Westinghouse flux switching device. NS: Permanent magnet. G H: Steel arms. A B: Keepers or work-pieces. C D E F: Coils*

### 7.2.6 Applications using induced magnetism

A permanent magnet may be used to induce attraction between other iron parts. A good example is a reed switch, sketched in Fig.7.17. The two reeds are held apart by their springiness, but are made to come together when they are magnetized by a distant permanent magnet. The obvious mode of action is for this coming together to make an electrical contact, but if desired switches in which a magnet breaks a contact can be made. Reed switches can be made to operate at various flux densities, but those I have seen were operated by flux densities between 0.001 and 0.01 T (10 to 100 G). Usually the reeds themselves are made of a semi-hard material so that they do not separate until the applied field has been appreciably reduced. The range of field for which the switch remains closed can be varied by the designer. Reed switches are offered with various characteristics, and the user should be careful to choose one suitable for his purpose. Often the switch is operated by the approach of a magnet, but there are other possibilities. Both the switch and magnet may be fixed, and the switch may be operated by the passage of some steel object which depending on the arrangement increases or reduces the field experienced by the switch. Reed switches are employed in many automated industrial processes.

A magnetic clutch in which a suspension of magnetic particles in a liquid is congealed by the field of a magnet is another example of a device operated by induced magnetism.

## 7.3 REPULSION

The laws of magnetic repulsion are similar to those of attraction. Two magnets with like poles approaching repel each other with the same force with which they would attract each other, if they retained the same strength with unlike poles approaching. Yet until recently magnetic repulsion had no applications and was even difficult to demonstrate.

As two permanent magnets approach each other they may induce changes in their respective magnetizations. If they are attracting both magnets become stronger, but if they are repelling both become weaker. If two old steel-magnets with unequal strength are brought together with like poles facing, weak repulsion may be observed at first, but as the magnets come closer this may change to attraction because the stronger magnet reverses the magnetization of the weaker. This reversal or reduction of magnetization is much less with the Al–Ni–Co alloys, and decreases with increasing coercivity. It is practically absent in those ferrites and $RCo_5$ compounds for which $\mu_0 H_{ci}$ exceeds $B_r$.

Repulsion can take place between two permanent magnets, two

electro-magnets or a permanent magnet and an electromagnet. A permanent magnet can also promote repulsion between two pieces of iron but cannot repel one piece of soft iron. Iron pole-pieces may be attached to permanent magnets that are being used for repulsion, but they are less effective than in applications involving attraction.

Many people have tried to devise systems involving complete levitation of an object without contact with any other body. There is a very old theorem attributed to Earnshaw, early last century that states that an object cannot be suspended in stable equilibrium by purely magnetic or electrostatic forces. It is very tempting to try to disprove this theorem and a number of patents have been granted to inventors who claim to have done so. A good review of some of the possible and impossible suggestions is given by Geary (1964).

At one time when I had only seen a statement but no proof of Earnshaw's theorem, I tried to disprove it myself with completely negative results. Eventually I delved into the ancient literature and was completely convinced by the mathematical soundness of the theorem. The mathematical arguments are examined in the next section, together with an explanation of why complete levitation is possible by other means.

### 7.3.1 Earnshaw's theorem and levitation

Let $\psi$ be the scalar magnetic potential at a point, then according to the well known equation of Laplace

$$\frac{d^2\psi}{dx^2} + \frac{d^2\psi}{dy^2} + \frac{d^2\psi}{dz^2} = 0 \qquad (7.7)$$

In order that a magnetic pole should experience no force and have no tendency to move, it is necessary that

$$\frac{d\psi}{dx} = \frac{d\psi}{dy} = \frac{d\psi}{dz} = 0 \qquad (7.8)$$

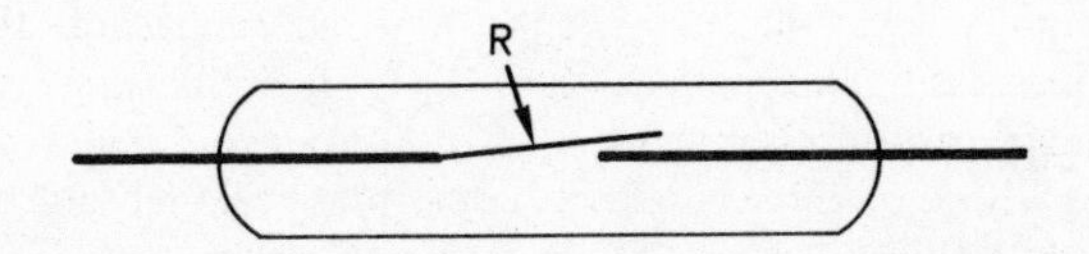


*Fig. 7.17. Reed switch. R: Reed of springy semi-hard magnetic material*

Equation 7.8 is the condition for a maximum or minimum in the potential $\psi$. At a potential maximum the pole is in unstable equilibrium. Only at a potential minimum is the equilibrium stable. The condition for a potential minimum is that its second derivative is positive. If two of the second derivatives, say $d^2\psi/dx^2$ and $d^2\psi/dy^2$are positive. Equation 7.7 shows that the third $d^2\psi/dz^2$ must be negative. Therefore the pole must be unstable with respect to movement in at least one direction. The theorem has been derived for one point pole, but it should apply to a magnet as a whole.

There is an assumption made in this theorem that the strength of all the magnets remains unchanged. If it were possible for a magnet to become stronger as it approached another magnet that repelled it, Earnshaw's theorem would no longer forbid complete levitation. Actually if the strength of magnets changes at all it does so in such a way as to increase the instability. Induced magnetism reduces the strength of a repelling magnet and increases the strength of an attracting one.

There are however a number of magnetic phenomena outside the realm of magnetostatics that make complete levitation possible. Some of these phenomena involve eddy currents. An alternating field produced by an alternating current induces eddy currents in any nearby conductor. These eddy currents produce a repulsive force between the conductor and the coil carrying the eddy currents. The eddy currents and force of repulsion increase as the conductor approaches the coil. As a consequence it is possible to design a coil, which when it is supplied with AC, levitates a conducting body. A practical application of this form of levitation is melting certain reactive metals without allowing them to be in contact with any crucible material.

Any form of magnet moving relatively to a conductor induces eddy currents in the latter that might help to promote a suspension system.

A permanent magnet will float over a sheet of superconducting material. If the magnet approaches closer to the superconductor it induces great currents which create a field that repels the magnet.

Diamagnetism is a very weak effect rather akin to superconductivity. Rather remarkably there are a number of claims to have promoted complete levitation by means of a magnet and a diamagnetic material. I imagine these claims describe laboratory demonstrations rather than practical applications.

### 7.3.2 Bearings

Magnetic bearings are at present one of the most important applications of magnetic repulsion. They are used particularly in watt-hour meters, such as are used to measure the consumption of electricity in both industry and the home. There is no attempt to secure complete levitation. The weight of the moving part is supported by magnetic repulsion, but sideways movement is prevented in the conventional manner. Provided the adjustment is correct the sideways restraint requires little force, and introduces little friction.

A simplified sketch of ferrite magnets supporting a rotor by repulsion is shown in Fig.7.18. Many millions of meters are made with this kind of suspension, although there is an alternative American design using attraction between an outer magnet of bonded Al–Ni–Co and an inner magnet of Cunife shown in Fig.7.18(b).

An important design consideration is that small changes in the strength of the magnets, such as result from temperature changes, should not displace the equilibrium position of the rotor too much. This requirement is met by a design that causes the force of repulsion to fall away quickly with distance. Ferrite magnets are commonly used, although high coercivity Al–Ni–Co has the advantage of a smaller temperature coefficient. Rare-earth-cobalt magnets appear ideal for high performance magnetic bearings.

The main advantage claimed for magnetic bearings in watt-hour meters is not so much that they reduce the initial friction as that they prevent wear and an increase of friction and low reading of the meter after prolonged use.

Several types of bearing that give complete levitation by means of a combination of permanent and electromagnets have been described. The principle of an early one manufactured in the USA is explained

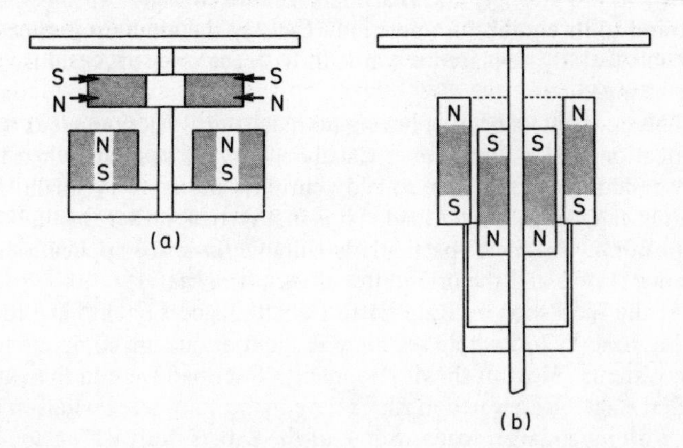


*Fig. 7.18. Magnetic flotation of rotor of watt-hour meter. (a) Using repulsion between ferrite magnets; (b) using attraction.*

with the help of Fig.7.19. Radial displacement of the rotor, that is movement in any direction at right-angles to the axle is restrained by ring magnets on the rotor and fixed concentric ring magnets of larger radius. This system of permanent magnets is quite unstable to axial displacement of the rotor. If the rotor is displaced by a small distance in a direction parallel to its axis, an increasing force tending to move it in the same direction arises. Eventually some part of the rotor magnet will be close to an unlike pole on the stator and repulsion between rotor and stator will be replaced by attraction. The slightest radial displacement of the rotor will be accelerated and the rotor will be found stuck to the stator in an oblique position, as predicted by Earnshaw's theorem.

To prevent the initial axial displacement of the rotor, a soft ferrite disc C on the rotor is situated between a pair of fixed electromagnets A and B. If nothing else is changed the system is still unstable, and if C moves towards one of the electromagnets say A, it is attracted more strongly by A and less strongly by B, and so continues moving until it touches A.

The electromagnets A and B are, however, fed with AC. As C moves towards one of the electromagnets it effectively increases the self inductance in its coil and reduces the current flowing therein. At the same time the self inductance of the coil of the other magnet decreases and its current increases. A special power supply containing capacitors is used, and is adjusted so that the coils on A and B almost resonate with the supply. At resonance the current in a circuit is much enhanced, as in a radio tuned to the frequency of the transmitter.

Now it is arranged that as C moves towards A and away from B, the tuning of the circuit A is worsened and that of B improved. Thus A attracts C less strongly and B attracts it more strongly. In this way C is restored to its equilibrium position. Exactly the same process occurs if C is accidentally displaced towards B, with the roles of A and B interchanged.

This contactless bearing having no mechanical friction needs no lubrication, and is therefore especially useful in a vacuum, when the only residual losses are due to eddy currents and are very small. The bearing can carry a bigger load if it is in a vertical rather than a horizontal position. It appears particularly suitable for space applications where $g$ is zero and the orientation does not matter.

At the Workshop on Rare-Earth-Cobalt magnets held in Dayton during June 1976 a whole session was spent discussing complete levitation systems. Most of the developments described were in the experimental stage. Rare-earth magnets have made complete levitation feasible with much larger loads. Some of the papers dealt with associated problems such as motors or methods of damping out resonant oscillations, in addition to the actual problem of levitation.

Most systems use permanent magnets to provide radial stability, but there has been a move towards using them in the attractive rather than the repulsive mode. There are many different arrangements, but the basic principles are shown in the simplified Fig.7.20. Three permanent magnet rings are magnetized axially in the same direction. This arrangement obviously gives a strong restoring force against any radial displacement of the central ring, but at the best it is in unstable equilibrium against axial movement. To keep this ring in the central position requires a sensing mechanism, an amplifier and an electromagnet that usually acts on another rotor permanent magnet. Theoretically if the rotor is positioned precisely in the equilibrium position, no current need flow in the electromagnet to keep it there. In practice the power required by the electromagnet is small, unless some external disturbance displaces the rotor.

Various types of sensing mechanisms are suggested for detecting the position of the rotor. They include photoelectric devices, magnets acting on Hall elements and magnets acting on coils. The last mentioned are sensitive to the movement rather than the position of the rotor.

Completely levitated bearings are particularly useful for high rotational speeds from 20 000 rpm to over 100 000 rpm. The system has to be stable against movement as a whole of the rotor in an axial or a radial direction, and also against angular deflection of its axis of rotation. Resonant frequencies of oscillation in any of these modes cannot always be completely excluded, and any such vibrations must be damped out, without introducing a braking action. The final structure may thus be much more complicated than is suggested by Fig.7.20.

The need for gyroscopes to maintain the orientation of space satellites for many years with the minimum expenditure of power and no wear has been a great incentive to develop magnetic bearings. It is suggested that large rotating wheels may also be used for energy storage in satellites. Solar energy is normally used, but energy may need to be

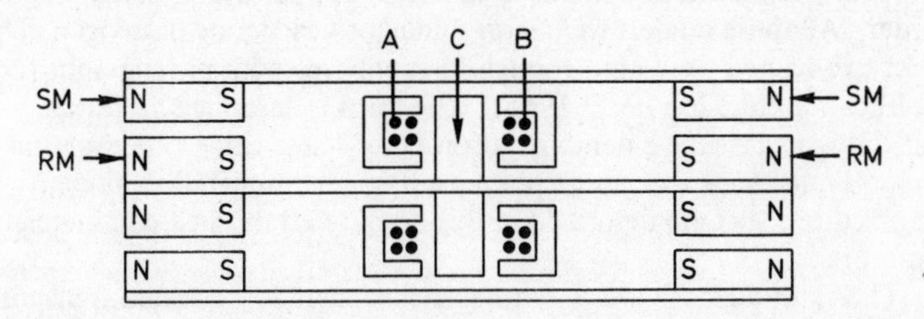


*Fig. 7.19. The Mag-Centric bearing. SM: Stator permanent magnets. RM: Rotor permanent magnets. AB: Similar electromagnets. C: Disc of soft magnetic ferrite*

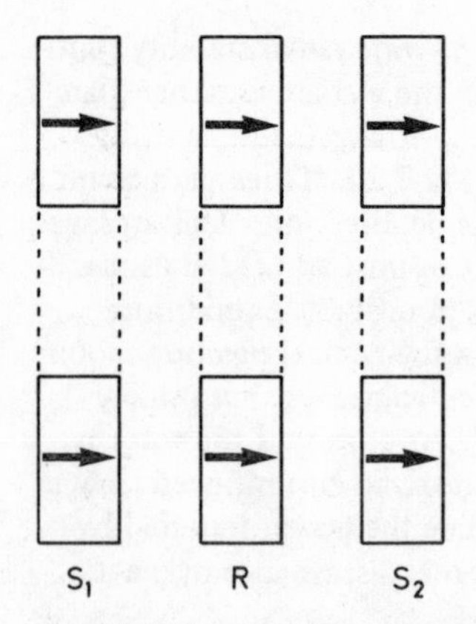


*Fig. 7.20. Magnetic bearing. Central rotor magnet R is supported radially by stationary magnets $S_1$ $S_2$, but is in unstable equilibrium against axial displacement*

stored if the satellite passes into the shadow of the Earth. Energy wheels are said to be more efficient than batteries. On the ground there are many requirements for high speed rotation in the textile industry. If magnetic bearings can be developed for this application, considerable quantities of rare-earth magnets will be required.

### 7.3.3 Magnetically levitated transport

At the present time there is great research activity in several countries into transport systems in which the vehicles are supported by magnetic forces instead of wheels, but the systems that are being actively supported do not use permanent magnets. Instead there are superconducting magnets in the vehicle reacting with eddy currents induced in a conducting track; DC electromagnets in the vehicle attracting a steel track with some form of electronic proximity control to prevent contact, and AC coils in the vehicle or track inducing eddy currents in the other members.

Three models of a vehicle supported by repulsion between ferrite permanent magnets in the vehicle and track have been constructed by the Westinghouse Electric Corporation in Pittsburgh, G. R. Polgreen of Wendover, England and the Krupps Research Institute in Essen, in that order. All these models were large enough to carry one passenger. The first two named were kind enough to supply me with photographs for publication (McCaig 1967, 1968). The third is described by Baran (1972) who gave me a demonstration ride. These three experimental models must have used more permanent magnet material than many applications that are described in this book as established commercial enterprises.

Thus although the use of permanent magnets for levitated transport is not now being actively pursued, it seems worth examining the principles involved, and also the objections that have caused the work to be abandoned. In my opinion some of these objections are justified, but are not greater than ones that can be raised against more fashionable solutions, while others have arisen because of a misunderstanding or are not valid at all.

The principles of permanent magnet levitation are discussed at some length by Polgreen (1966). The first point to be observed is that because of Earnshaw's theorem, permanent magnets can provide lift, but not sideways stability. This is probably the most formidable argument against the permanent magnet solution. Some additional mechanical or other form of sideways restraint must be provided. Other systems, however, are also not self-sufficient. Superconducting eddy current support does not operate at low speeds, so an alternative support such as wheels must be provided. The electromagnet attractive system requires an alternative support for safety reasons, as in fact does any system that uses electric power for the support.

In the paper cited above Baran discusses the efficiency and economics of permanent magnet repulsion systems in some detail. The ideal solution seems to require a compromise between conflicting factors. An attempt to explain these conflicts is made with reference to Fig.7.21. The first choice that is necessary is whether to use a homopolar arrange-

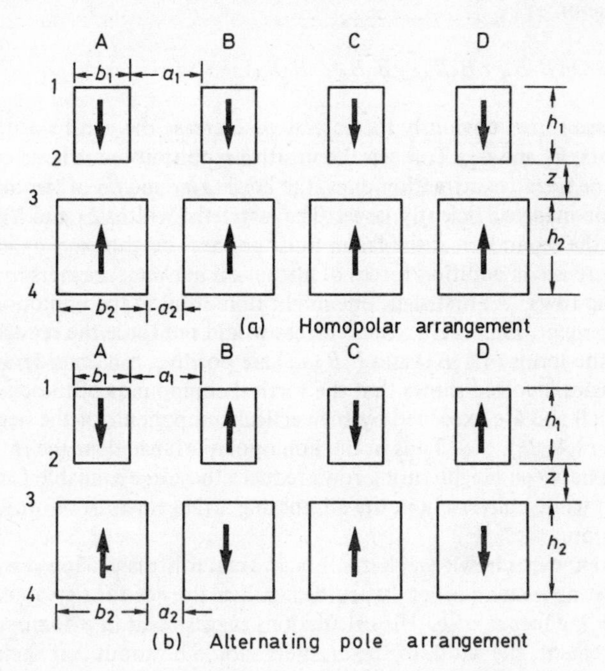


*Fig. 7.21. Magnets for levitated transport. (a) Homopolar arrangement. (b) Alternating pole arrangement*

ment as in (a) with the magnetization of all the magnets in the track in the same direction, or the arrangement shown in (b) with the magnetization alternating between consecutive magnet strips. In both cases the track magnets are long strips like railway lines, but probably with more than two strips. The magnets on the vehicle are also long strips with the same length as the vehicle. There must be the same number of strips on the vehicle as on the track; their magnetization must be mutually opposed so that both are homopolar or both are alternating, but the height and width of the vehicle and track magnets are not necessarily the same.

In Fig.7.21 the top and bottom surfaces of vehicle and track magnets are designated 1, 2, 3, 4 in that order, and the consecutive rows of magnets on both A, B, C, etc. If we consider one strip of magnets say A, the repulsion is a function of two repulsive and two attractive forces, or with obvious nomenclature, repulsion being counted positive:

$$F_A = F(A_2A_3 + A_1A_4 - A_2A_4 - A_1A_3) \tag{7.9}$$

If we consider the second row of magnets we obviously have another similar force:

$$F_B = F(B_2B_3 + B_1B_4 - B_2B_4 - B_1B_3) \tag{7.10}$$

These forces obviously increase as we increase the widths of the magnets, $b_1$ and $b_2$. To keep the positive repulsion terms large compared with the negative attraction ones, the heights $h_1$ and $h_2$ of the magnets must be made sufficiently large. The larger the widths $b_1$ and $b_2$ and the larger the separation $z$, the larger must be these heights.

There are in addition forces of attraction between magnets in neighbouring rows. At first sight one might think that in the homopolar arrangement, Fig.7.21(a), these forces would reinforce the repulsion since the terms $F(A_2B_3)$ and $F(B_2A_3)$ are positive, but consideration of the angles involved shows that the vertical compounds of these forces are small and are exceeded by the vertical components of the negative forces $F(A_1B_3)$, etc. Thus in the homopolar arrangement the interaction between neighbouring rows reduces the force available for levitation, while conversely in the alternating arrangement it reinforces the levitation.

Baran warns however, that although greater lift is produced by the alternating system, other difficulties such as the need for greater lateral restraining forces arise. His calculations suggest that in a homopolar arrangement, the width of the magnets should be about half their lateral separation. Experiments with his man-carrying model confirm the accuracy of his calculations.

Making the track magnets larger than the vehicle magnets slightly improves the pay load of the vehicle, but the more important object of substantially reducing the cost of the track is conversely achieved by making the height of the vehicle magnets greater than those of the track.

It is sometimes argued against levitation by permanent magnets that as they grow bigger they grow worse, until they are incapable even of supporting themselves. This is possibly a quarter truth to which a misunderstanding of a paper which I wrote years ago (McCaig, 1961) may have contributed. The argument is that the performance of any magnet system can be predicted from a small model, if all the dimensions are scaled up in the same ratio. Repulsion forces are proportional to the area (that is the square of the dimensions) while the mass is proportional to the volume or the cube of the dimensions. Thus the mass of the magnet appears to outstrip the force available. My paper implied, but some of those who have quoted it have missed the point, that the separation $z$ is also scaled up in this argument. The magnets that would have been unable to support their own weight would have been separated by a distance greater than is necessary in levitated transport. A further point is that as the vehicle becomes wider it is not necessary to go on increasing the width of the magnets. Instead, one increases the number of strips.

The question whether the cost of a ferrite track is acceptable depends upon the separation that is considered necessary. Baran's calculations suggest that if the separation does not exceed 2 cm the cost of the magnets for a ferrite track are a negligible fraction of the civil engineering cost of building a railway. At 3 cm the cost of the ferrite becomes a significant factor.

If rare-earth-cobalt magnets are used in the vehicle only, the separation $z$ can be increased and the cost of the track much reduced. Since the length of vehicles is so much less than the length of track this solution may be sensible. The use of ferrite or any other permanent magnets would of course involve an enormous expansion in the permanent magnet industry, the capital cost of which could only be justified by a sustained demand.

Baran does not deal with the use of steel in the magnetic circuits. If steel is cheaper than ferrite its use may be economically justified, but in repulsion devices the introduction of a given volume of steel is less effective than the introduction of the same volume of hard ferrite.

The present prejudice against the use of permanent magnets in levitated transport may not in the long term continue. They may well prove to have a useful role, possibly in conjunction with one of the other systems, all of which by themselves seem to have shortcomings. Repulsion between permanent magnets is not the only possibility. It appears that the superconducting systems are not expected to require enormous flux densities. Many model experiments to appraise the possible value of these eddy current devices have used permanent magnets to simulate the superconductors, and some improvement in permanent magnets may enable them to be used in the full scale applications.

### 7.3.4 Miscellaneous repulsion devices

The use of permanent magnets to give some assistance in holding dentures in place by repulsion between top and bottom plates has been suggested, and a fair number of such dentures have been used.

Magnetic repulsion is used in a number of amusing toys. These include dolls that seek a member of the opposite sex, and a dog with a ball or bone that always rolls away. A brake that is applied by magnetic attraction, and released by repulsion can be incorporated in electric motors that are required to stop quickly. Possible applications are in coil winders and tape recorders. The brake is a normal friction brake and is held on by attraction between a permanent magnet and iron. When the motor is switched on the iron is magnetized with a polarity that repels the magnet and releases the brake. The use of repulsion between an electromagnet and a permanent magnet should have other applications such as operating a valve or switch.

A bistable switch is shown in Fig.7.22. The permanent magnet NS may contact either of the iron pole-pieces A or B. The capacitor C is charged by a battery, and discharged by switch S through the coils A and B, the direction of the discharge being controlled by the reversing switch R. The magnet NS is thus repelled by A and attracted by B or vice versa, and can operate a switch or valve.

A sheet floater is an application in which repulsion is induced by permanent magnets between sheets of iron or steel. It is a common experience that it is not easy to remove the top sheet from a pile; the task can waste time and wear both tempers and finger nails. Provided the sheets are magnetic, permanent magnets can come to the rescue. Long U shaped magnets are used for the purpose, as shown in Fig.7.23, the number of magnets required depending on the size of the sheets. Repulsion makes the top sheet float, so that it can be easily removed. If it can be arranged that the plane of the sheets is vertical they can be separated by a smaller magnetic force.

## 7.4 INTERACTION BETWEEN MAGNETS AND CURRENTS

Permanent magnets interact with electric currents generating forces and torques, enabling us to convert electrical into mechanical energy. Loudspeakers, telephone receivers, electric motors and measuring instruments are important examples of this type of application. Sometimes the permanent magnet moves as in a synchronous motor and sometimes the current moves as in a DC motor. The current may flow in an air-cored coil, or it may have a core of soft magnetic material. In the latter case it is not always clear how far the magnetic flux acts directly on the current or through the agency of the core. Rather surprisingly the same result can sometimes be predicted by either assumption.

In the following pages the different devices are grouped under the title of the application, so that headings such as loudspeakers and telephone receivers, instruments and motors are used. Under each heading the different ways of applying permanent magnets are explained. The emphasis is on the function and design of the magnet, and space only permits the devices themselves to be described to the minimum extent necessary to explain the magnet design.

### 7.4.1 Loudspeakers and telephone receivers

In the vast majority of modern loudspeakers a magnetic field produced by a permanent magnet drives a moving coil. The coil carries currents which correspond to the sound to be reproduced. A good loudspeaker should convert as much as possible of the electrical energy supplied to the coil into acoustical energy, although the fraction that can be so converted is always small. A good loudspeaker is also required to respond faithfully to the signals, which it receives. It is difficult to eliminate

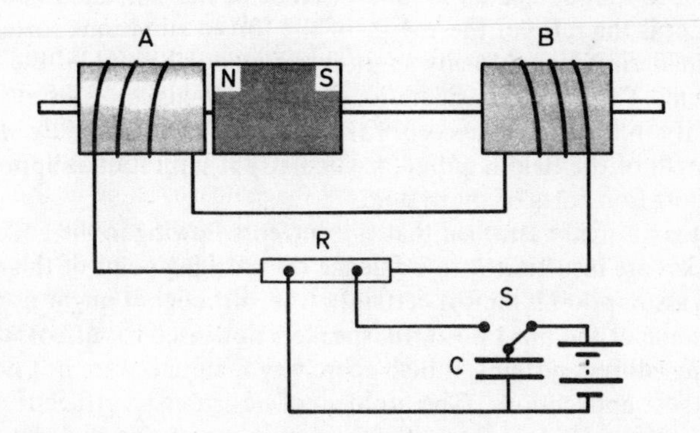


*Fig. 7.22. Bistable magnetic switch or valve. Permanent magnet NS may be in contact with either of the iron pieces A or B. Condenser C is charged by battery and discharged by switch S through coils wound on A and B. Direction of discharge can be reversed by switch R. Magnet is then repelled by A and attracted by B or vice versa*

*Fig. 7.23. Sheet floater. (Courtesy James Neill Ltd., Sheffield)*

natural frequencies of vibration that depend on the mechanical construction of the speaker, and cause it to resonate.

The larger the magnetic field in which the moving coil of the speaker is situated, the greater is the ratio of the electrical energy converted into sound to that dissipated in the resistance of the coil, and in addition the greater is the ratio of the energy of the forced vibrations corresponding to the desired signal to any spurious resonant vibration of the mechanical system. Thus both considerations require the highest possible flux density in the gap. The cost of the magnet increases rapidly with the strength of the field required, particularly if saturation is approached in the soft iron parts of the system.

It is normally assumed that the currents flowing in the coil of the speaker are insufficient to influence the working point of the magnet. This assumption is almost certainly true, although it might not have been for some of the most powerful speakers now used for discotheques and public address systems, if high coercivity materials were not now used for these applications. The problem of designing an efficient magnet system for a loudspeaker is thus regarded as ensuring that the magnet operates at its $(BH)_{max}$ point, and the leakage factor is kept to a minimum. Designs are often compared on the assumption that the flux in the geometrical gap is useful and the remainder is leakage. The warning given in Section 6.5 that the useful flux is actually any flux that cuts the moving coil while it is operating is particularly relevant in this connection.

Actually the coil is usually a little longer than the geometrical gap, which is itself not very precisely defined, if the flux passes from a top plate of definite thickness to a long centre core. If the coil extends outside the gap to places where the flux density is small, the flux cutting the coil does not alter as the coil vibrates. If the flux does alter, spurious harmonics may be produced.

In Fig.7.24(a) the magnet is in the centre and has the same diameter as the inside diameter of the gap; it is ground and acts as one pole. In (b) the magnet is still in the centre but an iron pole-piece is provided. To increase the flux density in the gap it may be necessary to make the magnet with a greater diameter than the gap as in (c) or taper the magnet as in (d). Tapering the magnet has been criticized for creating a smaller flux density in the base of the magnet than in the end near the gap, and such a situation is indeed inefficient. However, in a cylindrical magnet the flux density must be at least a little less near the gap than near the base plate, so at least a slight taper may be justified to produce uniformity of flux density.

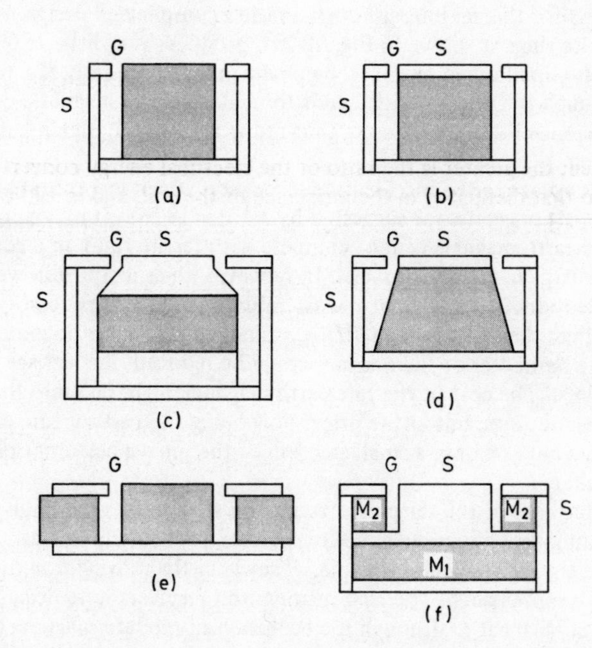


*Fig.7.24. Loudspeaker arrangements. Permanent magnet material shaded. S: Steel. G: Radial gap*

The above designs are all suitable for centre core magnets of anisotropic Al–Ni–Co alloys. Sometimes columnar or semicolumnar alloys are used. Larger loudspeakers usually have the permanent magnet outside, and an iron centre core as shown in Fig.7.24(e). At one time, rings of alloy magnets were used for this purpose, but ferrite magnets are now nearly always preferred. The high remanence anisotropic grade (material F2) is generally the most suitable. Its outer dimensions are always much greater than the internal diameter. The outer shape may be circular, but is not necessarily so. Some speaker manufacturers prefer square magnets, possibly with slots to accommodate clamping screws. It is claimed that the leakage factor is significantly reduced by making the magnet protrude slightly outside the top and bottom plates as shown in the diagram.

The large size of ferrite magnets is no disadvantage in hi-fi speakers, which must be large for other reasons, but is less welcome in compact portable radios. Although the efficiency of the system with large ferrite rings compares favourably with that of systems with alloy centre-core magnets, the leakage flux of the former is mainly outside the magnet, and can interfere with the operation of television tubes. Alloy centre-core magnets are therefore often preferred for the loudspeakers of television sets for this reason, although a rather complicated design using two ferrite rings as shown in Fig.7.24(f), produces very little external field. The small upper ring, $M_2$, helps the main ferrite ring, $M_1$, to drive flux through the gap, and also tends to neutralize any magnetic potential difference in the outer core, thus reducing the stray fields outside the system.

An unusual type of ultra-thin loudspeaker, which incorporated rare-earth-cobalt magnets was suggested by Narita (1976). It may appear that rare-earth magnets cannot compete with ferrite rings on a cost basis for large hi-fi loudspeakers. If, however, the substitution were made the reduction in size of the assembly would be considerably greater than the ratio of the $(BH)_{max}$ values of the magnetic materials. Since the perimeter of the magnet would be reduced, the leakage would also be less. The cost of the rare-earth magnets might be more than that of the ferrite rings, but at the prices now being forecast for rare-earth magnets would be only a small fraction of the price a customer pays for the speaker.

The purpose of the telephone receiver is the same as the loudspeaker but in miniature. Some telephone receivers, particularly in Germany are made on the moving coil principle. Elsewhere the moving iron diaphragm is more usual. The first moving iron receivers were designed about 100 years ago, although the evolution of modern magnets has resulted in great changes. Loudspeakers with moving iron diaphragms have also been made particularly in the early days of radio broadcasting, and occasionally since.

The principle of the Bell type receiver is quite interesting, and can be understood with the help of Fig.7.25. The diaphragm D is attracted by the permanent magnet M, round which a coil C is wound. At first sight the use of a permanent magnet rather than a soft iron core is surprising. If, however, a soft iron core were used, and a signal of constant frequency and amplitude were applied, the current in the coil would oscillate between $+I_s$ and $-I_s$, during each cycle. As a result a magnetizing force that would oscillate between $+H_1$ and $-H_1$ would be produced. Now the mechanical force on the diaphragm is proportional to $H^2$, say $kH_1^2$ in one case and $k(-H_1)^2$ in the other. Both these forces are attractions of the same magnitude. Thus the diaphragm experiences a force of attraction $kH_1^2$ twice per cycle, and it oscillates with a frequency twice that of the imposed signal.

Now consider the situation when a permanent magnet is imposing a magnetizing force $H_2$ greater than $H_1$. When the field due to the signal is in the same direction as that due to the magnet, the magnetizing force is $H_1 + H_2$ and the mechanical force

$$k(H_1 + H_2)^2 = k(H_1^2 + 2H_1H_2 + H_2^2) \tag{7.11}$$

and when the two fields are opposed the force is

$$k(H_2 - H_1)^2 = k(H_2^2 - 2kH_1H_2 + H_1^2) \tag{7.12}$$

Thus the force of attraction fluctuates by $\pm 2kH_1H_2$ each cycle. The permanent magnet ensures that the frequency of oscillation of the diaphragm is the same as that of the input signal, and also increases the amplitude of the oscillation, since $2H_1H_2$ is much greater than $H_1^2$.

The original steel magnets are quite well suited to this design. They

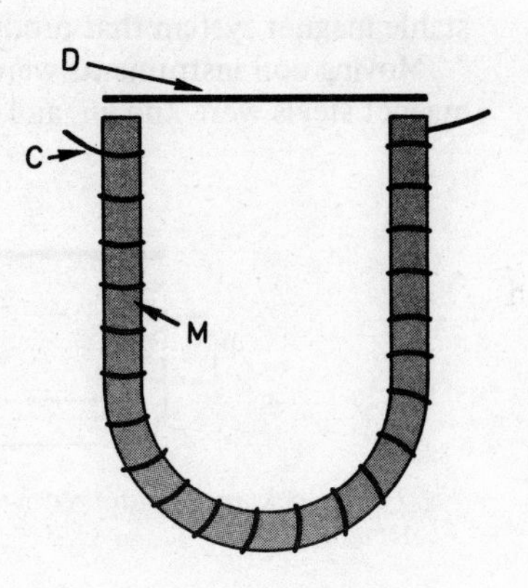


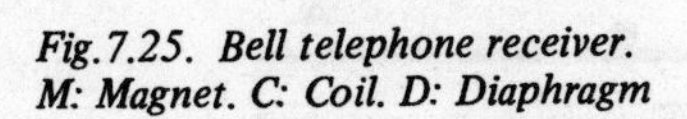
*Fig.7.25. Bell telephone receiver.*
*M: Magnet. C: Coil. D: Diaphragm*

have a good $B_r$ giving a high value of $H_2$ and a good recoil permeability $\mu_r$, ensuring a high value of $H_1$.

Modern Al–Ni–Co allow the magnets to be made much smaller, but their small recoil permeability is a disadvantage, which is overcome in the rocking armature design illustrated in Fig.7.26. There is a single small permanent magnet M in the centre and the coils are wound on the two poles $P_1$ and $P_2$. The current sets up a see-saw like motion in the armature with the same frequency as the signal.

### 7.4.2 Moving coil instruments

Moving coil galvanometers, ammeters, voltmeters and fluxmeters are an important and long established application of permanent magnets. The moving coil is situated in a radial air-gap in which flux is maintained by a permanent magnet. The coil experiences a torque which is proportional to the flux density and the current in the coil. Naturally the torque also depends on the number of turns and the dimensions of the coil.

The coil is deflected until the electromagnetic torque is balanced by a restoring torque produced by springs or the suspension of the coil. The deflection of the coil is measured by a mechanical pointer or an optical lever.

Obviously it is important that the flux density in the gap should not change with time. It is also necessary that any variation of the flux density with position of the coil should be completely predictable. If a uniform scale is required the flux density should be independent of the position of the coil and the reading of the instrument.

As the coil usually passes a very small current, it exerts negligible armature reaction on the magnet and a static design as in a loudspeaker is called for. The task of the magnet designer is to produce an efficient stable magnet system that produces a uniform field in a radial gap.

Moving coil instruments were first made when only low coercivity magnet steels were known, and the only way of generating the magneto-

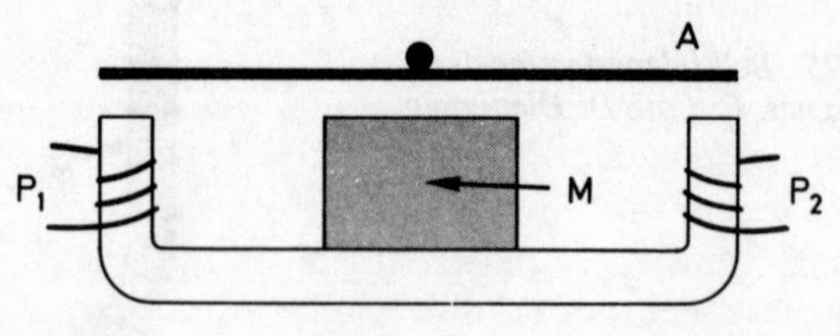


*Fig. 7.26. Rocking armature receiver. M: Magnet. $P_1$,$P_2$: Poles with coils. A: Armature*

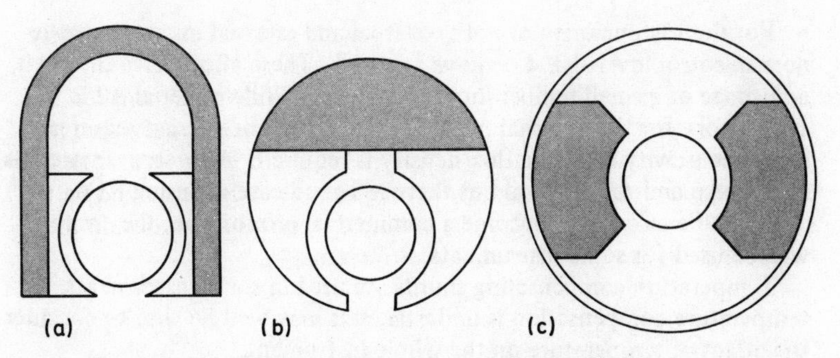


*Fig. 7.27. Instrument magnets with external permanent magnets*

motive force necessary to drive the flux through the gap was to use a very long magnet, such as the U shaped one in Fig.7.27(a). As permanent magnets improved innumerable designs evolved, but Figs.7.27(b) and (c) indicate the general trends. Increasing coercivities have permitted shorter magnets to be used. A design such as that in Fig.7.27(c) with the permanent magnet close to the gap minimizes leakage, but it is not very good to have permanent magnet material actually adjacent to the gap. Permanent magnets are less homogeneous than soft magnetic materials, and soft iron pole-pieces tend to smooth out irregularities in the gap flux density that might otherwise arise from material inhomogeneities. This precaution has not always been observed in cheap instruments.

In recent years even higher coercivities have enabled the permanent magnets to be placed in an internal position as shown in Fig.7.28. In this illustration (a) shows a complete magnet assembly, and (b), (c) and (d) the evolution of centre core magnets. This centre core design made possible by modern materials such as A3, A4 and A5 is cheap and efficient in the use of permanent magnet material. It is very compact, and has the advantage that the permanent magnet is well protected from outside influences.

Initially some manufacturers attempted to use a diametrically magnetized cylinder without pole-pieces as in (b). This design violated the principle of separating the permanent magnet from the gap, and difficulties were encountered in the scale-shape. Some experiments were made on the effect on scale shape of the precise orientation of the preferred direction of the anisotropic Al–Ni–Co discs. As a result manufacturers are sometimes asked to mark this direction accurately. We carried out experiments on the best way of determining this direction and our conclusions are summarized in Section 8.4.5.

The next step was to use a disc of permanent magnet material, but add iron pole-pieces as in Fig.7.28(c). One of the latest designs uses a rectangular permanent magnet and segmental pole-pieces as in (d).

For the vast majority of both external and internal instrument magnets anisotropic Al–Ni–Co alloys are used. These alloys have the great advantage of a small temperature coefficient. While material A2 is satisfactory for the external magnets, the centre core design requires a higher coercivity if a large flux density is required. At present, materials such as A6 and A7 are used but there is a good case for using $RCo_5$.

If Cr–Fe–Co alloys become a commercial proposition, they may well be used for some instruments.

Temperature compensating shunts are used in some instruments. If temperature compensation is undertaken it may well be wise to consider the effect of temperature on the whole instrument.

One type of instrument that does not necessarily use a radial field is the long scale instrument, the magnet of which is shown in Fig.7.29. The coil is threaded round one of the split iron rings. This type of magnet also requires a high coercivity material.

Instrument magnets of all kinds are usually designed so that when fully magnetized they are too strong. They are then 'knocked down', that is, partially demagnetized, preferably by an alternating field, so that the instrument has the correct calibration. A process of this nature is the only way in which just the correct strength of magnet can be obtained, and it also stabilizes the magnet against stray fields or any other adverse circumstances.

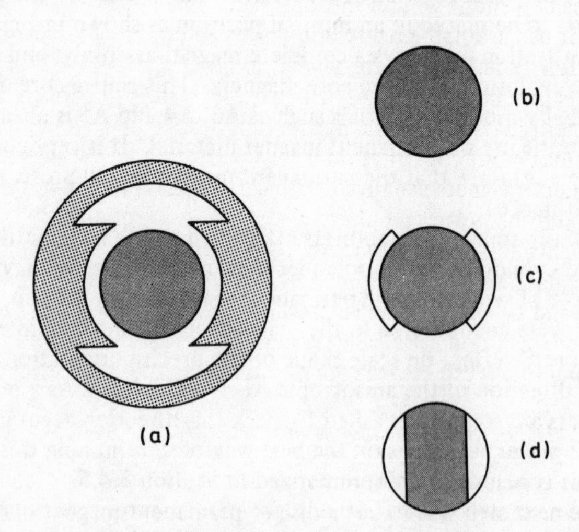


*Fig. 7.28. Centre core instrument magnets. (a) Complete assembly. (b), (c) and (d): Development of centre core magnets*

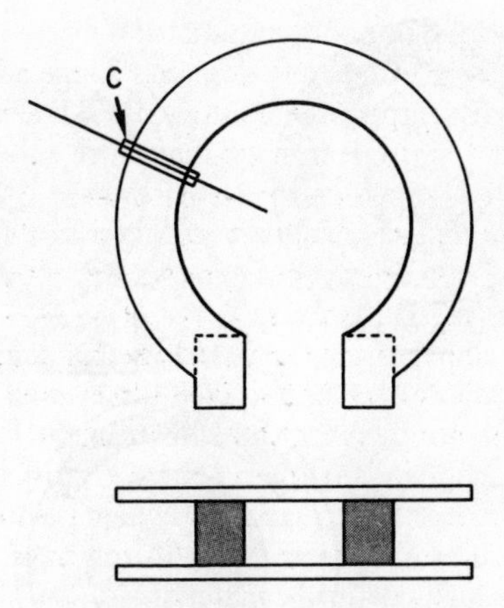


*Fig. 7.29. Magnet for long scale instrument. C: Moving coil*

### 7.4.3 Electric motors

Until recently DC permanent magnet motors were constructed with permanent magnet stators, replacing the iron stators with field coils of traditional DC machines. The rotors were fed through commutators in the usual manner. These permanent magnet motors have the characteristics of a shunt wound motor. Many small motors are made of this type for use in small battery appliances and in motor-cars. In recent years, ferrite segments have tended to replace Al–Ni–Co permanent magnets in such motors. Al–Ni–Co alloys are still used in some large permanent magnet motors, particularly for use in rolling mills.

An inside-out construction is now sometimes used with DC motors. Permanent magnets usually of rare-earth cobalt are embodied in the rotor, and the power supply to the stator is commutated.

Synchronous motors also use permanent magnets in the rotor. Such motors have been used for many years for low-power devices such as clocks and timing mechanisms, and recently their use has been extended to other small domestic gadgets such as mixers and shavers. DC, synchronous and hysteresis motors are discussed in separate sections.

#### *7.4.3.1 DC permanent magnet motors*

Various types of stator were developed to use Al–Ni–Co alloys in DC motors as shown in Fig.7.30, but the most common design today is Fig.7.30(c) with ferrite magnet segments. The highest quality ferrite magnets for use in motor segments is sintered and anisotropic intended

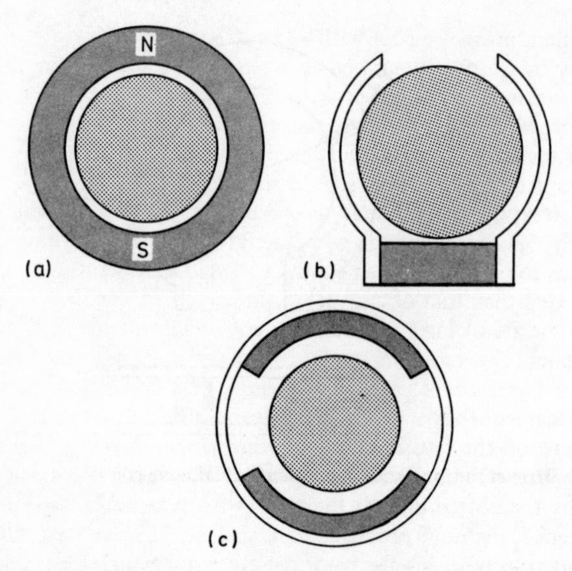


*Fig. 7.30. DC motor magnets. (a), (b) Alloy (c) Ferrite*

for radial magnetization. Cheaper motors may use isotropic or plastic bonded material.

Ireland (1968) describes the design of small DC motors, using ferrite segments, in considerable detail. The book deals with the design of motors as a whole, and not just the permanent magnets. The instructions are easy to follow, although in somewhat antiquated units, and they deal only with the empirical design of small motors with ferrite segments.

Gould (1972) considers the design of permanent magnets for DC motors generally. He brings out a number of useful facts about the behaviour of magnets in motors. His treatment is intended to enable the utility of different materials to be compared, but the design methods he suggests are possibly a little more difficult to follow, and certainly cannot be easily summarized.

In the discussion of electric motors, the flux produced by the stator is often considered to act on the current flowing in the conductors of the rotor or armature. If these conductors are sunk in slots of an iron rotor, little flux, if any, actually cuts the conductors. It is an interesting but academic question, whether the conductors push the armature or the iron poles push the conductors.

The armature conductors and armature iron themselves produce flux, but this flux cannot itself exert any force directly on the armature. In a permanent magnet motor with a ferrite segment stator, the flux produced by the rotor may exceed that produced by the stator. The flux density measured by a Hall probe inserted into the gap between stator and rotor, while the motor was running, would bear no relation to that exerting torque or force on the rotor. It would not even indicate the direction in which the rotor should rotate.

The armature reaction may, however, alter the flux produced by the stator, by demagnetizing the permanent magnets in the stator of a motor, which has to be treated as an example of dynamic working, rather more complicated than that of a lifting magnet.

With the aid of Fig.7.31 the influence of a simple armature winding on a segment motor is examined. At the centre of the segment the armature exerts no influence on the magnet. The whole permanent magnet segment shown in the figure tends to drive flux in one direction into the rotor, the return path being through another segment not shown. The armature winding tends to drive flux in closed paths, such as that shown by the dotted line. If there are $N$ turns per unit angle each carrying current $I$, the mmf produced and available to drive flux round the path shown, comprising the rotor windings in an angle $2\alpha$ is $2\alpha IN$. A small fraction of this mmf is used to drive flux through the gap and iron parts, but the remainder produces a magnetizing force in the ferrite segment. At A there is in consequence a magnetizing force $H_a$, while at B there is a similar demagnetizing force. $H_a$ is a little less than $NI2\alpha/L_m$, where $L_m$ is the thickness of the magnet segment.

Next we must consider the influence of such additional fields on the performance of the magnets.

In order to follow the changes that occur in the permanent magnets it is necessary to use the intrinsic curve as explained in Section 6.8; the

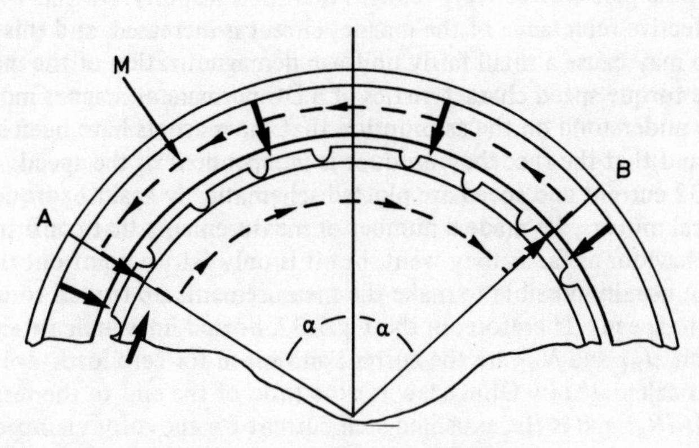


*Fig.7.31. Armature reaction on stator magnet. M: Magnet*

largest demagnetization is at the trailing end of the segment, B, and this will be caused to work on a lower recoil line. Once the motor has been used at its maximum load (in both directions if it is intended to be reversed) no further demagnetization can occur. Each part of the segment has a working point that moves up and down a recoil line. Although some parts of the segment may be working on lower recoil lines than others, the recoil permeability is approximately the same for all. When the motor is switched on, for any part of the segment that is subject to a demagnetizing force there is another that is subjected to an equal magnetizing force. Thus the average flux density and polarization of the segment are the same when the motor is running as when it is at rest, unless the motor is subjected to a bigger load than has ever previously been experienced.

Gould has pointed out that some protection against demagnetization of the trailing tip, when an overload is applied, can be provided by putting a few turns round the stator magnet. These turns are connected in series with the rotor winding and are arranged to provide a magnetizing force. This protection is only justified in the larger type of motor with alloy magnet stator, and appears difficult to arrange if the motor has to be reversed.

The limitation of the demagnetization of segment type stators to the trailing edge has been corroborated experimentally. In larger motors steel pole-pieces are sometimes needed to concentrate the flux, although it is recognized that in an efficient magnet assembly with low leakage pole-pieces are a disadvantage.

If pole-pieces are fitted, the net demagnetization due to an overload is probably reduced, but is no longer confined to the trailing edge. The flux produced by the armature windings now probably passes through the pole-piece without entering the magnets. This circumferential flux in the pole-pieces effectively reduces their permeability. In this way the effective reluctance of the magnet circuit is increased, and this change may cause a small fairly uniform demagnetization of the magnets.

The torque speed characteristics of a DC permanent magnet motor can be understood on the assumption that the magnets have been stabilized, and that the flux they produce is independent of the speed. In Fig.7.32 current and speed are plotted schematically against torque for a typical motor. We made a number of measurements that confirmed this behaviour as far as they went, but it is only fair to point out that it was not usually possible to make the measurements up to stall torque and zero speed. Therefore, in the Fig.7.32, dotted lines indicate extrapolation. $I_{n1}$ and $N_{n1}$ are the current and speed for zero load. $I_s$ is the current calculated by Ohm's law for the ratio of the emf to the resistance $(E/R_a)$ and is the expected stall current for the rotor clamped. $T_s$ is the extrapolated torque for this condition. The lines can also be extrapolated backwards to obtain a negative torque $T_i$ for zero current. This is a loss torque and the corresponding value of the speed $N_t$ is the speed at which the imput emf is equal to the back emf induced in the rotor.

From Fig.7.32, the power and efficiency values plotted against torque in Fig.7.33 can be calculated. Input watts follow the same line as current in Fig.7.32, since input voltage is constant. Resistance losses are proportional to $I^2$ and equal input watts at stall. If the loss torque $T_i$ is independent of speed, the mechanical losses $W_m$ are proportional to speed. The maximum value of $W_0$, the output power, occurs when the torque is about half the stall torque, but the maximum efficiency occurs at a lower torque and current.

The linear relation between torque and speed confirms that the effective flux produced by the stator magnets is independent of current and load, but this linear relation can only be expected after the magnets have been stabilized.

The design of the permanent magnets must take account of the expected armature reaction. Ireland suggests an arbitrary rule, that the trailing tips of the magnets should not be taken beyond a point at which the polarization $J$ falls to 0.8 $J_r$. This rule is probably only useful for the type of motor and ferrite magnets that he was using. In fact $J =$ 0.9 $J_s$ is now suggested for rare-earth magnets. Gould considers that it is desirable to maximize the working product of the magnets after they have been exposed to the most adverse conditions. Although realizing that the demagnetization is confined to the trailing tips, he concludes that the effect is much the same as would be obtained if one fifth of the armature ampere-turns were acting on the whole magnet, and gives data for a number of magnet materials appropriate to this assumption.

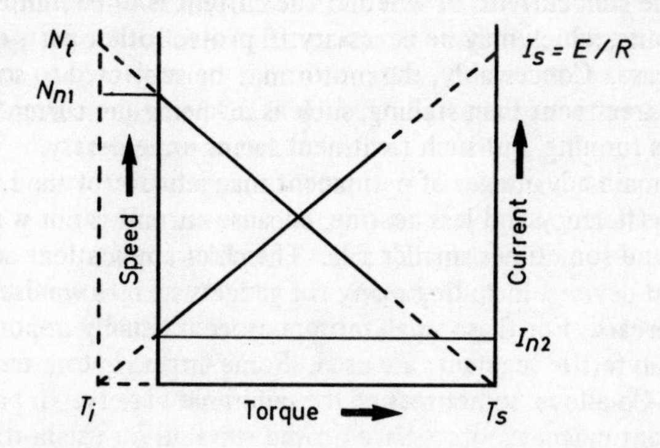


*Fig. 7.32. Speed and current plotted against torque for DC motor with stabilized permanent magnet stator*

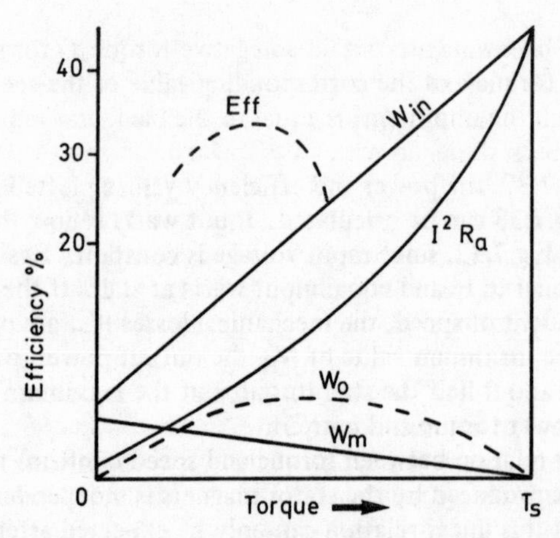


*Fig. 7.33. Input ($W_{in}$), output ($W_0$), ($I^2R$), mechanical losses ($W_m$) and efficiency calculated from Fig. 7.32 and plotted against torque*

The reader who wishes to use either of these methods of design is referred to the works by Ireland (1968) and Gould (1972) already quoted. The maximum demagnetizing force that the magnet experiences is to some extent within the control of the motor designer and user. We substituted alloy magnets into a motor designed for use with a wound field and were unable to find much demagnetization. The manufacturers produced a much greater demagnetization, the difference being due to our power supply being unable to supply the stall current $E/R_a$ although quite capable of running the motor in the optimum power condition.

Thus it is necessary to decide whether the motor is intended to withstand the stall current, or whether the current is to be limited by fuses or cut-outs, which may be necessary to protect other parts of the circuit in any case. Conceivably, the motor may be subjected to some more adverse treatment than stalling, such as reversing the current while the motor is running, but such treatment seems unnecessary.

The main advantages of permanent magnets over wound fields are greater efficiency and less heating, because current is not wasted in the stator, and sometimes smaller size. The chief applications are battery operated devices, including many for gadgets such as windscreen wipers in motorcars. For these small motors, price is usually important and the cheap ferrite segments are used. Some larger motors are made with Al–Ni–Co alloys, sometimes of the columnar variety. In particular permanent magnet motors have proved suitable for use in rolling-mills. The low temperature coefficient of flux density of these alloys, compared with the temperature coefficient of resistance of copper means that the permanent magnet motors are less affected by overheating.

Although the majority of permanent magnet motors have a radial field and are not unlike the equivalent motors with a wound field, some motors have been designed with an axial field, possibly with a printed circuit stator.

Rare-earth magnets have made many new designs possible for permanent magnet motors. Noodleman (1976) examines these possibilities and compares them quantitatively. Similar rectangular blocks may be used in the conventional manner in a motor stator as shown in Fig.7.34(b), or on the outside of a rotor in an inside-out construction as shown in Fig.7.34(a). In the latter design the current in the stator must be subject to commutation. This commutation may be performed by brushes in the conventional manner, provided the current is first taken into the rotor by means of slip rings. There are now also several

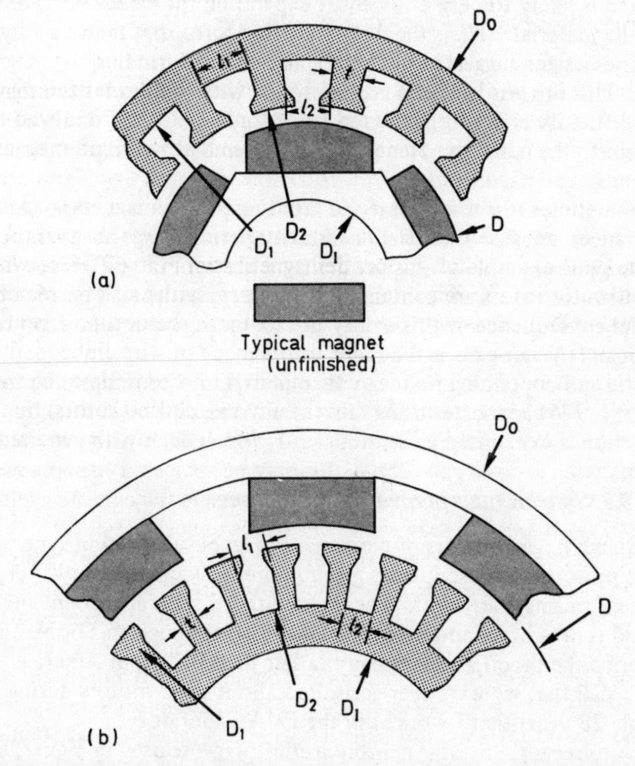


*Fig. 7.34. Inside-out and conventional permanent magnet motor configurations using rectangular magnets in salient pole concstruction. (Noodleman, 1976, courtesy Inland Motor Division, Radford, Maryland)*

possible methods of making brushless motors with electronic commutation. A photograph of the rotor of an inside-out machine is shown in Fig.7.35.

One measure of the performance of a motor is the factor $K_m$, which is defined as the torque divided by the square root of the input watts. Noodleman finds that $K_m$ is 16% greater for the inside-out than for the conventional construction. If the motor is to operate without forced cooling, the advantage of the inside-out design is even greater, because the winding cools naturally if it is on the outside.

In Fig.7.34 the magnets for both constructions are made from similar rectangular blocks which are internally ground for the conventional design and externally ground for the inside-out design. It has been argued that the inside-out design makes the magnets thinner at the trailing edges, where the armature reaction is greatest, and the magnets are therefore more vulnerable to demagnetization. Whether this demagnetization is likely to be serious, must depend on the values of $H_c$ and $H_k$ for the material. ($H_k$ is the demagnetizing force that reduces $J$ by 10%.)

The designs suggested by Noodleman require grinding the magnets in situ. This process has been accomplished with premagnetized magnets, but difficulty arises from the tendency for the material removed to adhere to the magnets. Hence there is a demand for virgin magnets that can be magnetized in situ.

Sometimes it is unnecessary to grind rare-earth magnets to close tolerances, because their large coercivity permits large air-gaps. A remarkable example of this possibility is shown in Fig.7.36, in which three motor rotors are contained in one rectangular air-gap. Another useful consequence of this ability of rare-earth magnets to drive flux through large air-gaps is that light rotors made of wire embedded in plastic and containing no magnetic material may sometimes be used. Voigt (1976) advocates using tangentially magnetized rotors; this construction is explainted in Section 7.5.1, which deals with generators.

### *7.4.3.2 Synchronous permanent magnet motors*

Although large numbers of permanent magnets are used in synchronous motors, the descriptions given in previous English books on permanent magnets are brief. The most detailed general account that I have found is in Schüler and Brinkmann (1970), although this book may describe the practice in Germany rather than Britain or America. I do not recall that we were ever consulted about these motors during the whole 28 years that I worked in the PMA Laboratory.

Until recently most synchronous motors were used in clocks, and timing devices that required little torque or power. There was little incentive to ensure that the magnets were efficiently designed. The currents in the stator coils were low and exerted little demagnetizing force on the magnets. The problem that received most attention was probably that of starting.

If a permanent magnet rotor with any number of pairs of poles is placed symmetrically inside a stator, having an equal number of poles, and fed with single phase AC, the magnet will rotate in either direction if it is once started, but will not start spontaneously when the power is switched on. It is not absolutely necessary that the rotor should be a permanent magnet to operate in this way. Similar behaviour can in certain circumstances be induced in a soft iron rotor of suitable shape. If the motor does operate it does so with a number of turns per second determined by the mains frequency divided by the number of pole pairs. Clocks in which only a slow rotation is desired may have a considerable number of pole pairs.

Many clocks were made that needed to be started by hand, possibly with a pawl to prevent the clock running backwards. Most electric clocks are now self-starting, although it is debatable whether this is an advantage. If there has been a temporary power supply failure, one is likely to notice if a clock has stopped completely, but be misled if it has restarted say a quarter of an hour slow. Timing mechanisms used to programme industrial operations and domestic equipment, such as washing machines, must usually be self-starting.

The more sophisticated devices use what is in effect an additional eddy current induction motor or even a hysteresis motor (see Section 7.4.3.3). Such auxiliary motors require a rotating magnetic field. A

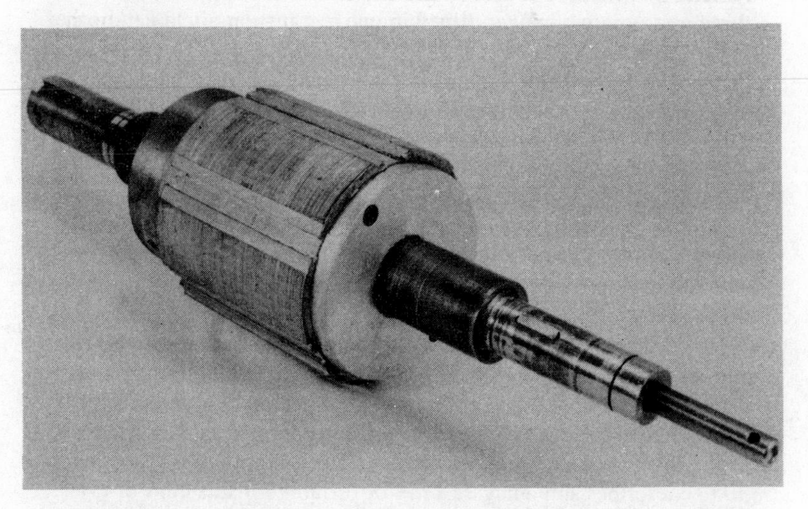

*Fig. 7.35. Permanent magnet rotor. (Noodleman, 1976, courtesy Inland Motor Division, Radford, Maryland)*

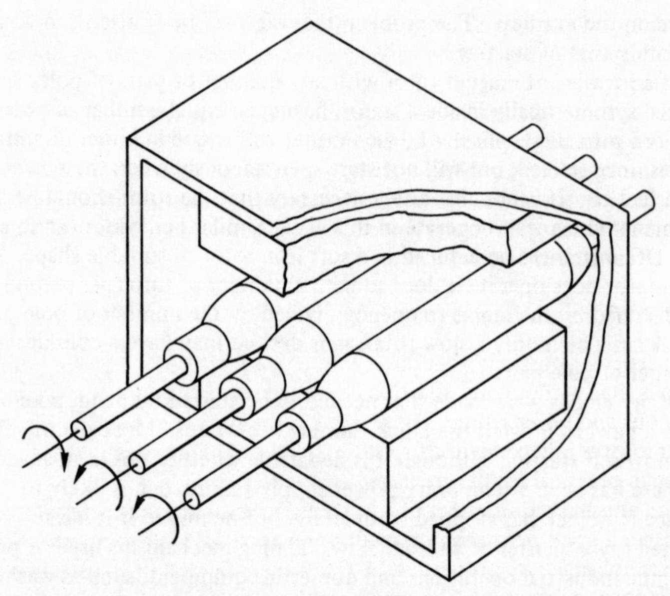

*Fig. 7.36. Multiple armature single air-gap motor. (Jaffe and Herr 1976, courtesty the Singer Co., Elizabeth, NJ)*

genuine rotating field requires a two- or three-phase supply, and is only likely to be available in industrial equipment. An approximation to a two-phase supply, giving a slightly imperfect rotating field can be produced by two pairs of coils at right angles. One pair is powered directly from a single-phase supply, and the other pair is powered from the same supply through a capacitor. The currents in these two pairs of coils differ in phase by approximately 90°.

A simpler starting device requires only a single stator fed by single phase AC, but has a rotor, which is constructed in some asymmetrical manner, such that its natural position of rest when the power is off differs from that when it is on. When the power is switched on, the rotor starts to vibrate, and if the design factors are correct the vibrations build up until rotation commences.

Rotors for synchronous motors often used steel magnets. Sintered and bonded Al–Ni–Co have been used. Isotropic barium ferrite is now often recommended, while some grade of Vicalloy is a possibility.

Synchronous motors are now being incorporated into rather more powerful domestic equipment such as shavers and mixers. Schemmann (1973) describes and analyses a motor suitable for this kind of application. It is fed by single-phase AC, and is self-starting by setting up vibrations as described above. The analysis shows that the motor runs smoothly only with certain combinations of voltage and load. Smooth running may not be very important for some of the domestic devices for which this type of motor is suggested. It is possible that full details of some of these more powerful synchronous motors have not yet been published. By the time this book is in the hands of the reader more information may be available, at least in the patent literature.

In recent years there has been considerable interest in brushless motors in which the rotor is a permanent magnet and the stator is fed with DC switched electronically.

### 7.4.3.3 *Hysteresis motors*

The particular property that distinguishes hysteresis motors from other types is that without any additional arrangements they are both synchronous and self-starting. Theoretically, the torque of an ideal hysteresis motor is independent of speed up to the synchronous speed, although practical results are usually rather less simple.

The unusual requirement of a hysteresis motor is for a material that is neither a good permanent magnet nor a good soft magnetic material, but has intermediate properties, so that it may be described as a semi-hard magnetic material. The semi-hard materials of today were often the permanent magnet materials of 50 years ago.

The electrical excitation of a hysteresis motor is fed into a stator which produces a rotating magnetic field, or at least a pseudo-rotating field, as described in the previous section.

The operation of a hysteresis motor can be explained by energy considerations, and because SI units give magnetic energies directly in $J$, they make discussion of this subject particularly simple.

The rotor of a hysteresis motor is subject to a rotating magnetic field. If the rotor is held still, a certain amount of energy, $W_h$, is converted into heat in each cycle of the rotating field because of magnetic hysteresis. If the rotor turns against a torque $T$, work $2\pi T$ is done for each revolution. If $W_h > 2\pi T$ the armature magnet accelerates until it rotates with the same speed as the field. If $2\pi T > W_h$ the rotor remains still and is heated.

There has been considerable discussion about the magnetic hysteresis energy, and the way it varies with the exciting field. The most familiar form of magnetic hysteresis is the ordinary alternating hysteresis, that can be obtained by measuring a complete loop of $B$, in tesla, against $H$, in $\mathrm{Am}^{-1}$, and measuring the area enclosed. (The procedure in CGS units is less simple, because if $B$, in gauss, is plotted against $H$, in oersted, it is necessary to divide the area of the loop by $4\pi$ to obtain the hysteresis loss in erg.) A typical family of loops, obtained under cyclic conditions with different maximum magnetizing forces is shown in Fig.7.37. In measuring such inner loops care must be taken to see that

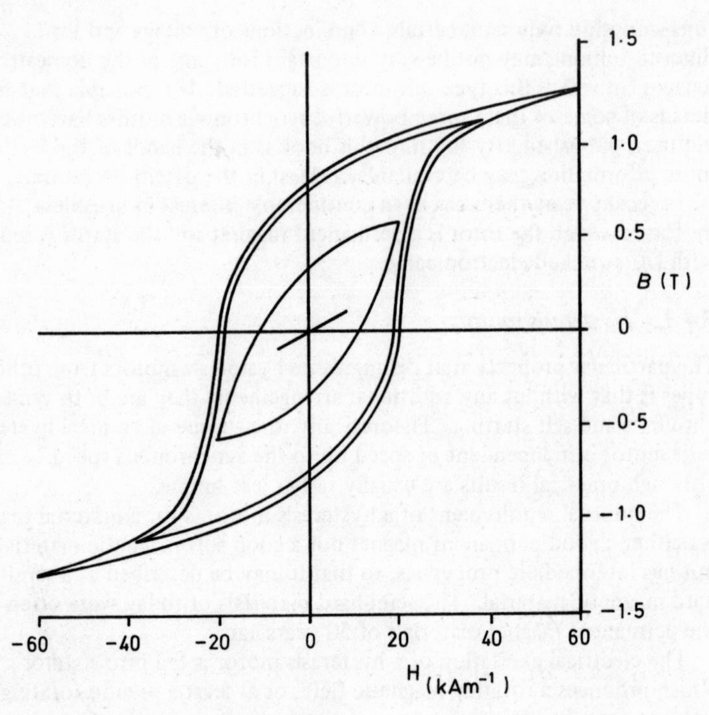


*Fig. 7.37. Inner loops for 35% cobalt steel*

the maximum magnetizing force is the same for the positive and negative branch of each loop. In Fig. 7.38 the value of $W_h$ is plotted against the value of $H_m$, the maximum magnetizing force for each loop. The value of $W_h$ rises at first slowly then more quickly and finally approaches a maximum value asymptotically.

As the magnetic field in a hysteresis motor does not alternate but rotates, suggestions have been made that rotational rather than alternating hysteresis may be involved. The rotational hysteresis of a material can be measured by rotating a disc (or ideally an oblate ellipsoid) about its axis in a magnetic field. The rotational hysteresis $W_r$ is the mean value of the torque multiplied by $2\pi$. Figure 7.38 also shows the typical behaviour of the rotational hysteresis plotted against the magnetizing force, which should be corrected for the self-demagnetizing field of the specimen. The rotational hysteresis rises to a maximum and then falls again to zero at infinite magnetizing force. The maximum of the rotational hysteresis is not necessarily the same as that of the ordinary alternating hysteresis, and may even be a little higher. There are, however, only a few published measurements of rotational hysteresis.

If the performance of a hysteresis motor is determined by alternating hysteresis, the output will increase asymptotically to a maximum as the current supply is increased. Resistance losses in the stator windings will increase as the square of the current. Thus the efficiency of the motor defined as output watts divided by input watts will also rise to a maximum and then decrease. Thus it is important that the input current should produce a field that causes the motor to work somewhere close to the maximum in the $W_h, H_m$ curve.

If rotational hysteresis predominates, the adjustment of magnetizing force to the optimum properties of the material is even more critical. In this case the output itself, and not only the efficiency, passes through a maximum and falls to zero.

Most hysteresis motors use a very thin ring of hysteresis material. The ring may be mounted on non-magnetic material as in Fig.7.39(a) or on soft magnetic material as in Fig.7.39(b). In (a) the direction of the flux in the ring is mainly circumferential, while in (b) it is radial. With both constructions, although the ring rotates, it appears that the flux within it alternates and the design can be safely based on measurement of the normal alternating hysteresis. Possibly, if a hysteresis motor were designed with a solid cylindrical rotor or a very thick ring rotational hysteresis would predominate.

To design an efficient hysteresis motor the exciting magnetic field should be matched to the properties of the rotor material so that the ratio of the area of the hysteresis loop to the maximum magnetizing force is large. For all materials this condition is fulfilled somewhere near the knee of the magnetization curve, but for some materials the ratio is more favourable than for others.

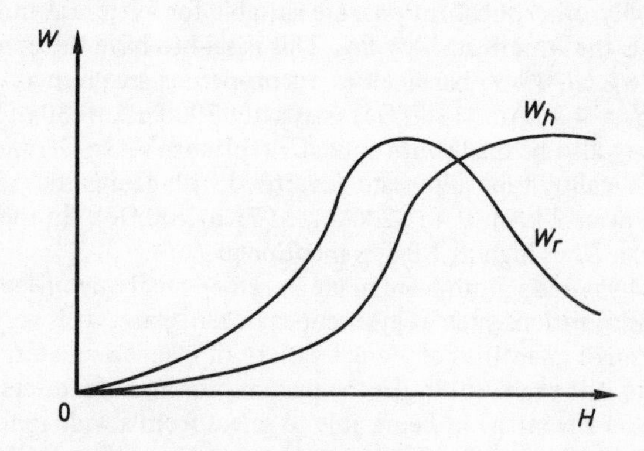


*Fig. 7.38. Typical curves of alternating hysteresis, $W_h$, and rotational hysteresis, $W_r$, plotted against H*

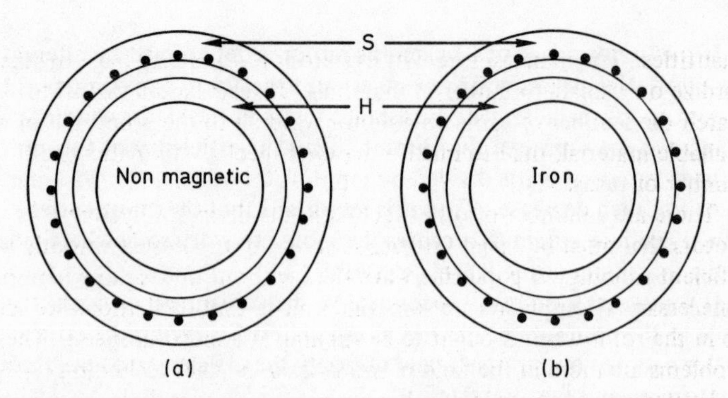


*Fig. 7.39. Hysteresis motor. S: Stator. H: Hysteresis ring (a) mounted on non-magnetic core and (b) on iron core*

A material with a square loop such as anisotropic Al–Ni–Co alloys appears ideal, but difficulties in making and heat treating thin rings of such material have restricted their use. Although higher powers can be obtained with high coercivity materials it is difficult to design stator windings capable of producing high enough fields to make use of such materials.

In the UK manufacturers of hysteresis motors have tended to use various grades of cobalt steel. Often the steels are tempered to reduce their coercivity slightly. It is possible that in this state the hysteresis loops are a little more square and the properties are as a result more suitable for use in hysteresis motors, but the main object of the treatment is probably to match the material properties more closely to the stator field.

Various modifications of Vicalloy usually with a lower V content, and possibly other substitutions, are suitable for hysteresis motors. One example is the American alloy P6. This is said to have the composition 45% Co, 6% Ni, 4% V, balance Fe. Its properties are given as: $B_r$ = 1.42 T, $H_c$ = 4.8 $\text{kAm}^{-1}$ (60 Oe) resistivity 3000 $\mu\Omega$m (30 $\mu\Omega$cm). Alloys may also be made with some Cr replacing V. In Germany, a whole range of Vicalloy type alloys are advertised with coercivities ranging from as low as 2 $\text{kAm}^{-1}$ to 32 $\text{kAm}^{-1}$ (25 to 400 Oe). In the low coercivity type, $B_r$ as high as 1.8 T is mentioned.

Many hysteresis motors are made in rather small quantities for specialized applications such as gyroscopes in aerospace. It is very expensive to make small quantities of a great variety of magnetic materials tailored to suit each motor. In the past, motor manufacturers have enjoyed the advantage of being able to select from a wide range of steels made for other purposes. These other requirements for steels are rapidly disappearing and the production methods are not geared to making small quantities. The manufacturers of hysteresis motors may have to standardize on a smaller number of materials. It seems more logical to match the number of turns on a motor winding to the properties of an available material, than to make a material specially to match the number of turns.

There are a number of other factors in the design of hysteresis motors that must be taken into account in order to ensure smooth efficient running. A good sine wave field without undesirable harmonics is necessary. Complaints are sometimes made that oscillations are set up in the rotor when it ought to be running at a uniform speed. These problems are more in the field of motor than magnet technology.

Hysteresis couplings in which a permanent magnet drives another member made of a semi-hard material can be made. Compared with synchronous couplings, described in Section 7.2.4, they can transmit a rather small torque but as with hysteresis motors they can both accelerate from rest and run synchronously. The requirements for the hysteresis material are similar to those for hysteresis motors.

## 7.5 INDUCED ELECTROMOTIVE FORCE

The ability of a magnetic field to induce an emf and an electric current in a conductor moving relative to it, is the basis of many applications, which are often complementary to those based on the force existing between a magnet and a current. Many pieces of equipment can work in reverse: if an electric motor is driven it can generate a current, and a loudspeaker or telephone receiver can act as a microphone. Often, however, the equipment does not operate equally efficiently in both directions.

An expression for the emf induced in a coil is easily written down. If a coil of $N$ turns is threaded by a flux of $\phi$ weber

$$E = -N\frac{\mathrm{d}\phi}{\mathrm{d}t} \tag{7.13}$$

If the coil has a core of area $A$ m$^2$ and is cut by a uniform flux density $B$ T, Equation 7.13 becomes

$$E = -NA\frac{\mathrm{d}B}{\mathrm{d}t} \tag{7.14}$$

(in CGS units these equations are unchanged, but $E$ is given in EMU, and its value has to be multiplied by $10^{-8}$ to obtain a value in volts).

The significance of the minus sign in the above equations is that if a current flows as a consequence of the induced emf, its direction is such as to oppose the change of flux.

Induced currents and potentials in closed coils are unambiguous, but there is a little confusion about the circumstances in which an emf can be induced and observed in a single conductor. I remember seeing a physics examination question, which asked the student to calculate the emf induced in the axle of a train moving in the Earth's vertical field. Such an emf cannot be observed in the train, because an equal and opposite emf is induced in the leads that connect the ends of the axle to any instrument. In theory at least the emf can be measured between the railway lines that do not move relatively to the field.

The Faraday homopolar dynamo consists of a conducting disc that rotates in a magnetic field parallel to its axis of rotation. Brushes make contact with the axle and circumference of the disc and are connected by stationary wires to stationary equipment permiting an emf to be measured. The Faraday dynamo has been suggested as a source of high-current, low-voltage power, but I personally have seen it used only in laboratory demonstrations and in a piece of equipment, now practically obsolete, for measuring the demagnetization curves of permanent magnets. One may ask whether the dynamo would still work if the disc were still and the magnet rotating. The answer is that no emf could be detected because the rotating magnetic field would cut the whole of the stationary circuit and induce no net voltage.

Permanent magnets and coils are used for many applications in several different ways. A coil may be moved relative to a fixed magnet, a magnet may be moved relative to a fixed coil, or both magnet and coil may be fixed, and the flux cutting the coil may be altered by changing the reluctance of the magnetic circuit by moving other steel parts. These three principles recur in different applications. The arrangement that follows is to describe the applications under the headings: generators, magnetos, microphones, pick-ups, detection and counting devices.

### 7.5.1 Generators

A conventional DC motor with a permanent magnet stator generates DC if it is driven by another machine, although for efficient working it is usually necessary to alter the position of the brushes. A permanent magnet alternator with the winding on the stator is at first sight the converse of a synchronous permanent magnet motor, but it is in fact a more flexible type of machine. The synchronous motor can run only at a speed linked to the frequency of the power supply, which is not usually adjustable, whereas the corresponding alternator can be driven at any desired speed and produce an alternating supply at any desired frequency. Both the synchronous motor and the rotating magnet alternator have the advantage of dispensing with brushes and commutators or slip rings, but the alternator possesses this advantage with fewer attendant disadvantages.

There are some limitations to rotating large permanent magnets at high speeds, because of the mechanical unreliability of permanent magnets. A third type of machine with both coils and permanent magnets in the stator and a rotor that varies the reluctance has found considerable applications.

In a generator, as in a motor, the permanent magnets are subjected to varying working points because of the demagnetizing forces produced by the currents generated. This aspect of the operation of generators is discussed by Schüler and Brinkmann (1970) and Parker and Studders (1962). The latter give the simpler account. They point out that in a generator the voltage is proportional to the total flux and the current is proportional to the magneto-motive force supplied by the permanent magnets. However, on closer examination they conclude that only a small fraction of the mmf supplied by the magnets is used in this way, most of it being needed to drive flux through the gap. This conclusion seemed to hold in an experimental generator we made some years ago, the magnets suffering little or no demagnetization in use. The difference between a motor and a generator in this respect is that there is no limit to the current that can be driven by an external source through a stalled motor, except for the point at which the motor burns out, whereas the short circuit current of a generator is limited by what it can itself produce.

The way in which the voltage of a generator varies with load is called the regulation, and the regulation of a generator with permanent magnet excitation differs significantly from that of one with separate electrical excitation. The terminal voltage of any source of emf falls with the current taken from it, because of the potential drop over its internal impedance. The voltage of a generator depends on its speed and the flux cutting its coils. If a generator is excited by a constant current flowing in its field coils, then as a current flows in the output coils, a back magneto-motive force reduces the flux cutting these coils and the output voltage is reduced. Permanent magnets respond to a back mmf to a degree by increasing the mmf they produce. This compensation cannot be complete, but it does make permanent magnet generators to some extent self-regulating. In some circumstances this self-regulating behaviour of a permanent magnet generator is an advantage, but in other circumstances the ease with which the output of an electrically excited generator can be manipulated by controlling the field current is preferred.

One type of generator that has to produce scarcely any power is the tachometer generator. A small generator is connected to a voltmeter calibrated in units of speed. Provided the permanent magnet is stable the voltage is proportional to speed. The use of alloy magnets with their low temperature coefficient is desirable for this application.

A long established application of permanent magnet alternators is for cycle lighting. Multipole rotors are desirable for two reasons. The frequency of the output increases with the number of poles. The emf induced is proportional to $d\phi/dt$ and is therefore greater for a given flux divided over a large number of small coils alternating with a high frequency. The frequency of the output is of course the product of the number of pole pairs and the number of revolutions per second of the rotor. A large number of pole pairs produces the voltage necessary to give light while the cycle is being ridden at a walking speed.

The voltage must also not rise to such a high value that the light bulbs are destroyed when the cycle is ridden at a high speed. Here again a high frequency output helps. The impedance (effective resistance for AC) of a piece of equipment such as an alternator is $(R^2 + (2\pi nL)^2)^{1/2}$, where $R$ is the ohmic resistance, $L$ is the self-inductance and $n$ is the frequency of the AC. If the frequency $n$ is sufficiently high so that $2\pi nL \gg R$ the current output of the machine is not increased by a further increase in speed and the light bulb is protected from damage.

Cycle generators were originally designed with multipole Al–Ni–Co magnets that were not considered very easy to make or efficient to use. Multipole ferrite rings were a definite improvement when they became available.

Electrical generators for cars have had a different history. Cars are always provided with a battery which requires a DC supply to charge. Until recently rectifiers were expensive and a dynamo, i.e. a machine with a commutator was used. Battery chargers for cars were developed before good permanent magnets were available.

Now it has been mentioned that wound-field machines require regulation of the stator current supply. This regulation once it has been provided can without much extra trouble be made more complicated and do more than just maintain a constant current or voltage. It has become customary to regulate charging systems in cars so that the charging current is much reduced when the battery is fully charged.

Today it would be possible to design an efficient and economical permanent magnet alternator, which combined with a rectifier would start to charge a battery at low speed, and would not produce an excessive current at high speed. It would not be impossible but it would be expensive to regulate the charging current according to the state of charge of the battery.

This difficulty is put forward as an explanation why permanent magnet chargers are little used in automobiles. One suspects that although over-charging the battery is undesirable, it is not really disastrous, and inertia against change in a mass-production industry, where any innovation requires large capital expenditure is equally important.

Although ferrite rings are very suitable for magnetization with multiple poles, larger machines can be made more compactly with anisotropic Al–Ni–Co alloys. One simple method of using such material is in the form of an axially magnetized cylinder with steel poles in the form of teeth bent over as shown in Fig.7.40. Any convenient number of teeth can be used. The 'Cobalt in Industry' supplement of 'Cobalt' (1973), No.1, contained a brief description of a French design that used high coercivity Al–Ni–Co (Material A4), which seemed to be an elaboration of this idea. This machine was intended for use in milk delivery vans.

The magnetic circuit with teeth bent over the outside of the permanent magnet does not appear very efficient from the point of view of flux leakage, and we tried experimentally the design shown in Fig.7.41. The rotor, Fig.7.41(b), consists of 8 Alcomax III (Material A2) trapezium shaped blocks, interspersed by 8 similar mild steel blocks. These are all mounted on a steel backing plate, which is rotated about an axis perpendicular to its plane. The rotor was separated by a small gap from the stator, which had 16 laminated poles wound with coils connected in

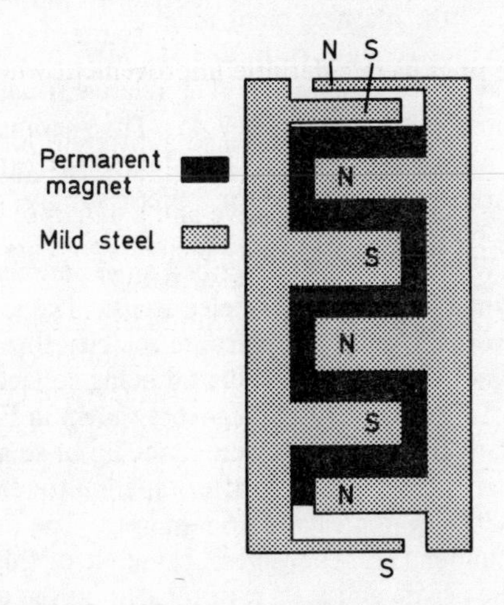


*Fig. 7.40. Rotor with imbricated poles (side view). Cylindrical slug of permanent magnet material magnetized axially and fitted with mild steel end-plates of larger diameter. The imbricated or interleaved teeth project alternately from these plates to form the poles of the rotor. The magnet can just be seen between the gaps in these teeth*

series as shown in (c). This construction, in our opinion, worked satisfactorily.

Bailey and Richter (1976) discuss the principles of permanent magnet generators in considerable detail, and describe a projected design for a machine capable of generating 195 kVA at 155 V and 1·200 Hz. The speed of rotation of this generator, which is required for use in aircraft, is specified as 12 000 to 21 000 rpm. A diagram of the rotor with flux paths is shown in Fig.7.42, and a photograph in Fig.7.43. The permanent magnets are sintered rectangular blocks of $SmCo_5$, tangentially magnetized in alternate directions, so that flux is concentrated by feeding from two magnets into one pole-piece. To enable the rotor to remain intact at high speeds, the magnets and rotor are contained in a shrink ring, but to avoid leakage the shrink ring is made with a composite structure of magnetic and non-magnetic parts to cover the magnets and pole-pieces respectively. These parts are welded into a single ring shown in Fig.7.44. Some form of containment is required for permanent magnets in any high speed rotor. Tangential magnetization would also be possible in motors.

A somewhat similar generator requirement was presented to us some years ago. An account of our proposed solution is included in a paper presented at Dayton (McCaig, 1976). In this suggestion the rare-earth magnets would have been trapezia contained in discs of light metal. Flux would have been fed into fixed stator teeth from both sides, the direction of magnetization of the magnets being axial.

Variable reluctance generators up to 1215 MW for use as excitors in power stations have been described.* The relative situation of the magnets, coils and rotor are shown in Fig.7.45. The Alcomax III (Material A2) magnets are magnetized in situ. The slots in the rotor and the stator windings are shown arranged for 3-phase supply, but other arrangements are possible.

Kennedy and Womack (1964) described an experiment in the magnetohydrodynamic generation of electricity. The ionized gases from a rocket motor were blown at a supersonic velocity through the gap of a permanent magnet, the current generated being collected by suitably placed electrodes. The permanent magnet is shown in Fig.7.46. The large central rectangular structures were made up of smaller blocks of Columax (Material A2col). The gap after tapering for flux concentration was 133 x 380 mm in area and 36 mm long. The flux density in this gap was just under 1 T. The general principle of this design with the magnets in the centre and steel return paths on the outside has been used for other applications such as nuclear magnetic resonance.

* *Cobalt* (1974), No 3. 'Cobalt in Industry' Supplement.

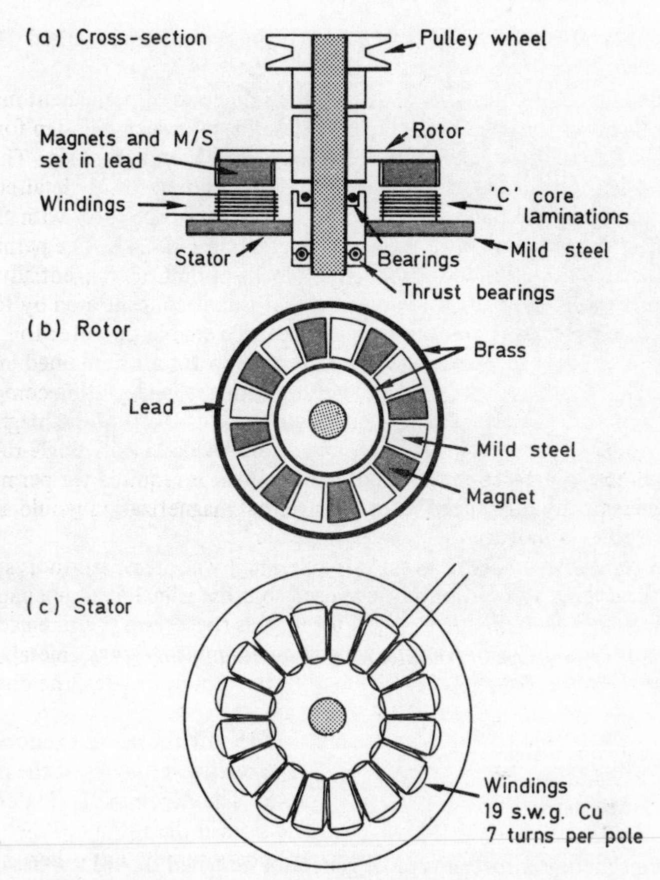


*Fig. 7.41. Construction of axially magnetized alternator*

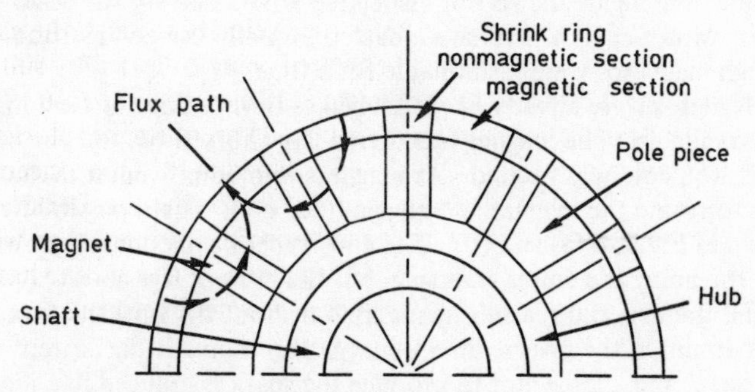


*Fig. 7.42. Layout of generator rotor and flux paths (Bailey and Richter, 1976, courtesy General Electric, Schenectady, NY)*

*Fig. 7.43. Photograph of generator rotor (Bailey and Richter, 1976, courtesy General Electric, Schenectady, NY)*

### 7.5.2 Magnetos

The magneto is a special type of generator designed to give a series of high voltage pulses for the ignition of the gas mixture in an internal combustion engine. Magnetos are used in motor cycles and for any internal combustion engine that is not associated with a battery for other purposes. Motor cars, which have a battery, usually use coil ignition, although magnetos were fashionable for a time.

Magnetos were already needed when only very inferior steel magnets were available. The magnet was then a large horseshoe, and the coil on a soft iron core was rotated. As magnets were improved, it became possible to rotate the magnet and have a fixed coil. These possibilities are shown in Fig.7.47(a) and (b). The emf reaches a maximum just as the flux threading the coil is reversing, but the current lags about ¼-phase behind the emf and reaches its maximum about the time the flux reaches a maximum in the reverse direction. At this moment the current circuit is broken, and a large emf to produce the spark is induced in a many turn secondary coil.

Magnetos can also be operated by the reluctance effect (Fig.7.47(c)), but this method does not provide such a large flux change.

### 7.5.3 Microphones

A moving coil microphone is similar to a loudspeaker in reverse. The movement of the coils in a microphone is usually much less than that in a loudspeaker, and the coil of a microphone is usually shorter than the gap, whereas as previously explained that of a speaker is usually longer. Instead of a many turn coil a single ribbon of aluminium may be stretched across a rectangular gap, rather like the arrangement of a string galvanometer. The output emf of a microphone is proportional to the product of the flux density, the total length of wire or ribbon and the velocity set up in it by the sound wave.

The problem presented to the magnet designer is a fairly conventional and simple one of providing a sufficient flux density in the gap. The currents involved are very small, so that there is no question of armature reaction and the circuit is completely static.

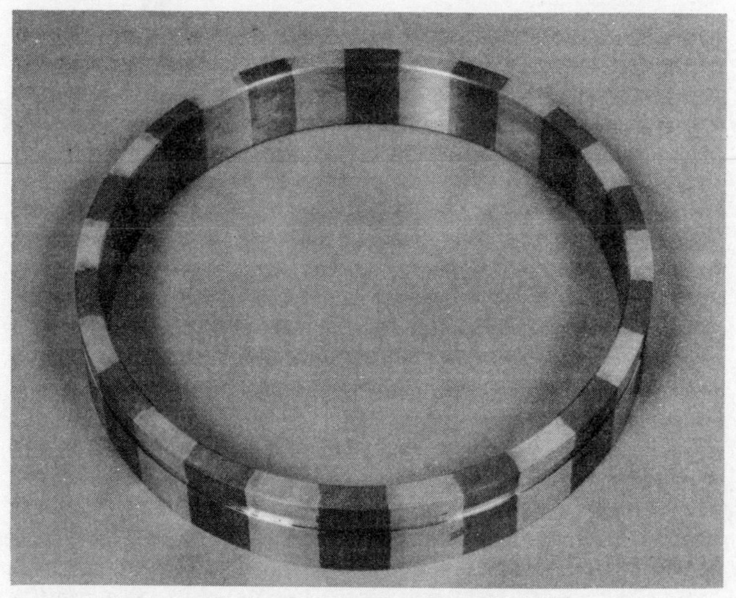

*Fig. 7.44. Shirnk-ring of welded magnetic and non-magnetic parts (Bailey and Richter, 1976, courtesy General Electric, Schenectady, N.Y.)*

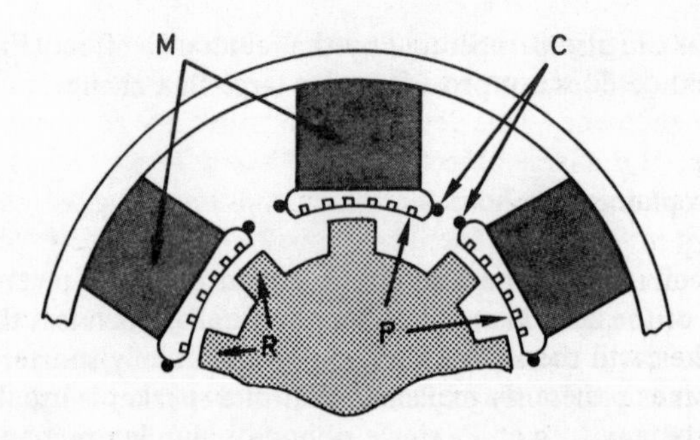


*Fig. 7.45. Variable reluctance generator. M: Magnets. C: Coils to magnetize magnets in situ. P: Three phase winding. R: Rotor teeth create varying reluctance*

An alternative type of microphone is more like the converse of a telephone receiver. The small movements of a magnetic diaphragm vary the reluctance between it and the pole pieces of a permanent magnet with coils wound round them.

### 7.5.4 Gramophone pick-ups

In a gramophone pick-up the movements imparted to the stylus by the grooves in the disc may be communicated either to a coil which moves relatively to a fixed magnet, or to a magnet that moves relatively to a fixed coil. Small bars of alloy or ferrite magnets are used in the moving magnet type.

Stereophonic sound systems are increasingly popular. The vibrations of the stylus are in two directions, differing by a right-angle, and these are detected by two coils also at right angles as shown in Fig.7.48. An Al–Ni–Co magnet weighing as little as 30 mg is used. Rare-earth cobalt magnets appear very suitable for this application.

### 7.5.5 Control and counting

Control and counting may be performed by a magnet on the moving part, which induces a pulse in a fixed coil as it passes. This pulse may be amplified and used for control and counting operations. We used this method once for revolution counting, a small piece of rubber-bonded ferrite providing an adequate magnet. There are may different types of operation and situation in which this principle may be used. The method is of course in competition with other methods in which magnets actuate Hall probes or reed switches.

Whiteley (1976) and Hollitscher and Whiteley (1976) describe a multipurpose speed sensor, which makes use of a number of different magnetic phenomena. This device is used with large DC motors in rolling mills. It is a large object weighing nearly 60 kg. Speed is measured by an analogue tachometer, which embodies an axially magnetized stator shown in Fig.7.49. There are actually two such rings on opposite sides of a multi-turn rotor set in epoxy, and containing no magnetic material. Each magnet that can be seen in Fig.7.49 consists of two pieces of rare-earth cobalt. As the equipment is likely to operate in a hot environment, these magnets are temperature compensated by iron–nickel shunts.

Digital signals are provided optically, but an overspeed switch uses two magnetic principles. An eddy current disc (Section 7.6) rotates in the gap of a magnet built of a U-shaped piece of steel with two rare-earth magnets, one on each side of the disc. This U can be seen in Fig.7.50; it moves about a pivot, and is attached to a snap action switch. This latch assembly also has teeth which act as a keeper to another U-shaped

*Fig. 7.46. Magnet for magnetohydrodynamic generator. (Courtesy Swift Levick and Sons, Sheffield)*

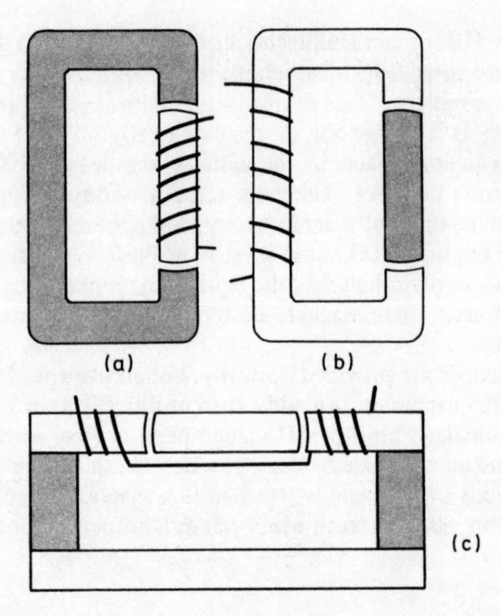


*Fig. 7.47. Magneto constructions. (a) Horseshoe magnet and rotating coil. (b) Rotating magnet. (c) Reluctance type with coil and magnet fixed*

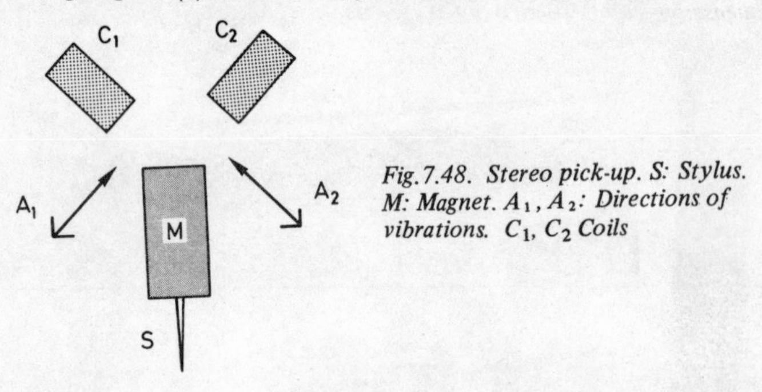


*Fig. 7.48. Stereo pick-up. S: Stylus. M: Magnet. $A_1$, $A_2$: Directions of vibrations. $C_1$, $C_2$ Coils*

magnet. When the speed of the eddy current disc reaches a certain value, the eddy current drag becomes sufficient to overcome the force between the second U magnet and the teeth. At this point the assembly turns and the snap action switch operates, giving the overspeed signal. This device is claimed to be more reliable than purely mechanical devices for the same purpose. One difficulty with mechanical overspeed switches is that if they are rarely operated, because the plant is working correctly, they may fail, when they are required, due to dirt or lack of lubrication, in a manner that is common to many mechanical systems that are rarely used.

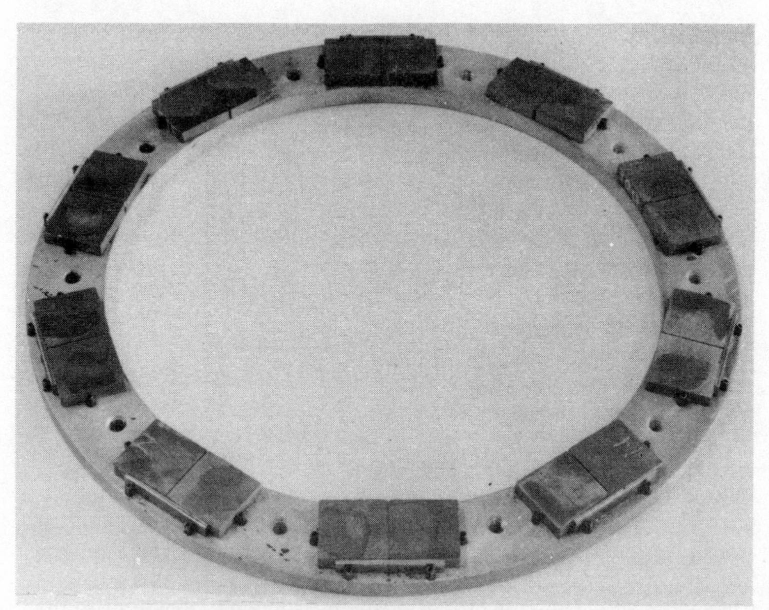

*Fig. 7.49. Axially magnetized ring forming stator of tachometer. Each magnet is made from two blocks of rare-earth cobalt, and is fitted with a temperature compensating shunt. (Courtesy Canadian General Electric, Peterborough, Ontario)*

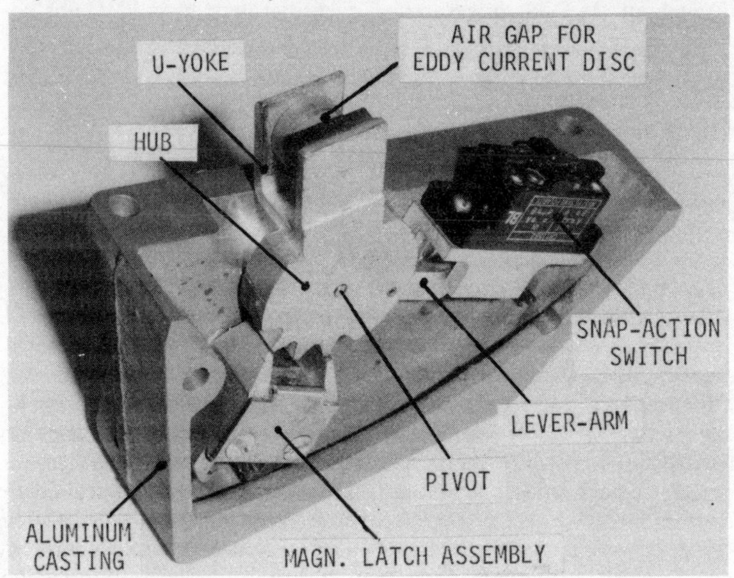


*Fig. 7.50. Inside of overspeed switch. (Courtesy Canadian General Electric, Peterborough, Ontario)*

## 7.6 EDDY-CURRENT DEVICES

Eddy-current devices combine the power of magnets to induce currents in conductors moving relatively to them, described in Section 7.5, with the power of a magnet to exert a force on a moving conductor carrying a current described in Section 7.4. That is to say, the magnet first induces a current and then exerts a force on the conductor in which the current is induced.

If a conductor moves in the vicinity of a stationary magnet it experiences a braking action. A magnet moving in the vicinity of a conductor free to move may set it in motion, but there is always some slip between the motion of the magnet and the conductor. The velocity of the driven member is always less than that of the driving one. Permanent magnet eddy-current devices include brakes, damping systems and drives.

Most of the eddy-current devices used in the past have been fairly small, and have involved small velocities; thus the theory is often developed for small velocities only. There appear to be possibilities that permanent magnets could be used in more powerful eddy current applications and we performed some experiments to investigate this possibility some years ago. Our conclusions, including some modifications to the theory at high velocities, were published by Gould (1968). Most of this theory holds in SI or CGS units provided the system is used consistently throughout. In CGS units electrical quantities must be in EMU not amperes, volts, and ohms for example. No statement of units is made unless a divergence arises.

Consider as in Fig.7.51 a conducting sheet of thickness $z$ and resistivity $\rho$ moving with velocity $v$ past $n$ alternating poles of pitch $\frac{1}{2}p$; each pole has a transverse width $a$, and a breadth in the direction of motion $b$. The flux density in the sheet is assumed to have a uniform value, $B_g$, under the poles and to be zero elsewhere.

In the sheet under the poles there is a transverse emf

$$E = vaB_g \tag{7.15}$$

The emf gives rise to currents within the sheet under the pole that return by closed paths outside the pole. If the resistance of the part of these paths under the pole is a fraction $k$ of the total closed path, the resistance of these paths is

$$R = a\rho/bzk \tag{7.16}$$

Therefore current per pole is

$$I = E/R = kvbB_g z/\rho \tag{7.17}$$

The force per pole = $IaB_g$ and with $n$ poles of alternating polarity the force $F$ is given by

$$F = nkvabB_g^2z/\rho \tag{7.18}$$

$$= cv \tag{7.19}$$

where

$$c = nkabB_g^2z/\rho \tag{7.20}$$

The above formulae apply to low speeds and can be found in many books. The value of $k$ depends upon the particular geometry, but a value of 0.3 is fairly usual.

If the speed increases, the inductance of the eddy current path takes the current loops outside the magnetic poles before the maximum current is attained. Gould assumes that the variation of force with speed can be described by a formula

$$F = \frac{cv}{(1 + v^2/v_m^2)} \tag{7.21}$$

$v_m$ is the velocity at which the force is a maximum. By differentiating it can be shown that the maximum value of $F$ is $\frac{1}{2}cv_m$. According to Gould

$$v_m = p\rho/(12\pi mkbz) \tag{7.22}$$

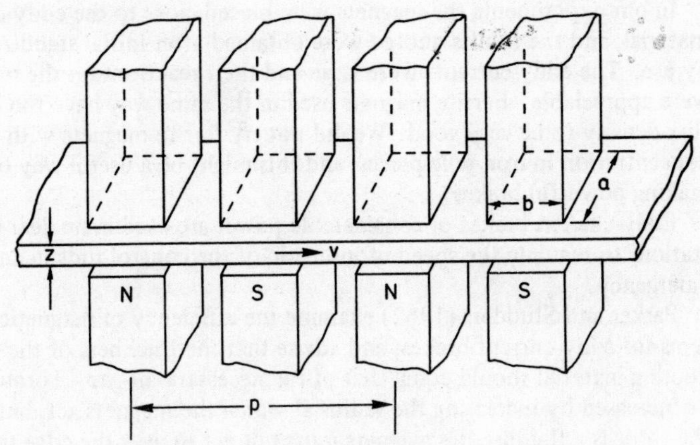


*Fig. 7.51. Diagram of brake, showing symbols used. (From Gould, 1968)*

the new term $m$ is a constant that takes into account the self-inductance of the current paths, but which in practice can only be determined experimentally.

We tried a considerable number of magnet configurations, the results of the experiments being summarized by Gould (1968). These arrangements included horseshoe magnets that were swung over the rotating disc. In another arrangement magnet blocks were fixed on a mild steel backing plate rather like the drive magnets described in Section 7.2.4. This device acted rather like a disc brake, except that the magnets did not quite touch the rotating disc. The rotating disc for this experiment was made of soft iron, but preferably had a thin copper facing in which the main currents were induced. The copper facing greatly reduces the speed at which the maximum braking force is exerted, but has less influence on the magnitude of this force.

The curve which displayed the largest force reported by Gould is shown in Fig.7.52. It was obtained with a concentric brake, in which 24 Hycomax IV blocks (material A5) of total weight about 1 kg were fixed on the periphery of a steel cylinder. The brake was operated by inserting this member into a concentric hollow steel cylinder with copper facing on the inside. A theoretical curve according to Equation 7.21 was fitted to the experimental curve at velocities up to $v_m$. At higher velocities the experimental forces do decrease, but rather less quickly than predicted by the theory. The brake can be partially applied by partial insertion.

Some heavy vehicles are fitted with electromagnetic eddy current brakes, mainly for use in mountainous districts. The work described above was intended to show that permanent magnets could beneficially replace electromagnets.

In our experiments the magnets were placed close to the eddy-current material, and the results quoted were obtained after initial stabilization by use. The eddy-currents were large and their reactions on the magnets were appreciable. Ferrite magnets used in the same way have too low a flux density to be very good. We did not try ferrite magnets with flux concentration in iron pole-pieces, and this might be a useful way of making powerful brakes.

Eddy-current brakes of considerable power are used in nuclear power stations to regulate the speed of insertion of the control rods in an emergency.

Parker and Studders (1962) examine the efficiency of magnetic systems for eddy-current brakes, and advise that the thickness of the conducting material should equal that of the necessary air-gap. Torque can be increased by increasing the radius at which the magnets act, but if the rotor is a flat disc the magnets must not act so near the edge that the return paths for the eddy-currents are restricted.

The watt-hour meter is one of the oldest and most common uses of eddy-current brakes. The electrical arrangements of a watt-hour meter provide a torque that is proportional to the product of the current times the voltage or the watts consumed. This torque must drive something which has a resistance to motion that is a constant linear function of speed. This condition is difficult to satisfy by mechanical devices but, at low speeds so long as Equation 7.18 applies, it is fulfilled by the eddy-currents induced in a rotating aluminium disc by one or more permanent magnets. For many years a rather special type of steel horseshoe magnet was always used, but this design is now obsolete. Many different designs are in use, but generally one or more small Al–Ni–Co magnets drive the flux through the disc at least twice. Good stability is the most important quality of magnets for this application.

Many speedometers use eddy-currents. A small magnet rotates in a copper or aluminium cup or near a conducting disc. The cup or disc is restrained by springs as in an instrument, and its angular deflection is proportional to the eddy current torque, and hence to the speed.

Many laboratory instruments such as balances can be damped by a conducting sheet moving in the gap of a permanent magnet. Moving coil instruments are often damped by eddy-currents induced in their own coils. The degree of damping depends upon the external resistance.

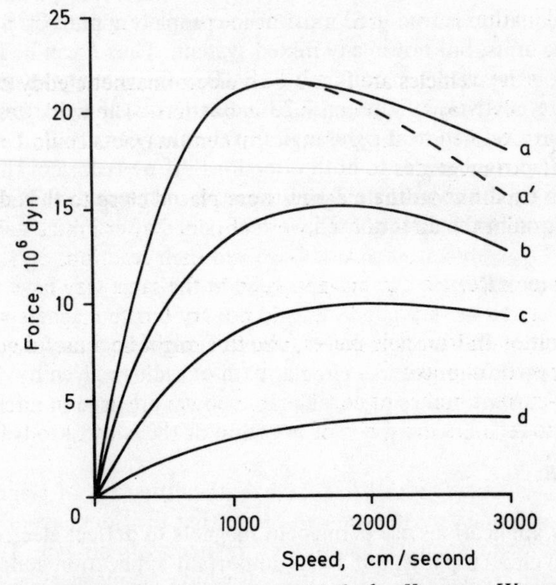


*Fig. 7.52. Force vs. speed curves for concentric brake, Hycomax IV magnets. a: Magnets fully in cylinder (a′: theoretical). b, c, d: Magnets ¾, ½ and ¼ in cylinder, respectively. (From Gould, 1968)*

Sensitive instruments such as galvanometers are critically damped for one particular value of the external resistance. If the resistance is much lower than this critical value the instrument is too sluggish, and if the resistance is much higher they continue to swing for a long time.

A more complicated system of damping uses a permanent magnet to induce an emf in a coil. This emf is amplified and used by negative feed-back to exert a force on the same or another magnet to oppose the motion. When really good damping without sticking is required this method is excellent.

## 7.7 ACTION OF MAGNETS ON FREE ELECTRONS AND IONS

This section is concerned with the action of magnetic fields on moving electrons or ions in the gaseous phase, often but not invariably in a high vacuum. The subject of electrons and ions in the liquid phase is considered in a later section.

The fundamental interaction of a magnetic field and a moving charge is expressed by the vector relation

$$\boldsymbol{F} = -q\boldsymbol{v} \times \boldsymbol{B} \tag{7.23}$$

The equation is true in SI units or in completely pure CGS electromagnetic units, but not in any mixed system. Thus $\boldsymbol{F}$ can be in newton or dyne, $q$ the charge in coulomb or EMU, $v$ the velocity in $\text{ms}^{-1}$ or $\text{cms}^{-1}$, respectively. Equation 7.23 indicates that if a charged particle moves with velocity $\boldsymbol{v}$ at right-angles to the flux density $\boldsymbol{B}$, it experiences a force $\boldsymbol{F}$ at right-angles to both directions. If we represent that third direction by the coordinate $Z$ and the mass of the particle is $m$, its acceleration in the direction $Z$ is

$$\ddot{Z} = -q\boldsymbol{v} \times \boldsymbol{B}/m \tag{7.24}$$

As a result of this acceleration it follows from elementary mechanics that the particle moves in a circular path of radius $r$ given by

$$r = \frac{mv}{qB} \tag{7.25}$$

Many applications use permanent magnets to deflect electrons or ions into circular paths, but a very important application concerns charged particles moving in the direction of the magnetic field. Suppose such a particle happens to have a velocity component perpendicular to the field. It will experience a force at right-angles to both the velocity and the field. Thus the particle experiences a circumferential force and begins to follow spiral paths around the field direction. The circumferential velocity is also at right-angles to the field, and by following the directions of these forces and motions with the left-hand rule, described in elementary books on electricity and magnetism, you can verify that the result is a force directing the particle back to its original path

If a beam of electrons is moving in a given direction a solenoid or a permanent magnet can be used to focus the electrons, so that any tendency of the particles to spread out is counteracted, and they are all brought to a focus at a point. A solenoid or permanent magnet can thus be made to act like a lens for a beam of electrons or other charged particles, and for a particular beam has a definite focal length.

The field configuration that can be produced by any permanent magnet system cannot be made quite the same as that produced by a solenoid. Figure 7.53(a) shows a solenoid, and Fig. 7.53(b) shows the axial flux density it produces. Suppose a tubular permanent magnet is used in an attempt to imitate the field. The result is shown in Figs. 7.54 (a) and (b). The field falls gradually to zero outside the ends of the

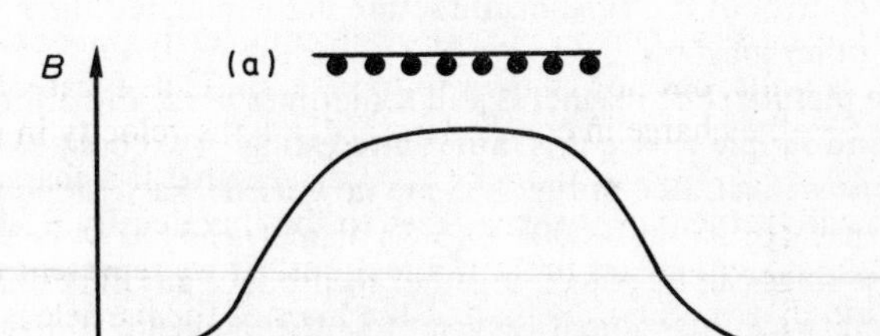


*Fig. 7.53. Flux density on axis of solenoid.*

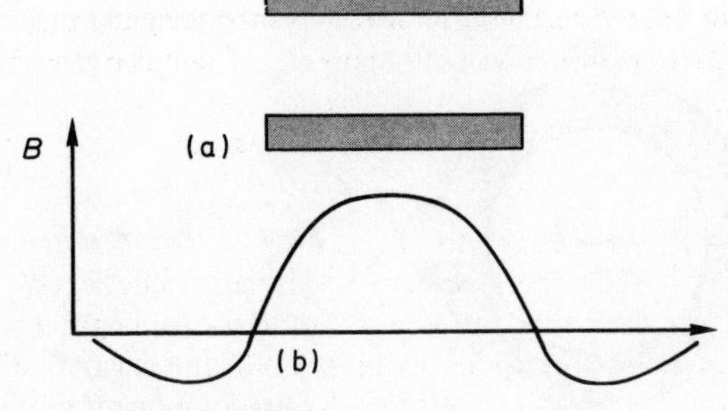


*Fig. 7.54. Flux density on axis of tubular magnet.*

solenoid, but the field produced by the permanent magnet reverses direction outside each end of the magnet. Fortunately the focusing power does not depend on the direction of the field, and many focusing devices deliberately use alternating fields.

One simple device in which a permanent magnet acts upon electrons and ions moving in free air is a blow-out for switches. Switches that break DC currents in inductive loads are particularly liable to produce arcs when the circuit is broken. A small magnet placed near the contact deflects the electrons and ions away and blows out the arc.

Most applications in which magnets act on moving particles do so in an enclosure. Sometimes the magnet may be situated in the enclosure. Cast Al–Ni–Co can be heated to 500°C and is not too difficult to degas for use in a vacuum. Sintered magnets may be more difficult to degas.

Often the magnet must be situated outside the vacuum enclosure, and then the gap tends to be rather large. If a large flux density is also required, the magnet itself must also be large. Some of the largest and most expensive magnets are made for this type of equipment.

At one time or another magnets have been used in television receivers for focusing, separation of ions that would damage the screen, and many other purposes. Some of these purposes were to correct for faults in the manufacture of the tube. Electrostatic focusing is now used, and better control of the tube manufacture has eliminated the need for many of the other magnets.

The picture shift magnet is still sometimes used, and although a very weak and simple device it is rather interesting. Two bonded ferrite magnets as illustrated in Fig.7.55 are now used. Each alone produces a flux density of about 0.0008 T (8 G) diametrically at its centre. Two of these rings can be placed together to produce any flux density between twice this value if they act together and zero if they are opposed. For intermediate values the two rings are arranged with their magnetization directions making a suitable angle. The pair of magnets can then be turned together to produce the resultant field along any diameter. A

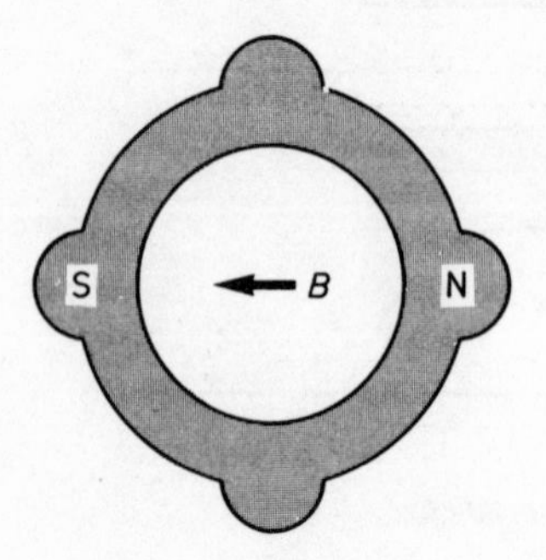


*Fig. 7.55. Picture shift magnet*

pair of these magnets are placed on the television tube, and if the picture is not quite central owing to some inaccuracy in manufacture, the orientation of the rings can be adjusted to correct the fault. The protuberances on the rings are for ease of manipulation and have no specific magnetic function.

Although magnetic focusing is no longer used in television, there are other needs for focusing magnets such as electron microscopes, linear accelerators, klystrons and travelling wave tubes. The last two named electronic devices have electron beams that need to be focused, while they are in some way modulated. They replace ordinary valves and transistors at very high frequencies and sometimes at very high powers.

The focusing magnet may be a single hollow cylinder or ellipsoid. The flux density inside such a magnet is always less than $\mu_0 H_c$, and when high flux densities are required, high coercivity alloys such as A4 are often used.

Alternating fields are produced very successfully with ferrite magnets with alternating polarity and fitted with pole-pieces, as shown in Fig.7.56. Radially magnetized magnets have been suggested and may have been tried but present difficulties. A rectangular tube with block shaped magnets as shown in Fig.7.57 is being used. In space applications

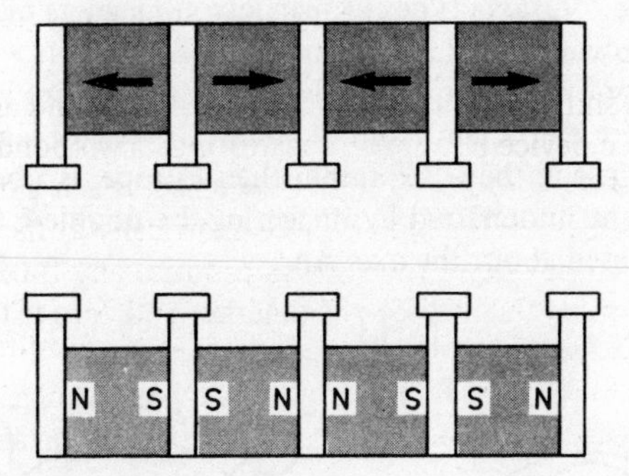


*Fig. 7.56. Alternating field focusing magnet*

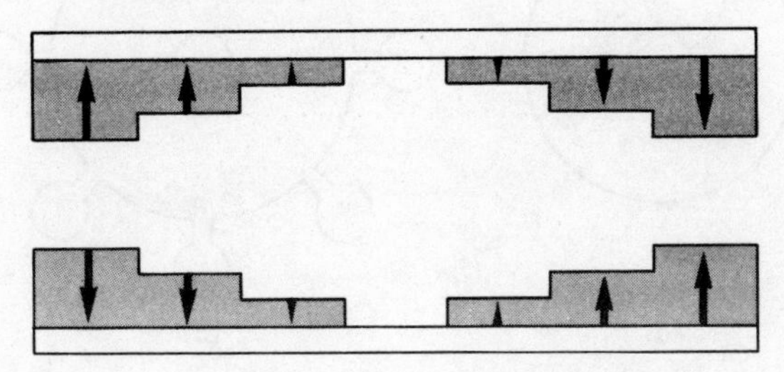

*Fig. 7.57. Focusing magnet made up from rectangular blocks*

such as communication satellites, it is most important to reduce size, and at one time a considerable amount of PtCo was being used. $SmCo_5$ is both cheaper and better than PtCo and this application was a powerful incentive to develop this material. In some of the applications the operation of the apparatus heats the magnets and the equipment has to be put out of action for long periods to allow the magnets to cool. This important economic consideration has led to the great interest in the temperature stability of $RCo_5$ magnets mentioned in Section 5.5.6.

Probably the best known large component in which a permanent magnet produces a magnetic field at right angles to the direction of motion of a beam of electrons is the magnetron. In the magnetron electrons emitted by a cathode are accelerated towards an outer concentric anode as shown in Fig.7.58(a). The cathode may have a more complicated shape with cavities as shown in Fig.7.58(b). A magnetic field, perpendicular to the plane of the paper, constrains the electrons to move in circular paths and the magnitude of the field in relation to the voltage is adjusted so that the electrons normally just fail to reach the anode. The permanent magnet technologist is concerned with providing this field.

The need for magnetrons during the war was so great that the anisotropic Al–Ni–Co material A2, invented in 1940, was almost immediately pressed into use. Various types of magnetron magnets made in this material are shown in Fig.7.59. Some of these magnets weigh over 50 kg and the need for a fairly large flux density in a large gap led to many designs that were inefficient from the point of view of leakage. A further type in use in the USA, rather than Europe, is a bowl shape, which can best be understood by imagining the double-E shape of Fig.7.59(c) rotated about the axis AA.

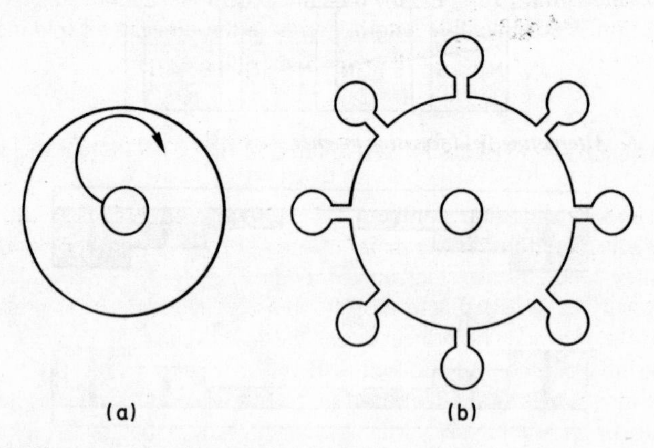


*Fig.7.58. Magnetron, (a) showing electron paths, (b) showing cavities*

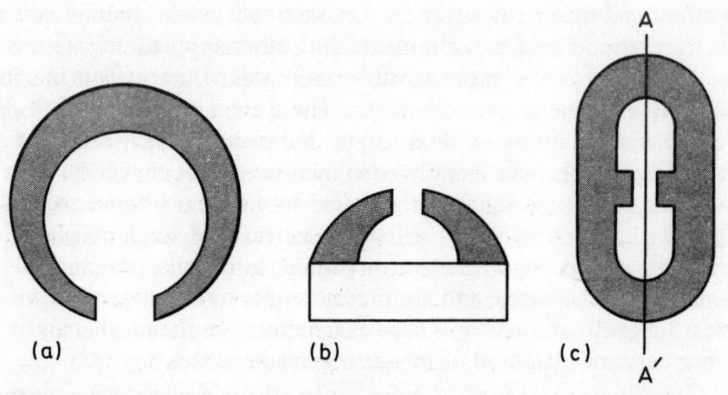


*Fig. 7.59. Magnetron magnets, (a) single C-shape. (b) double-horn, and (c) double-E*

Other cross-field applications include the mass spectrometer, used to separate and identify isotopes and a number of other pieces of laboratory equipment. In an ion pump what is effectively a large diode with a reactive metal used for the electrodes is inserted in the space to be evacuated. A crossed magnetic field causes the electrons to follow closed paths. They thus have a greater probability of colliding with residual gas molecules and ionizing them. The ions are then attracted to one of the electrodes and absorbed by them, acting as a getter.

The omegatron uses the oscillations of ions in a high frequency electric field and a steady magnetic field as a means of identifying them at a pressure of $10^{-5}$ to $10^{-11}$ torr. As it requires flux densities of 0.3 to 0.5 T over a considerable volume, very large magnets are required. Schüler and Brinkmann (1970), who incidently discuss the subject of this section at considerable length, compare the omegatron to a mass spectrograph.

## 7.8 MAGNETIC FIELDS IN SOLID AND LIQUID MEDIA

There are a considerable number of phenomena and applications that can be considered under the main heading of this section. Some of these may generally use electromagnets, but permanent magnets may occasionally be involved, and more such use of permanent magnets may arise in the future. The subject is divided in the remainder of this section into a number of sub-sections.

The most obvious applications are those such as saturable reactors and saturistors in which a permanent magnet alters the characteristics of a magnetic circuit. Next come electromagnetic effects such as the

Hall effect and magnetoresistance. Considerable use of these effects is made in magnetic field measurements, but other applications are possible. It seems however more sensible to consider these effects in connection with magnetic measurements. There are a number of optical effects in which a magnetic field rotates the plane of polarization of light. These effects have already been mentioned in connection with domain studies, but also have other uses. Magnetic resonance studies require the interaction of an oscillating electric field and a magnetic field. They may occur with electron moments in ferromagnetic and paramagnetic substances, and are the basis of several electronic devices. Nuclear magnetic resonance is used as a method of chemical analysis and as an accurate method of measuring magnetic fields.

### 7.8.1 Permanent magnets in magnetic circuits

Permanent magnets can be used to magnetize other permanent magnets (see Section 8.1), and in the field treatment of Al–Ni–Co and Cr–Fe–Co alloys (Sections 4.2 and 4.3). Polarized relays that remain in one or more different positions after receiving a temporary signal usually use permanent magnets for the latching effect, and are employed in railway signalling, telecommunications and many other branches of technology. There are many variations in detail, some of which are described by Tyack, in Hadfield (1962) and Schüler and Brinkmann (1970). Only the broad principles are explained here. The permanent magnets may be contained in the moving or stationary part of the component.

In Fig.7.60 the permanent magnet moves and the action is obvious, the direction of the current in the coil C determines which pole of the magnet comes down in contact with the yoke, and of course remains in contact after the current has ceased. Many more complicated devices are in use – some are described in Hadfield (1962) and Schüler and Brinkmann (1970).

Chokes are often used to control alternating current by their impedance arising from their self-inductance, as this method of control avoids the losses associated with rheostats. The self-inductance of a choke depends upon its permeability. If the core of the choke is saturated by a unidirectional mmf its permeability falls, and a greater AC can flow. The control may be performed by a DC in a coil, and the whole component is then known as saturable reactor. Tyack, in Hadfield (1962), suggests that the control could well be effected by a permaneht magnet.

Shepherd and his co-workers have studied the use of permanent magnets in saturistors. Shepherd recommends Alger (1965) as an introduction to the subject. If a closed yoke is constructed partly of a square loop permanent magnet material such as A2 or F2 a coil wound round

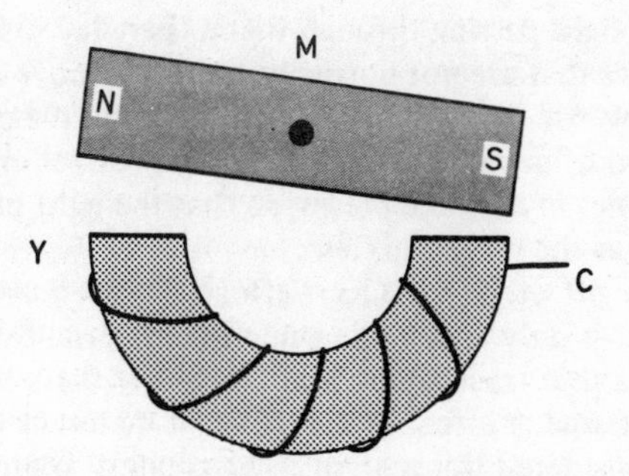


*Fig. 7.60. Polarized relay. Y: Yoke. C: Coil. M: Pivoted magnet*

the yoke does not have a very large self-inductance and impedance until the current in the coil produces a magnetizing force greater than that of the coercivity of the permanent magnet. The permeability of the magnet then increases considerably and with it the impedance of the device, the purpose of which is to provide a safety limit for the current in the circuit.

In one suggested application permanent magnets are incorporated into the rotor of an electric motor and provide a safeguard against excessive starting currents.

The design of saturistors requires a knowledge of the family of hysteresis loops of a saturistor material for various maximum magnetizing forces. It has also been suggested that the operation of a saturistor might be controlled by superposing a unidirectional magnetizing force. Although the family of hysteresis loops of a saturistor material obtained by standard DC measurements is often used in the design of saturistors, it is necessary to point out that eddy-currents and magnetic viscosity may modify these loops somewhat at the operating frequencies.

### 7.8.2 Magneto-optical effects

Two magneto-optical effects, the Faraday effect and the Kerr effect have been known for a long time. The Kerr effect is a small rotation of the plane of polarization of light when it is reflected from the surface of a magnetized material. Ordinary light can be polarized when it is passed through a Nicol prism or a piece of polaroid. As was described in Section 3.8.1 the Kerr effect has been extensively used in research on the domain structure in magnetic materials, including a considerable amount of recent work on permanent magnet materials.

A number of substances such as sugar solutions rotate the plane of polarization of light passing through them. Faraday discovered that many substances that are not normally optically active in this way acquire this power if they are placed in a powerful magnetic field. The classical method of demonstrating the Faraday effect employs an electromagnet with holes in the pole pieces, so that the light path is in the same direction as the magnetic field.

The Faraday effect, like the Kerr effect, has been used in the study of domains, but is only applicable when the material can be obtained in the form of a thin transparent film. Recently there has been much interest in the Faraday effect in thin films of materials that would not normally be considered transparent. The greatest interest has centred round a new class of magnetic garnets, but experiments have also been performed with some recognized permanent magnet materials such as barium ferrite, MnBi and PtCo. Some of the interest has been a purely academic study of domain processes, but there is also interest in the use of such films for computer memory stores. The desirable properties for this application appear to be a film with permanent magnet properties at room temperature that can be removed by moderate heating for the purpose of erasure.

Although I have not been able to find any practical examples, it is obvious that magneto-optical effects could be controlled by permanent magnets, and one can envisage control devices operated by permanent magnets, a material that is magneto-optically active, polarized light and a photo-electric cell.

### 7.8.3 Magnetic resonance

If an electron or an atomic nucleus with a magnetic moment is acted upon by an alternating electric field in one direction and a steady magnetic field in a direction at right angles, the moment vector precesses around the steady field. At some frequency determined by the strength of this magnetic field, the mass, charge and magnetic moment of the particle and sometimes other interactions, such as crystal fields in a solid, resonance occurs.

Resonance can occur in nuclei or in electrons in paramagnetic or ferromagnetic substances. Magnetic resonance is a research tool in atomic physics, a sophisticated but now routine method of identifying atomic nuclei and therefore of chemical analysis, and if known nuclei such as protons in water are used, a very accurate method of measuring magnetic fields. Magnetic resonance has also been suggested as a method of measuring magnetocrystalline anisotropy constants, but those requiring these constants for the study of permanent magnets have made little use of this method.

Although some magnetic resonance experiments can be performed with quite small fields, flux densities of 0.5 to 2 T are frequently required. The higher the field the greater the resonance frequency. In the Earth's field the frequency is in the audio range.

Both permanent magnets and electromagnets are used to produce fields for magnetic resonance experiments. If electromagnets are used the frequency can be kept constant and the field slowly varied to find the resonance point. The control equipment even to maintain a constant field with an electromagnet is quite costly. If a permanent magnet is used the field can be kept constant with comparatively simple precautions, and the frequency can be slowly changed to find the resonance value.

Spatial homogeneity of the field to one part in $10^6$ or better and a similar constancy with time is often required. Permanent magnets of the anisotropic Al–Ni–Co alloy, often of the columnar type A2col., are used. The magnets may contain a considerable weight of permanent magnet alloy and even greater weight of iron. In appearance they do not look unlike a conventional laboratory electromagnet without the coils.

To secure the high homogeneity of field demanded, the most obvious first step is to make the area of the gap much greater than that of the sample. The pole-pieces may have a diameter of say 10 cm for a sample of volume 1 $cm^3$, although the area of the pole-pieces can be reduced by the skilful use of shims. Permanent magnet material is notoriously inhomogeneous and must be separated from the gap by well homogenized good quality iron. An air gap of 1 or 2 mm between the magnet and the pole shoes, or within the pole-shoes, although it slightly reduces the gap flux improves the homogeneity. Incidently, practically the only way of telling whether the required homogeneity has been achieved is by resonance experiments themselves. If the field is not homogeneous the resonance is not sharp. To avoid time variations the equipment should be kept in a thermostatically controlled room.

There are a number of electronic devices that may be considered in conjunction with magnetic resonance, because they depend on electron moments precessing around a magnetic field. In these devices a radio wave is polarized, as in polarized light, as it passes through a piece of magnetized material, usually a soft ferrite. Often the ferrite is magnetized by an alloy or hard ferrite permanent magnet.

If the direction of magnetization coincides with the direction of propagation of the wave, a rotation of the plane of polarization (Faraday effect) occurs. In other applications crossed-fields may be used. By taking advantage of selective absorption of certain waves near a resonant frequency, non-reciprocal devices which transmit certain waves in one direction only, can be constructed.

These polarized ferrite components are used largely in connection with wave guides. The magnetic field needed depends upon the frequency. The fields required vary from about 0.01 to 1 tesla. There is thus a demand for magnets of many different sizes. Sometimes temperature compensation is necessary, possibly for the soft ferrite rather than the permanent magnet.

### 7.8.4 Biological and chemical effects

There are of course many medical and veterinary applications of permanent magnets, in which the magnets are used for conventional purposes such as attraction or control, but in unusual situations connected with living bodies.*

Certain biological influences of magnetic fields have been demonstrated. Slight distortions of vegetable growth have been observed probably due to the concentration of paramagnetic salts by a strong field. There have also been claims of changes in the growth rate of mice living in a cage situated in the gap of a large magnet. These effects involve periods of several weeks in a powerful field, and cannot be regarded as evidence for the value of wearing a small magnet round one's neck or wrist. It is however reasuring that most of the folk lore connected with the influence of magnets suggests that they are beneficial.

Permanent magnets are used for the purpose of preventing scale in hot water systems when the water is hard. This application is widespread and is known in Russia as well as the Western countries. Surprisingly the magnets are placed upstream of the places where scale is expected rather than in those places themselves. It is suggested that in some way the magnets promote a form of crystal growth that does not adhere to the walls of the boiler or hot water system.

Again there is some scepticism about the process which is difficult to explain. There is one effect that a magnetic field has on water that persists for several seconds. After the momentary application of a magnetic field the proton magnetic moments precess about the Earth's field with a frequency in the audio range for several seconds. This precession takes about a quarter of a minute to decay, and is also very sensitive to certain impurities, which this effect is used to detect. Although one would not really expect the precession of nuclear moments to have much influence on crystal structure, the time scale of the phenomenon is about that required to account for the claims. Demand for magnets for this application appears to be increasing.

* The papers presented at a symposium on the Application of Magnetism in Bioengineering were published in IEEE Transactions on Magnetics, Mag.-6, June (1970).

# Chapter 8

# Magnetizing, demagnetizing and testing

Magnetizing and demagnetizing have been grouped in the same chapter as testing, because it is often necessary to magnetize and demagnetize in order to test. Testing involves the determination of magnetic quantities such as flux density, the measurement of the hysteresis loops of magnetic materials and the estimation of the fitness of finished magnets for the purpose for which they are intended. In connection with the last process a decision between 100% testing and statistical quality control is often necessary.

## 8.1 MAGNETIZING

According to long standing tradition a magnetizing force 3 to 5 times the coercivity is required to magnetize a magnet. The lower value is sufficient for materials that have square loops such as Al–Ni–Co alloys that have been treated in a magnetic field. The higher figure is desirable for isotropic materials. If the magnet is not contained in a closed circuit of soft iron, extra magnetizing force must be provided to overcome the self-demagnetizing field of the magnet. Very large magnetizing forces are needed to magnetize the new rare-earth-cobalt magnets, but these values fall within the above rules, provided it is understood that coercivity implies the intrinsic coercivity $H_{ci}$.

Magnets that are used in a magnetic circuit, e.g. loudspeaker magnets and instrument magnets usually require to be magnetized in situ. This necessity is a considerable complication; it can be avoided with high coercivity ferrite and rare-earth-cobalt magnets.

Simple bar-magnets may be magnetized by a larger permanent magnet, a solenoid or an electromagnet. The use of other permanent magnets is only suitable for magnetizing small magnets, because of the large force needed to separate large permanent magnets. If a permanent magnet is used for magnetizing, its poles should be covered with a thin layer of non-magnetic material to facilitate this separation. Care must be taken with the way in which the magnetized magnets are withdrawn as there is a danger of partial demagnetization or distortion of the flux paths as they are pulled away from the larger magnet.

Electromagnets may be powered by conventional battery or rectifier sources of DC. High powers and elaborate cooling are necessary for continuously operated solenoids capable of magnetizing modern magnets. Magnetization is often performed with small coils or solenoids energized by large current pulses of short duration.

The most common source of energy for pulse discharges is a bank of condensers. These are quite commonly made to store energies up to 10 000 J. There seems to be some difference of opinion, whether it is better to use a few hundred volts and a large capacity or a few thousand volts and a smaller capacity. The more quickly the capacitors are discharged the higher the current and magnetizing force. The large capacitors required can be provided at a reasonable price and within a reasonable volume by means of electrolytic capacitors. An oscillating discharge must be avoided in many capacitors of this type. At one time it was thought necessary to prevent such oscillations by making the ratio of ohmic resistance to inductive impedance sufficient to ensure critical damping of the circuit. By using an ignitron or similar device as a switch that can pass current in one direction only, this undesirable constraint on the circuit can be avoided.

Large magnets cannot be magnetized by a pulse of short duration, because eddy-currents prevent the flux reaching the centre of the magnet.

Another method of providing a short pulse of high current is to isolate one half-cycle from the AC mains. Equipment for this purpose, using ignitron valves, was originally designed for the purpose of spot welding. It was found suitable for many magnetization processes and we and many of our members used it successfully for many years. It is remarkable that a pulse of over 1000 A can be taken from the mains for one half-cycle without blowing a 30 A fuse.

The limit to the current that can be provided by this type of apparatus seems to be set by the resistance of the mains. For the best results the equipment should be connected across two phases of a three phase supply, and be situated as near the main fuse box as possible. The most efficient use of the apparatus requires that the impedance of the magnetizing coils should approximately equal that of the mains.

We regularly obtained flux densities of 5 T in a coil about 3 cm diameter. We increased this value ot over 10 T in a smaller coil cooled in liquid nitrogen. The ohmic resistance of copper at the temperature of liquid nitrogen is about one sixth of that at room temperature, but the result of cooling a coil that is satisfactory at room temperature may be disappointing, because the self inductance may predominate at low temperatures. To obtain the best results in a low temperature coil, we found it necessary to wind it with relatively thin wire. In this way we reduced the outer area of the coil and therefore its self inductance. At room temperature the resistance was too high, but at the temperature of liquid nitrogen it was just right.

Half-cycle mains equipment is much cheaper than capacitor banks of equivalent power. It is however much less adaptable, as all the magnetizing coils have to be designed for a pulse of just 0.01 s with 50-cycle mains. Capacity banks can be made much more powerful than half-cycle equipment, but only at ten or more times the cost. The half-cycle mains equipment is less useful on the low voltage supplies common in the USA than on the 240 V supplies that are standard in the UK.

Recently half-cycle mains equipment based on solid state switching has become available. The last major piece of equipment that we acquired was a magnetizer of this type. After an initial fault which was quickly rectified by the manufacturer, it worked satisfactorily for the short time that we ourselves remained operational.

Other methods of producing a large pulse of current include a DC generator that is suddenly shorted and a DC pulse transformer. In the latter a large yoke of high quality soft magnetic material is provided with a primary circuit, which is supplied with DC. Owing to the large self inductance of the circuit, the time required to saturate the yoke may be several seconds. Small gaps are usually left in the yoke to ensure that the flux collapses when the primary circuit is broken. A large pulse is then induced in the secondary magnetizing circuit. There are several interesting points about the storage of energy in such a transformer which are discussed in Section 6.10.

To magnetize C-shaped magnets the best method is often to pass a very high current through a single conductor threading the C. Such a conductor may be connected to a pulse producing source through a step-down transformer.

Superconducting solenoids are occasionally used to magnetize rare-earth-cobalt magnets.

Magnetizing assemblies often sets special problems. Multipole magnets in particular often require special jigs designed for the particular magnet. Large DC motors and generators are sometimes provided with special coils that are left on the stator magnets. These coils are intended to carry only a short pulse and are nothing like the size that would be required for continuous excitation.

Assemblies such as loudspeaker magnets are often magnetized in the gap of a large electromagnet. During this magnetization the flux in the iron return paths is in the opposite direction to that after the magnet is removed for use. There is no disadvantage in this procedure provided there is sufficient flux in the electromagnet to ensure that the permanent magnet itself is saturated.

## 8.2 DEMAGNETIZATION

Magnets may need to be demagnetized for convenience in transit, to prevent them picking up ferromagnetic powder or as part of the testing process. There are three main methods of demagnetization: a gradually reduced alternating field, a single application of a carefully adjusted demagnetizing field and thermal demagnetization. The first mentioned method is the most satisfactory and the most widely applicable.

In the alternating field method the magnetizing force must start at a value appreciably greater than the coercivity and be gradually reduced with each alternation until it reaches a very low value. For a rough demagnetization 3 or 4 alternations may be sufficient, but when the specification for demagnetization is very strict as many as a hundred alternations may be needed.

Mains power may be used to demagnetize small magnets. The demagnetizing unit may be solenoid or a small electromagnet with a laminated core of transformer steel. The magnet to be demagnetized may be gradually withdrawn from the solenoid or gap of the electromagnet. We found with this method that for good demagnetization the withdrawal process should occupy about two seconds which, with 50 Hz mains, implies about 100 field cycles. Probably the time could be reduced if one knew at exactly which distance it was necessary to withdraw the magnet slowly. Alternatively one can leave the magnet in position and reduce the current slowly by means of a variable voltage transformer.

It is difficult to supply sufficient power to use mains frequency to energize a large demagnetizer. Difficulties may also arise with eddy-currents in a large magnet. To demagnetize a large magnet requires a special power supply that operates at a lower frequency. An effective but very tedious method of demagnetizing large magnets is to use DC gradually reduced by rheostats and alternated by hand operation of a reversing switch. When only a few reversals are required an oscillating capacitor discharge has been tried. It is claimed that a unit using high-voltage capacitors can roughly demagnetize rare-earth-cobalt magnets.

It is usually not too difficult to demagnetize ordinary magnets, so that no flux densities greater than about 0.003 T can be detected in their vicinity. Sometimes specifications as low as 0.001 T are made and these are difficult to achieve, even in materials with only minimal permanent magnet properties such as tool steel. The anhysteresis phenomenon may have to be taken into account. If an object is subjected to a reducing alternating field whilst it is also exposed to a small steady field it may actually be magnetized. The domains are switched backwards and forwards by the alternating field, and should be left randomly in all directions. If there is a small underlying direct field the domains tend to be left predominantly in its direction. Even the Earth's magnetic field may be sufficient to exert the anhysteretic effect, and consequently one is sometimes advised to perform demagnetization in the east–west direction.

If it is necessary to demagnetize a long bar that cannot all be accommodated in the demagnetizer at the same time, the part that is outside the demagnetizer may promote anhysteretic magnetization in the part that is inside.

In order to demagnetize a body by the application of a single demagnetizing force, it is necessary to adjust its value to that of the remanent coercivity $H_r$ as shown in Fig.8.1. The flux density must be taken to a negative value from which it recoils to zero. The value of $H_r$ is usually about 5 to 10% greater than $H_{ci}$, although in a few materials with a special kind of structure it may be greater. Measurements of $H_r$ are sometimes made in pure research on magnetic structures. If a magnet is to be demagnetized on open circuit a demagnetizing force slightly greater than $H_r$ must be applied. These details may help the engineer to understand the process and choose suitable equipment, but practical demagnetization by means of a direct field is more likely to be carried out by trial and error.

Thermal demagnetization can be safely used on ferrites, as they are the only class of permanent magnets that are not metallurgically damaged by heating above their Curie point. The Curie point of barium and strontium ferrite is about 450°C, but it must be remembered that ferrites are bad conductors of heat and the magnets must be left in a furnace at a considerably higher temperature for some time to ensure that they are demagnetized completely.

Because of the difficulty in demagnetizing and remagnetizing rare-earth-cobalt magnets, thermal demagnetization is sometimes tried. Something similar to the original heat treatment must be repeated if the magnets are not to be spoilt, and it is unwise for anyone to attempt thermal demagnetization of these materials without prior consultation with the manufacturer. More study of the properties of thermally demagnetized rare-earth magnets is desirable.

Besides complete demagnetization partial demagnetization to secure stability or the correct calibration of an instrument is frequently necessary. Any of the above three methods may be used but now in a controlled manner. If possible a reducing alternating field is to be preferred, as it leaves the magnets in the most stable condition. An exception may be in connection with temperature stability, as discussed in Section 5.5.

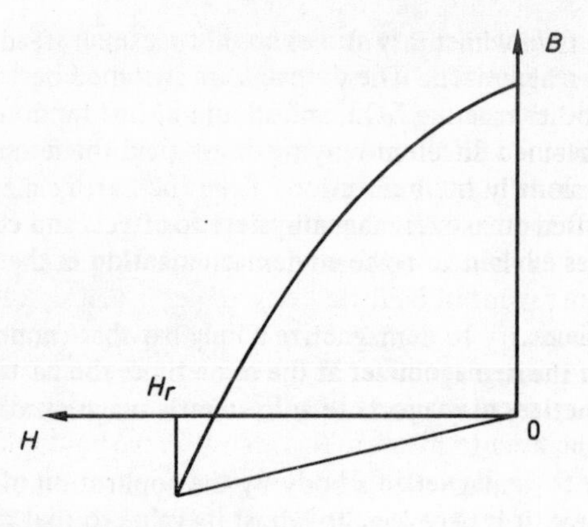


*Fig.8.1. Remanent coercivity $H_r$*

## 8.3 BASIC MAGNETIC MEASUREMENTS

The basic magnetic measurements include:

(1) the calculation of magnetizing force from the measurement of the current in a coil with a known number of turns

(2) magnetometer measurements that involve the measurement of the deflection of a magnet suspended in a magnetic field

(3) ballistic measurements involving a coil and ballistic galvanometer or fluxmeter

(4) various measurements involving solid state phenomena such as the Hall effect.

These basic methods of measurement are first described and then the way in which they are used to measure hysteresis loops and similar curves, and finally the way in which they are applied to assess finished magnets are considered. The calculation of $H$ needs merely a good ammeter and does not seem to need a separate section.

### 8.3.1 Magnetometer measurements

Some of the oldest type of magnetic measurement used magnetometers. The type of experiment in which the product of the moment of a permanent magnet and the Earth's field, or the field produced by a solenoid, determined by a vibration experiment, and the ratio of the moment to the Earth's field measured with a deflection magnetometer, is rarely performed in a permanent magnet laboratory. In 28 years we only had to attempt such experiments twice to measure the magnetic moment of some standard magnets. Although these experiments used to be given as an exercise to elementary students, they are exceedingly difficult to perform accurately. The worst place to perform them is a laboratory containing other magnets. On one occasion after two graduate assistants had obtained different results I resolved the discrepancies myself when I was alone in the building.

We also used an astatic magnetometer for a time for stability measurements, but as explained in the chapter on stability this method was abandoned in favour of ballistic measurements because of the effect of disturbing fields.

One type of magnetometer that can be successfully used for measurements on permanent magnets is the Silmanal type described in Section 7.1. It can be used to measure flux densities in small gaps or $\mu_0 H$ by the side of a magnet. We used it in permeameter experiments for a time, but care had to be taken not to expose it to magnetizing forces greater than 400 $kAm^{-1}$ (5000 Oe). The range of usefulness of this type of instrument could be greatly extended by the substitution of $SmCo_5$ for Silmanal.

### 8.3.2 Ballistic measurements

The most common methods of making magnetic measurements are ballistic ones. Originally these were made by means of a coil and a so-called ballistic galvanometer. If a coil is connected to such a galvanometer, and the flux cutting the coil is changed quickly, the first throw of the ballistic galvanometer is proportional to the product of the flux change and the number of turns in the coil.

Students are often taught to use galvanometers in the undamped state (or at least they used to be). This method is unnecessarily difficult, and it is just as accurate and far more convenient to use the galvanometer in the critically damped state in which, after a deflection, it returns to the zero position immediately but without overshooting. There is one particular resistance for the external circuit which gives critical damping and this is an important characteristic of a galvanometer. If the resistance is much greater the galvanometer is underdamped and continues swinging for a considerable time after being deflected. If the resistance is much less the galvanometer is overdamped and returns very slowly to rest. A very much overdamped galvanometer is just as inconvenient to use as one that is underdamped.

If one is willing to sacrifice some sensitivity a galvanometer can be arranged to be approximately in the critically damped state for a wide range of external resistances. The arrangement that secures this state is sketched in Fig.8.2. The sum of the resistances $R + S$ is that which gives critical damping. Provided $R$ is considerably greater than $S$ the

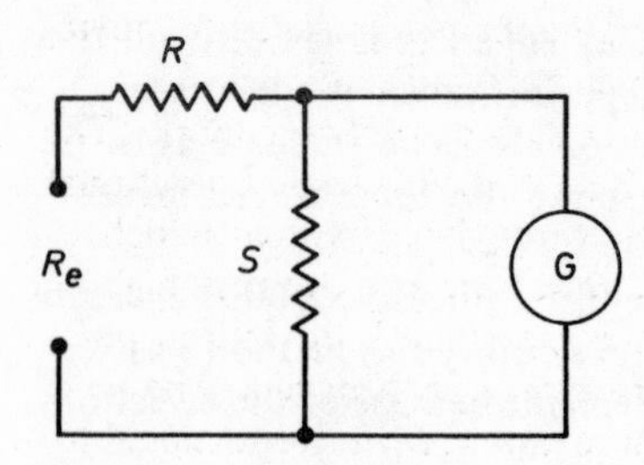


*Fig.8.2. Circuit for critical dampling damping of galvanometer. S is shunt that critically damps galvanometer G (R > S). $R_e$ is resistance of external circuit*

resistance seen by the galvanometere is not greatly affected by that of any external circuit, $R_e$.

Until 1950 or later ballistic galvanometers were still used and recommended by the National Physical Laboratory for the measurement of hysteresis loops, although fluxmeters had been known for a long time and were used for the production testing of the working points of finished magnets.

When a ballistic galvanometer is used the flux change must be completed quickly before the galvanometer coil has begun to move. Consequently a ballistic galvanometer cannot be used if the flux change is produced by switching a large electromagnet in which delays are produced by self-inductance and eddy-currents. The latter can be minimized by using a laminated yoke.

Moving coil fluxmeters are more robust than galvanometers. They are moving coil instruments that are very much overdamped. After a deflection the indicator which may in some models be a pointer and in others a spot of light remains practically stationary and can be read at leisure. Some auxiliary mechanical or electrical method must be provided to return the indicator to zero. The damping is effective only so long as the resistance of the external circuit does not exceed a value which in some of the older instruments was inconveniently low. Consequently the sensitivity of a moving coil fluxmeter, which is one or two orders of magnitude less than that of a ballistic galvanometer, cannot be compensated by using a coil with a large number of turns. Galvanometers continued to be used when a high sensitivity was required.

Some fluxmeters do have a suspension that exerts a slight torque and causes a very slow drift towards a zero that can be adjusted. It is preferable to adjust this zero to a point near the expected final position after the deflection rather than the starting zero. If the instrument is drifting near the zero point it is possible to note the position just before making the flux change or even wait until the reading is precisely zero. If on the other hand the instrument is drifting after the deflection it is sometimes difficult to decide when the deflection has finished.

Electronic fluxmeters can now far surpass the performance of moving coil instruments. The simplest type, which is described by Scholes (1970) consists of an operational amplifier with capacity feedback. The basic idea of the circuit is shown in Fig.8.3. A is a high gain operational amplifier. An input voltate, $E_i$, is applied through a circuit with high impedance and the capacity feedback $C$ must have an extremely high resistance. So long as an input potential is maintained, the output potential $E_0$ builds up and remains constant after $E_i$ is removed. The deflection of the output instrument that measures $E_0$ is proportional to the time integral of $E_i$. The instrument can be returned to zero by closing the switch S which may conveniently be a reed switch, as it too must have a good insulation.

Many additions may be made to this basic circuit to offset drift, to compensate for temperature changes and to improve and control the sensitivity. The best commercial and homemade instruments I have seen have one or two orders of magnitude less drift with the same sensitivity as an ordinary fluxmeter. Alternatively they can have one or two orders of magnitude more sensitivity with comparable drift. If one has attempted to make the sensitivity equal to that of the best ballistic galvanometer the drift becomes troublesome, but I have no doubt that at an increased cost modern electronics can improve on this performance.

An alternative method of integration, described by de Mott (1970), connects the coil to a solid state voltage-to-frequency converter, and this in turn to a digital counter. It is claimed that a stable digital integral of the input potential is given. It is also claimed that an absolute calibration in terms of the primary units the volt and second is possible, so that the need for secondary standards such as those of mutual inductance is eliminated.

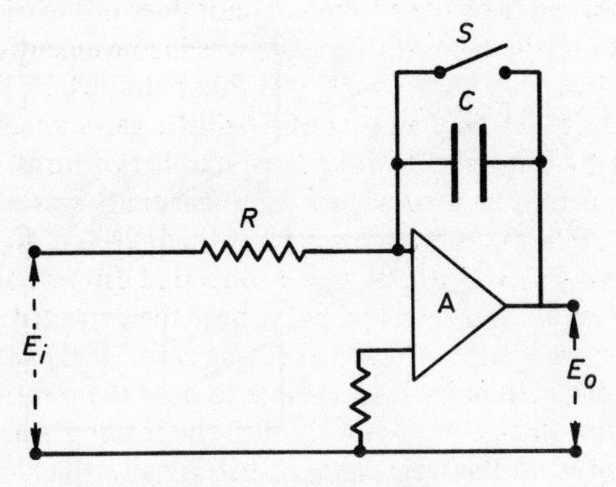


*Fig.8.3. Integrating circuit with operational amplifier*

I have not enjoyed the pleasure of using this elegant apparatus, but from the claims it appears very accurate but more expensive than the operational amplifier instruments.

Some instruments with a digital output can be set to read the maximum flux change, and so avoid errors due to pulling a coil off a magnet from a point that does not correspond to the maximum flux density.

### 8.3.3 Solid state and other devices

The Hall effect is very useful in magnetic measurements. It occurs in all conductors, but is largest in semiconductors. Suppose a current $I$ flows in a rectangular plate of a Hall element, as shown in Fig.8.4, and a flux density $B$ is applied perpendicular to the plane of the diagram, then an emf is generated perpendicular to both current and flux density as shown. The emf depends on the current $I$ and flux density $B$. In general the measured value of $E$ may not be strictly proportional to $B$, but by a suitable choice of the parameters of the external circuit, it can often be made virtually proportional over at least a limited range of $B$ values. Fairly elaborate precautions are needed to eliminate parasitic emf's.

Germanium, indium antimonide and indium arsenide are materials that have been used in Hall probes. Indium antimonide is very sensitive, but indium arsenide is now often preferred because it has the smallest temperature coefficient.

Simple Hall effect instruments can be made that are powered by a battery and have no other components than a network of resistances, switches and a sensitive moving coil instrument. It is difficult to make a portable instrument of this character that has a full scale deflection much less than 0.1 T. Better instruments are fed with AC and contain

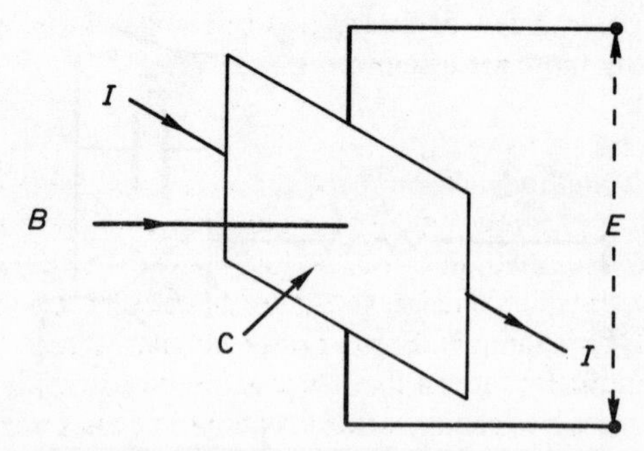


*Fig.8.4. Hall effect. C: Semi-conducting crystal. B: Flux density to be measured. I: Control current. E: Hall voltage.*

solid-state amplifiers. Such instruments can be made with full-scale multiranges varying from 3 to 0.0001 T, or less.

Hall probes can be produced in various shapes and sizes suitable for measuring fields in small gaps in which the field is in a transverse direction, the axial field in solenoids or the field parallel to the surface of a magnet.

The magnetoresistive effect is a change of resistance with applied field. It becomes quite large in some semiconductors and is one cause of non-linearity in Hall-effect instruments. It has been used for field measurements but its use does not appear to be very widespread.

The fluxgate principle deserves a brief mention. The characteristics of the soft magnetic core of a small transformer are modified by an external field. It has been developed into an exceedingly sensitive instrument by Institute Dr. Forster, of Reutlingen. It is extensively used in non-destructive testing, but is rather too sensitive for most tests connected with permanent magnets.

Nuclear magnetic resonance is a method of measuring flux densities with great accuracy and may be used as a standard.

## 8.4 MEASUREMENTS OF HYSTERESIS LOOPS AND DEMAGNETIZATION CURVES

Most measurements of hysteresis loops involve a coil and an integrating device to measure $B$. There are however other possibilities such as the Faraday disc method used for many years in the UK. $H$ may be calculated from the applied ampere-turns, measured by another coil system or by a solid state device such as a Hall probe. An important innovation is the recent replacement of the older point-by-point methods by continuous plotting. Minor modifications in the above methods, and various combinations of them, mean that there is a considerable variation in the equipment in use. A few of these types can be purchased as complete units, but many more are assembled by the user.

### 8.4.1 Ballistic methods

The most common method of measuring the whole or part of a hysteresis loop is one in which changes in $B$ are measured by means of a coil wrapped round the sample. Formerly the coil was connected to a ballistic galvanometer, now a fluxmeter is more commonly used. This arrangement measures changes and not absolute values of $B$, and therefore some base point must be established from which other values can be calculated. Zero flux density may appear an obvious base point, but often it is more convenient to use the value of $B$ corresponding to the maximum magnetizing force, or $B_r$ the flux density remaining after this has been applied and the magnetizing force is again zero. When I was a student and a young lecturer, practical physics courses in universities often included an experiment in which a hysteresis loop was measured by a step-by-step method. In this method a series of increments (positive or negative) in $B$ and $H$ were made starting from the base point, and the value of $B$ was obtained by summing these increments. This procedure is unsound and leads to large errors, because as well as random errors there are systematic ones in each change in $B$, due to the after effect known as magnetic viscosity. These systematic errors are small for one step, but are additive, and become large after many steps. It is desirable therefore to return to the base point after each step, or at least after a maximum of two steps.

A distinction must be made between the procedure necessary to measure an inner loop and an outer loop, which is one in which the maximum magnetizing force produces saturation. The measurement of inner loops is much the more difficult. Before starting the measurement the material should be put in a cyclic state by reversing the peak magnetizing force several times, and to return to a given base point on the loop, the material must be taken round the whole loop, great care being taken to ensure that the same peak magnetizing force is applied every time and in both directions.

The majority of measurements on permanent magnet materials are confined to the second quadrants of full loops and for these tests the cyclic state is not involved, and it is necessary only to return to positive saturation between each measurement. The precise value of the magnetizing force used for saturation is immaterial, provided it is sufficient.

To plot a hysteresis loop it is necessary to measure the magnetizing force $H$ as well as the flux density $B$. The simplest method to explain is the calculation of $H$ from the current and number of turns in the magnetizing coil. Although the method is now rarely used for measurements on permanent magnet materials it is the basis of the SI unit of $H$, and the procedure for determining a hysteresis loop can be conveniently explained in terms of this method.

A calculated value of $H$ is strictly correct for a thin homogeneous ring of material. It is difficult to make rings of permanent magnet material and even more difficult to provide sufficient ampere turns to magnetize them. Permanent magnets are usually tested in a yoke as shown in Fig.8.5. In practical equipment the yoke has an adjustable gap so that it can accommodate magnets of different lengths. The iron yoke has a magnetizing coil of $N$ turns, which is fed by DC from a battery or rectifier. The figure shows the current controlled by two variable resis-

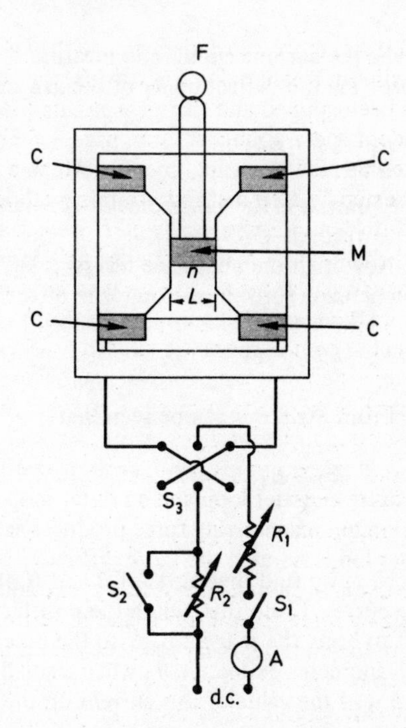


*Fig.8.5. Simple permeameter. M: magnet to be tested. C: Magnetizing coils, total N turns. n: Measuring coil, n turns. F: Fluxmeter*

tances $R_1$ and $R_2$, two on–off switches $S_1$ and $S_2$ and a reversing switch $S_3$. There is also an accurate ammeter, A. In fact the circuit needs to be more complicated, because it is necessary to measure and control currents that vary over a range of 100:1.

The magnet of length $L$ and cross-sectional area $A$ is shown with a closely wound coil of $n$ turns, connected to a galvanometer or fluxmeter. Fluxmeters are usually sold calibrated with a direct reading in weber or maxwell, but galvanometers need to be calibrated. The secondary coil of a standard mutual inductance must be inserted in the circuit for this purpose. While the actual measurements are being made this inductance must be replaced by an equivalent resistance. It may alternatively be left to the circuit, but it is then necessary to be sure that the inductance is designed and situated so that it does not pick up spurious signals.

Suppose first that we require only the second quadrant or demagnetization curve. The best reference point is $B_r$, that is the value of B after fully magnetizing, and switching off the magnetizing current. $B_r$ is called remanence in the UK and residual magnetization in the USA. To find $B_r$ apply a large magnetizing current; it is sometimes advisable to disconnect the galvanometer G while this current is switched on. Now break the current circuit and measure the change in flux density $B_1$ by observing the deflection $\theta_1$ of the galvanometer as the circuit is broken.

$$B_1 = k\theta_1/(nA) \tag{8.1}$$

In SI units $k$ is the galvanometer constant in weber per division (maxwell per division in CGS units).

Now with the aid of the reversing switch $S_3$ apply the same large magnetizing force in the opposite direction, and obtain:

$$B_2 = k\theta_2/(nA) \tag{8.2}$$

From Fig.8.6 it can be seen that

$$B_r = \frac{B_2 - B_1}{2} \tag{8.3}$$

Next we find the values of $B$ and $H$ of a point P on the demagnetizing curve. To do this switch the positive magnetizing current on and off to bring the magnet back to the base point $B_r$. Then observe the galvanometer deflection $\theta_3$ when a small negative current $I$ is applied and read the value of this current on the ammeter. The change in $B$ is

$$B_3 = k\theta_3/(nA) \tag{8.4}$$

The value of $B$ corresponding to the point P is

$$B_p = B_r - B_3 \tag{8.5}$$

The value of $H$ for this point is

$$H_p = -NI/L \tag{8.6}$$

In SI units Equation 8.6 gives $H$ directly in $\mathrm{Am}^{-1}$, but in CGS units

$$H_p = -1.257NI/L \text{ Oe} \tag{8.6a}$$

The procedure can be repeated with different currents to obtain further points on the demagnetization curve.

If a full outer loop is required the procedure can be continued with larger demagnetizing forces that take the magnet into the third quadrant. To obtain points between saturation $B_s$ and $B_r$ the switch $S_2$ is employed. The value of $R_2$ is preset to give the magnetizing force corresponding to Q. With $S_2$ closed the material is magnetized; $S_2$ is then opened to bring the working point to Q. The circuit is then broken and the change $B_4$ observed.

$$B_q = B_r + B_4 \tag{8.7}$$

It is important to follow the sequence of manipulations in the order described. The material must go from $B_s$ to Q and Q to $B_r$; it must not reach Q via $B_r$. Even greater care is required to work out the correct sequence when an inner loop is measured, and the cyclic state must be observed.

The shorter the permanent magnet the less satisfactory is the use of a calculated $H$, because a greater proportion of the applied mmf is used up in the air gaps and the yoke.

When the National Physical Laboratory was the ultimate authority on magnetic measurements, the method developed there by C. E. Webb for the measurement of $H$ used a rather complicated system of flat coils. In the language of CGS units acceptable at the time the field parallel to the surface of the magnet was continuous across the boundary. SI units introduce a logical difficulty because any coil, flat or otherwise, measures a flux density and not a magnetizing force. Perhaps the best we can say is that we measure $\mu_0 H$ just outside the magnet and use this quantity to calculate $H$ inside.

A more practical problem is that any ballistic method of measuring $H$ measures changes in $H$. As with $B$ measurements we must establish a base value. Now in practice the working value of $H$ of a zero current is not

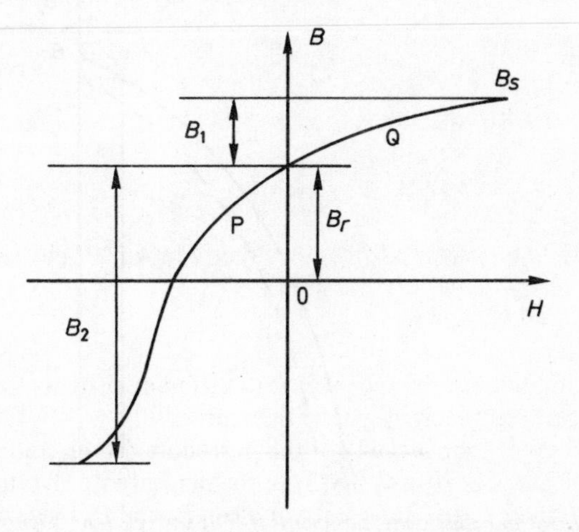


*Fig.8.6. Ballistic method of measuring hysteresis loops.* $B_r = \frac{1}{2}(B_2 - B_1)$

necessarily zero. I have observed with the same yoke and different magnets that the magnetizing force corresponding to zero current was for different magnets sometimes positive and sometimes negative. A negative value was usually associated with a magnet that because of bad or non-parallel surfaces did not make good contact with the surfaces of the pole-pieces. A positive value was usually observed on a short magnet, and was caused by the coercivity of the yoke material. It is not easy to find a yoke material that has a saturation polarization higher than Columax or certain grades of Vicalloy and at the same time a coercivity less than say 0.2% of a rather low coercivity Al–Ni–Co. As the magnetic length of the yoke may be 50 to 100 times that of the magnet it is easy to see that the permanent magnetism of the yoke may exert a significant magnetizing force. In general these two influences tend to neutralize each other, but when one is small and the other large, the positive or negative $H$ for zero current may approach 10% of $H_c$.

If a calculated $H$ is used the error is maximum at $B_r$, and tends to zero as the coercivity point $H_c$ is approached. The consequences of ignoring the error in the ballistic method of measuring $H$ is at first sight surprising. Suppose the demagnetization curve shown in Fig.8.17 is being measured, and suppose that after magnetization the working point is P rather than $B_r$. If this error is ignored in the ballistic method of testing the change in $H$ that reduces $B$ to zero is $QH_c$. The result is that the measured curve looks like the broken one in the figure. In other words the measured value of $H_c$ is too low.

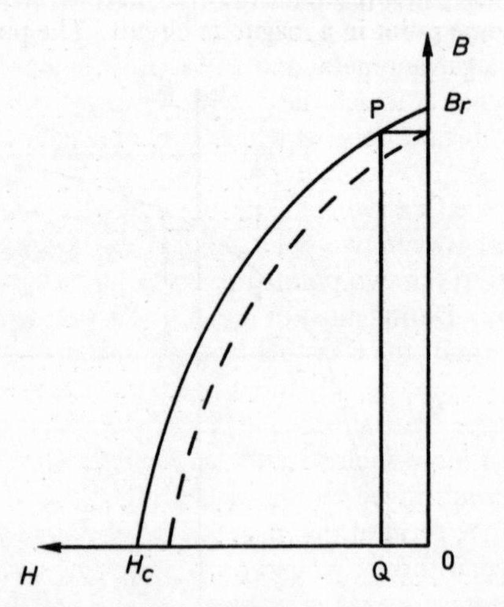


*Fig.8.7. Error in ballistic measurements if H at zero current is ignored. Broken curve incorrect. Full curve correct*

In theory one might measure the value of $H$ corresponding to the point P in the same way that was used to establish its base value of $B$. In practice this method is unsatisfactory. The magnetizing force required to magnetize a permanent magnet should be five times greater than $H_c$. One cannot read a ballistic galvanometer with much better than 1% accuracy, so that a value of $H$ depending on the difference in two such readings is inaccurate. A better method is to observe the galvanometer deflection when the flat coils are pulled off the magnet to infinity. A positive change in $H$ as a result of this operation implies that the $H$ value of the point P is negative and vice versa.

The NPL method used three flat coils at different distances from the surface. The object was to find by extrapolation the magnetizing force at the surface, errors due to the finite thickness of a coil being thereby eliminated. If reasonable precautions are taken to ensure that the gap diameter is appreciably greater than the dimensions of the magnet, and there is good contact between the magnet and the pole-pieces, it is my experience that the use of three coils is unnecessary.

Instead of flat coils a magnetic potentiometer may be used. The original Chattock potentiometer was a uniformly wound coil on a long former of uniform cross-section. This former could be straight or curved. In CGS units it was explained that the turns on an element of length $\mathrm{d}L$ measured $H\mathrm{d}L$ and the whole coil measured $\int H\mathrm{d}L$ over its whole length. $\int H\mathrm{d}L$ is the magnetomotive force or magnetic potential difference between the ends of the coil. The reader is left to consider how this explanation should be rendered in SI units.

The Chattock potentiometer could be used to measure the magnetic potential at some point in a magnetic circuit. The potentiometer was connected to a galvanometer and one end of the potentiometer was applied to a point at which the potential was required, while the other extended to a distant point at which the potential could be taken as zero. The whole potentiometer was then removed to a place of zero potential and the first swing of the galvanometer was read. The potential difference between two points could be found by applying the potentiometer to the two points in turn or by using two potentiometers simultaneously. During most of my career in permanent magnetism, the original form of the Chattock potentiometer was little more than a memory. It was occasionally mentioned, but I am certain that one never entered the Central Research Laboratory at the Permanent Magnet Association. A more sophisticated form of the Chattock potentiometer has recently been revived by Steingroever (1974).

An alternative form of the magnetic potentiometer consists of a small semicircular former wound uniformly with several layers of fine wire. The windings are taken as close to the flat ends of the former as possible. If these flat ends are placed on a plane side surface of a magnet

*Fig.8.8. Magnetic potentiometer with many turn winding on semi-circular former*

as shown in Fig.8.8, the device can be used instead of flat coils and does measure the field close to the surface. Absolute measurements of $H$ can be made by pulling the potentiometer off and changes can be measured during switching operations. Bars with a length to lateral dimension ratio of not less than 10 can be tested on open circuit with a magnetic potentiometer. This kind of potentiometer is usually satisfactory but very tedious to make. I would add that in my opinion, in terms of the distinctions made in SI units, a magnetic potentiometer like any other coil must measure $\mu_0 H$ in tesla and not any quantity in $Am^{-1}$.

For many years our standard tests were made by ballistic methods, a fluxmeter being used for $B$, and flat coils or a magnetic potentiometer with a ballistic galvanometer for $H$. Later we tried Silmanal magnetometers and Hall effect meters for the $H$ measurements. These instruments give a direct reading that automatically takes account of any residual magnetizing force at zero current. Hall effect meters are now widely used.

Discrepancies in testing permanent magnets arise not necessarily from defects in the apparatus or carelessness of the operator, but also from the characteristic inhomogeneity of permanent magnets. In some semi-columnar materials the inhomogeneity is deliberately encouraged.

The best columnar samples of the standard 25% Co Al–Ni–Co alloys have a $(BH)_{max}$ of about 68 kJ $m^{-3}$ (8.5 MGOe), whilst the equiaxed variety has a $(BH)_{max}$ of 43 kJ $m^{-3}$ (5.4 MGOe). It is not much more expensive and therefore commercially justified to make a semi-columnar material with $(BH)_{max}$ say about 50 kJ $m^{-3}$. This material is probably mainly equiaxed with a small volume of almost fully columnar material with much higher properties. The commercial grade of columnar material A2 col usually has a $(BH)_{max}$ about 85 to 90% of that of the best laboratory samples. This deficiency is just as likely to be caused by 10 to 15% of completely misoriented material, as by a general failure to obtain the best properties throughout the sample.

When tests are made on an inhomogeneous sample the results depend on how the tests are carried out, and particularly on the disposition of coils, Hall probes etc. Probably the best average for the whole sample would be given by a $B$ coil that was almost the same length as the sample, combined with flat coils that covered most of its surface.

Although Hall probes are very convenient many of them make measurements that are very localized. In these circumstances it is not always certain that they are best placed close to the sample. Although $H$ is likely to diverge from the mean value in the sample in some slow consistent way with increase of distance, close to the sample there may be large irregularities that are much more serious. It is wise to measure several curves with the Hall probe in different positions.

### 8.4.2 Direct plotting

Most plotting of demagnetization curves and hysteresis loops today is carried out automatically by means of a hysteresigraph. Several organizations will now supply the complete apparatus. The one used in the laboratory of the former PMA, with which I am consequently the most familiar, is described by Scholes (1968).

In this equipment $B$ is measured by a capacity feedback integrator as described in Section 8.3.2. This integrator must be sufficiently free of drift to produce no error during the period that is required for the measurement, say 1 min for a demagnetization curve, and perhaps three or four times longer for a full loop. $H$ is measured by a Hall probe and both measurements are recorded directly by an $XY$ plotter. The equipment has sufficient controls to make it direct reading in whatever units are chosen, but changes of scale can be made with the controls of the $XY$ recorder.

When a curve is being plotted directly $H$ must be changed gradually, not in sudden steps as in a ballistic method. Scholes solved this problem by means of a power supply to the yoke winding consisting of a rectifier and variable voltage transformer. The latter was driven by a small electric motor through a worm gear. If one is drawing the curve of a material with a square loop, $H$ can be changed fairly quickly until the knee of the curve is reached; the speed must then be reduced or $B$ will change more rapidly than the slewing speed of the recorder. Scholes met this requirement by reducing the speed of the motor. There are also electronically controlled rectifiers that can perform these functions. In some systems the rate of change of $H$ is controlled to ensure a constant rate of change of $B$.

The method of using $B_r$ as a base point is not convenient for automatic plotting, and for testing the new rare-earth-cobalt magnets that have to be premagnetized outside the yoke in one direction. Zero flux density can be used as a base point if a reading can be taken with the flux coil outside the yoke before the magnet is inserted. In the ballistic method dispensable close fitting coils are often wound directly on the sample with a minimum of insulation. A removable close fitting coil can however be made with a paper former. The paper is wrapped not too tightly round the sample and stuck to itself. The coil is wound over the paper and extra adhesive outside gives the coil some strength. With care such a coil can be used not indefinitely, but for a considerable number of times.

It is opportune to mention here that the leads to and from all coils to the instrument for any ballistic or hysteresigraph testing, must be kept close together. Any loop or open area left anywhere may pick up stray flux and lead to error.

For routine tests it is inconvenient to use a close fitting coil which must be specially made for every size of sample. The method of circumventing this difficulty used by Scholes, was to use a permanent coil compensated for air flux. The coil is larger than the sample and outside the coil is an additional compensating coil, which must have more area turns than the main coil. Both the main and compensating coils must be within the part of the gap in which the field is uniform. The main coil measures flux in the air space between it and the sample as well as that in the sample itself. By means of a potentiometer a signal from the compensating coil is placed in opposition to that from the main coil. This adjustment can be made to draw either a $J,H$ or a $B,H$ curve. For a $J,H$ curve the coil is placed in the gap on a non-magnetic former of the same length as the sample to be tested and, the compensating potentiometer is adjusted so that the recorder draws horizontal lines when $H$ is changed. To draw $B,H$ curves, the adjustment is made so that applying a demagnetizing force $H$ reduces the reading of B by $\mu_0 H$.

If desired the instrument can be calibrated so that to test a sample of given dimensions potentiometer readings can be read off from a graph or table to give direct readings of either $J$ or $B$.

An alternative method of measuring $B$ was devised by Steingroever and is supplied in commercial apparatus by his firm Elektrophysik of Cologne, and by O. S. Walker of the USA. In this equipment a coil is permanently inserted close to the surface of one of the pole-pieces, or both if desired. As little pole-piece material as possible is removed for this operation. This coil measures $B$ in the pole-piece, which is the same as that in the sample.

Only samples with an area greater than that of the pole-coil can be tested but there is no upper limit to the size of the sample which may even be greater than the size of the gap. The advantages of this method are that it is quick, because no change in calibration is needed for different samples, it allows large samples to be tested without destroying them and it also allows inhomogeneities over the area of a surface or from one end to the other of a magnet to be investigated.

The method does not have a very high accuracy and an alternative facility to use close fitting coils for accurate measurements is usually provided. Steingroever (1974) has also suggested a magnetic potentiometer built into the pole face for measuring *H*.

While *H* is often measured by means of Hall probes, coils and a second operational amplifier are another possibility. The *H* values that have to be measured in Al–Ni–Co samples give a signal ten or twenty times less than the *B* values, and an exceptionally drift-free integrator is necessary.

### 8.4.3 Miscellaneous methods

A brief reference to some older equipment for measuring demagnetization curves may be made. In the UK a rather remarkable piece of equipment was described in 1931 in a British Standard long since out of print. The sample was surrounded by a coil and inserted into an adjustable yoke. *H* was determined by the current in the coil, and *B* by the potential induced between the centre and circumference of a Faraday disc generator. This disc was made of iron, apart from the contact surfaces for the brushes, and was rotated at a constant speed, the gaps between the rotating disc and the yoke were kept as small as possible.

Adjustable resistance allowed the equipment to be direct reading for magnets of any length and cross-section. The equipment was used by most British magnet manufacturers and a few of their customers for many years, but only a very small number were exported.

A skilled operator could obtain good results, and the method was interesting in that it measured realistically what a magnet as a whole could do rather than the properties in a rather ill-defined part of the magnet. It is now almost obsolete, but one would like to think that one example of this apparatus will be preserved in a museum somewhere.

A number of methods in which *B* and *H* measurements are made by rotating or vibrating coils originated in Germany (DIN 50471), and have been widely used in Europe. In one such method a double yoke as shown in Fig.8.9 is used. With equal mmf's applied in the arms no signal is detected by the sensor at S. The balancing coil G enables a true zero value to be obtained. With a sample M inserted in one arm, S detects a signal proportional to the flux produced by M. *H* is measured by a vibrating coil by the side of M.

The Magnetic Materials Producers Association (1975) recommend a double yoke bridge method of measurement, similar to the German one, but with Hall probes replacing the vibrating or rotating coils. This equipment is claimed to enable hysteresis loops to be measured speedily, and to eliminate the errors that arise from drift in operational amplifiers.

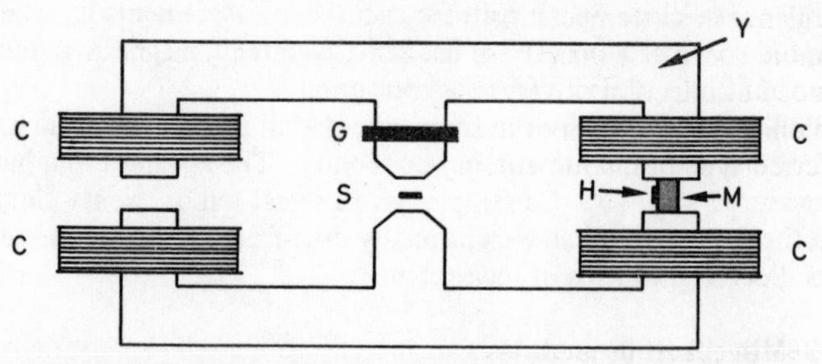


*Fig.8.9. Double-yoke testing apparatus. Y: Yoke. M: Magnet to be tested. C: Magnetizing coils. G: Balancing coil. S and H: Sensors for B and H measurements. (Vibrating coils or Hall elements)*

### 8.4.4 Standards and comparisons of accuracy

Since the National Physical Laboratory abandoned measurements of hysteresis loops, there has been a void in this type of measurement. The Central Research Laboratory of the former Permanent Magnet Association was recognized as a test house by the Ministry of Aviation and its various aliases, until December 1974. We did not feel justified by the number of tests we were asked to make under this arrangement to try to meet the much more rigorous standards of the British Calibration Service.

To satisfy the requirements of this service it is desirable to keep apparatus used for standard and routine measurements separate. On the other hand the NPL found that the personnel who do standard testing of permanent magnets need a supply of routine tests to keep them in practice. There is a possibility that the Magnet Centre recently formed at Sunderland Polytechnic may be able to assume this role.

The NPL is still able to check auxiliary standards such as standard inductances, search coils, magnets, solenoids and current measuring devices. A supply of these standards, which should be arranged to permit cross-checking is necessary for any more specific magnetic testing. An example of cross-checking is to compare the output of a mutual inductance with that of a standard solenoid and search coil. In the USA the National Bureau of Standards checks standards, but like the NPL does not normally measure hysteresis loops on high coercivity materials.

There has been considerable international co-operation in comparing testing methods and accuracy. Greppi and Buzzetti analysed tests on 18 samples made by 17 establishments in different European countries and reported to the Third European Conference on Hard Magnetic Materials. The full report was circulated to members of the conference, but only a brief synopsis was included in the Proceedings. The statistical spread of the results amounts to a standard deviation of 2 or 3%. At first glance measurements made by the German double yoke method are more consistent than those made by hysteresigraphs, but the latter included a collection of very different types of apparatus. It is my opinion that some form of hysteresigraph is likely to emerge as the preferred type of instrument.

### 8.4.5 Special problems of measurement

Rare-earth-cobalt magnets present special problems. The magnetizing force necessary to saturate them is much greater than can be produced in a yoke. They can however be premagnetized by means of a pulse or a superconducting coil, and then tested by the hysteresigraph method. It is difficult to measure $H_{ci}$ in a yoke if it is much greater than 1 T. Even if the yoke is capable of producing much larger fields the measurements become inaccurate because the pole-pieces of the yoke become saturated, and therefore effectively of low permeability. The field distribution around the test pieces changes gradually from that of a magnet in a yoke to that of an isolated magnet on open circuit. Sometimes the errors introduced are attributed to image effects. If one wishes to talk in terms of poles and images it would be more accurate to describe the problem as the disappearance of the images in soft iron pole-pieces, which no longer perfectly neutralize the poles on the ends of the magnet. Thus complicated corrections are necessary.

$H_{ci}$ can be measured on a small sample on open circuit if a solenoid capable of producing a sufficient demagnetizing force is available. A search coil is fixed in the solenoid and connected to a galvanometer. It is advisable to short circuit the galvanometer while large changes in the solenoid current are made. The sample is premagnetized and inserted into the search coil and solenoid with a small demagnetizing force applied. The sample is extracted and the galvanometer deflection noted. The process is repeated with increasing demagnetizing forces until the galvanometer deflection produced by withdrawing the sample is reduced to zero. The demagnetizing force calculated from the ampere-turns in the solenoid is then equal to $H_{ci}$.

A substitute for measuring $H_{ci}$ is to measure $H_r$, the demagnetizing force that after application and removal leaves the magnet demagnetized. $H_r$ can be measured by placing a magnet in the gap of an electromagnet, applying a measured demagnetizing force, removing the magnet and testing whether it is still magnetized. When it is no longer magnetized the demagnetizing force applied was $H_r$, which is usually about 10% greater than $H_{ci}$.

Some measurements on rare-earth-cobalt magnets have been made with a vibrating sample magnetometer, originally invented by Foner

(1959). To estimate $H$ in the specimen it is necessary as in all open circuit tests to calculate the self-demagnetizing field from the shape of the sample and its polarization. This correction is unnecessary for determinations of $H_{ci}$, because $J = 0$. If an electromagnet is used to provide the magnetizing force errors due to the change in the magnetic images arise as saturation is approached in much the same way as in a yoke test.

Measurements of the initial magnetization curve are often made by a ballistic method, starting with a demagnetized magnet. A small magnetizing force is applied and reversed. The corresponding change in flux density is double the value of $B$ corresponding to $H$. The procedure is repeated with increasing values of $H$, which is often reversed several times before taking each reading. Once a value of $B$ has been measured for one value of $H$, measurements for lower values of $H$ cannot be checked without demagnetizing the sample.

Recording hysteresigraphs make it easy to measure the initial magnetization curve with $H$ increasing steadily from the demagnetized state. A curve measured in this way often differs considerably from one obtained by the peak to peak method described in the previous paragraph and also depends on the precise method used to demagnetize the sample. This kind of measurement gives interesting information about domain processes, but the peak to peak method gives more reproducible information about the material that does not depend on the method of demagnetization.

In CGS units $\sigma$, the magnetization per unit mass was often measured. International Committees have until recently ignored this quantity, but happily have now chosen a definition which gives the same numerical value in SI as CGS units. The measurement of $\sigma$ is usually made by comparison with a standard sample of pure iron or nickel. It is particularly useful for measuring the properties of small samples, and can be used as a method of assessing the saturation magnetization of loose powders. Provided the powders consist of single crystals free to rotate, the measurements can be made with magnetizing forces much less than $H_{ci}$ and therefore much less than would be necessary to saturate solid samples. Consequently the measurement is very useful for monitoring the properties of ferrite and rare-earth cobalt products.

We used a method suggested by Klitzing (1957), and illustrated in Fig.8.10. A large permanent magnet with conical pole-pieces concentrates a large flux density in a small gap. In our case the flux density was about 1.9 T in a gap 12 mm diameter and 12 mm long. Coils were fitted around each conical pole-piece. Klitzing suggested two pairs of coils of different diameter. The outer coils were connected in series opposition to the inner pairs, and all were connected to a ballistic galvanometer. The outer and inner coils had the same number of turns.

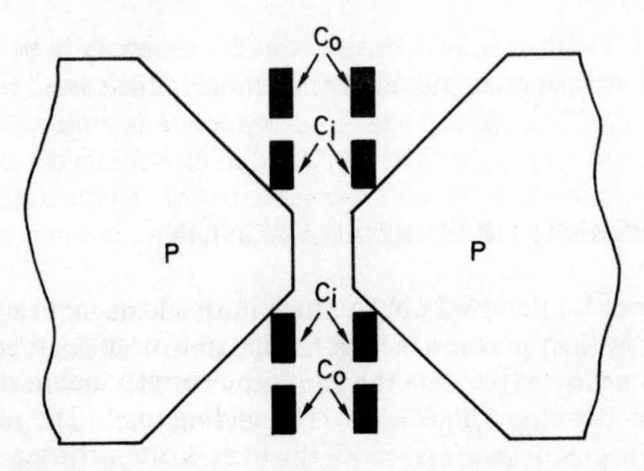


*Fig.8.10. Klitzing apparatus for σ. P: Pole-pieces of permanent magnet. $C_i$, inner coils in series opposition to outer coils, $C_0$*

The deflection $\theta_s$ of the galvanometer produced by withdrawing a standard sample of mass $W_s$ and magnetization per unit mass $\sigma_s$ was compared with that $\theta$ obtained with a test sample of mass $W$. Then the required value is

$$\sigma = \frac{\sigma_s \theta W_s}{\theta_s W} \tag{8.8}$$

According to Klitzing the deflection is produced partly by removing the field produced by the sample and partly by a change in the flux density produced by the large permanent magnet as a result of the change in its gap reluctance. He claimed that the latter effect depends on the shape of the sample, but is eliminated by having two pairs of coils in opposition.

Using two sets of coils in opposition greatly reduces the sensitivity, and I was unable to confirm the benefit claimed by Klitzing. As we had designed the coils to give the required sensitivity for normal samples, when they were connected in opposition, we continued to use them in this way, keeping in reserve the possibility of greatly increasing the sensitivity by disconnecting the outer coils, or even connecting them in series, for measurements on very small or weakly magnetic samples.

I found the response was not quite linear over a wide range of sample sizes and $\sigma$ values, and recommended that the standard should be chosen to give a deflection of the same order of magnitude as the test sample with the same galvanometer sensitivity. The value of $\sigma$ at room temperature and with a gap flux density of 1.5 to 2 T may be taken as 217.8 CGS or SI units for Fe and 55.1 CGS or SI units for Ni.* The

* For many years most writers quoted $\sigma$ for Ni as 54.6, but the value given here is due to Crangle and Goodman (1971) and is probably more accurate.

variation of $\sigma$ with gap flux density, which should really be corrected for the self demagnetizing field of the sample, is discussed by Hoselitz (1952).

## 8.5 ASSESSMENT OF FINISHED MAGNETS

The customer is interested not so much in the demagnetization curve of magnets as in their performance in his apparatus. A final test should simulate as nearly as possible the conditions in that apparatus and particularly the working $B/\mu_0 H$ ratio. The test may actually use the effect which the magnet is required to produce, such as the lifting force, or torque on a moving coil.

Often, however, a $B$ measurement is made by pulling off a coil or by means of an instrument such as a Hall probe. It is highly desirable to place the specimen in a magnetic circuit that simulates its operating conditions. A value that is to be considered satisfactory must be agreed between manufacturer and customer.

The testing may be hand operated if the numbers to be tested are not too large; it is then the responsibility of the operator to reject magnets that fail the agreed specification.

To keep the numbers of magnets to be tested within reasonable bounds, statistical quality control may be used. Table 8.1 shows examples of batch and sample sizes and the number of rejects that may be accepted by the inspector. If more rejects are found the whole batch must be tested. This particular table is intended to guarantee the customer not more than 1% of defective magnets.

**Table 8.1** SAMPLING FOR INSPECTION

| *Batch size* | *Sample Size* | *Acceptance number* |
|---|---|---|
| up to 25 | 5 | 0 |
| 26–50 | 8 | 0 |
| 51–90 | 13 | 0 |
| 91–150 | 20 | 0 |
| 151–280 | 32 | 1 |
| 281–500 | 50 | 1 |
| 501–1200 | 80 | 2 |
| 1201–3200 | 125 | 3 |
| 3201–10000 | 200 | 5 |
| 10001–35000 | 315 | 7 |
| 35001–150000 | 500 | 10 |

The above values are based on Table I and IIa (with AQL 1%) of Spec. DEF-131-A of December 1965 (HMSO)

Other attributes besides magnetic ones, e.g. dimensions may also be the subject of testing by statistical methods. It is unwise of the customer to insist on mechanical specifications that are not relevant to the operation of the magnets, since this must increase the cost.

Whether quality control or 100% testing is required depends upon the consequence of passing a faulty magnet. If the user will discover a faulty magnet at an early stage, for example in trying to calibrate an instrument, quality control is indicated. If a bad magnet will only be discovered by a fault in an expensive piece of equipment that will be costly to replace, 100% testing seems necessary.

Large numbers of magnets may be tested by automatic methods. The magnets are fed into the testing machine by a bowl feeder, and the test which may be by a coil, a Hall probe, a reed switch or some other device operates a mechanism that rejects any faulty magnets. Properly designed, such equipment is not only quicker but less liable to error than a human operator.

# Chapter 9

# Future prospects for permanent magnets

The last ten years have been noteworthy for intense activity in research and development in rare-earth-cobalt magnets. This work has achieved a wide range of permanent magnet materials with different properties that have good prospects of being commercially successful. An important factor in this success has been the close co-operation between academic and industrial scientists. Such co-operation existed between 1935 and about 1955, while the Al–Ni–Co alloys were being developed, but had been less apparent between then and 1966, which in retrospect was a relatively barren time in the history of permanent magnet research.

There are still many problems of detail to be solved in connection with the existing rare-earth-magnet materials. Perhaps the greatest need is to obtain consistent production properties. The ideal material should have a high value of $B_r$, a square loop to give good values of $(BH)_{max}$ and $H_k$ and a sufficient coercivity to guarantee resistance to demagnetization and good stability. The very high values of $H_{ci}$ that are found in some but not all production samples are for many purposes a disadvantage, since they make the magnets more difficult and expensive to magnetize and demagnetize.

An attempt to find permanent magnet materials with the very high magnetic moments and saturation magnetization values of some heavy rare-earth elements seems well worth while. At present these materials have low Curie points and are non-magnetic at room temperature. To increase the Curie points some means of increasing the exchange interactions is required. The task therefore seems to require co-operation between theoretical and experimental scientists. If the exchange interaction can be increased, the very narrow domain walls found in $Dy_3Al_2$ would presumably be lost. Although these narrow walls do produce high coercivities, our experience with $RCo_5$ and $R_2Co_{17}$ materials shows that they are not essential for this purpose.

A reasonable assessment of the future of rare-earth magnets lies somewhere between the extreme views that they will only be used in very specialized applications such as aerospace or that they will replace all other permanent magnet materials. There are, however, rather a large number of firms competing for the rare-earth magnet business. Although the demand for rare-earth magnets seems certain to grow, it is not so certain that all the firms at present offering them will continue to do so.

Already some rare-earth-cobalt magnet materials are being offered at a price that is no more in terms of cost per joule than that of some of the more expensive grades of Al–Ni–Co, containing 35% or more of Co. It is difficult to imagine that rare-earth-cobalt can ever compete for price in this way with alloys containing less than 25% Co, and even less probable that they can be as cheap as ferrites containing no Co. If it is just a question of the cost of the magnet, ferrite will almost certainly remain cheaper than rare-earth-cobalt. However, in an appliance such as a motor, the magnet often represents only a small fraction of the total cost. A motor containing a rare-earth-cobalt magnet can be many times smaller and lighter than a motor of the same power making use of ferrite. Manufacturers of many different products now believe that they can reduce their overall prices by substituting rare-earth for conventional magnets, even though the magnets themselves may be slightly dearer.

On the other hand, rare-earth magnets cannot yet be used up to such high temperatures as Al–Ni–Co alloys. They can also not be made with such a low reversible temperature coefficient without spoiling their other properties. One should also allow for competition from rare-earth magnets including a 'sailing ship effect' and improving the properties or reducing the price of existing materials. For example up to the present time columnar varieties of Al–Ni–Co alloys have usually been made in relatively small quantities as expensive premium materials. In the UK columnar magnets are usually cast into exothermic moulds. This method makes good magnets and can easily be operated on a small scale in any foundry, but it is in effect using a very expensive form of fuel. In the USA some firms use moulds preheated in a furnace. This method requires a foundry specially laid out for the purpose, with a large furnace for heating moulds close to the melting furnace. It may well be cheaper for large scale production. When one considers all the costs of making a magnet, including buying the raw materials, melting them in the correct proportions, performing the heat treatment in a magnetic field and subsequent tempering, fettling, grinding, inspection and testing, plus all the overheads and sales expenses, the single process of heating the moulds to 1200°C ought not to make an enormous percentage increase in the cost. Thus a comparison of the cost per unit of energy of rare-earth with Al–Ni–Co ought to be made relative to the columnar variety.

Another possibility is that alloys of Cr–Fe–Co may prove both cheaper and better than those of Al–Ni–Co. They use cheaper raw materials, have better mechanical properties and may be hot or cold worked to produce desired shapes and possibly also columnar textures. They have already been shown capable of yielding properties equal or better than the 25% Al–Ni–Co alloys (A2) with less than 20% Co and no Ni. So far they have not produced coercivities greater than about 80 $kAm^{-1}$ (1000 Oe), but there is no theoretical reason that values as high as any obtainable from 35% to 40% Co Al–Ni–Co alloys may not be found if they are sought. Research on these lines cannot be guaranteed to succeed, but it does not require expensive high technology, and could be carried out by any reasonably competent magnet manufacturer, or university department of metallurgy.

A reasonably balanced view is that rare-earth magnets, including a certain proportion of the bonded variety, will replace some but by no means all existing Al–Ni–Co and ferrite magnets. Al–Ni–Co magnets made loudspeakers with permanent magnets preferable to those with electromagnets. Ferrite magnets opened up new markets in small DC motors and hi-fi loudspeakers. Many examples of possible uses of rare-earth magnets were given in Chapter 7, and some of them were for mass production requirements.

For the present, permanent magnets seem to have been written off in research into levitated transport. The possibility of using rare-earth magnets in the vehicles and ferrite tracks could reduce the cost of the systems by a factor of two or three. Recent work on levitated bearings suggests that it is worth considering a combination of permanent magnet lift with electromagnetic drive and control. Perhaps some day permanent magnets will emerge as part of a cheap and safe system.

# Appendix 1

# Units

With the possible exception of the 1940s, when the war gave us a respite, no decade has passed recently without some major change being made in the internationally agreed definitions of magnetic units. I confidently expect further changes to be made in the future, but I have no means of foretelling what these changes will be. My reason for expecting further changes is because there are certain obvious practical inconveniences and philosophical contradictions in the SI system as it now stands.

A system of units is like a language, and I am faced with the problem of trying to write for readers, who understand two different languages. In Chapter 1 I stated what I intended to do; in this Appendix I explain in a little more detail the philosophical and practical reasons for these decisions. I am also allowing myself the luxury of expressing my own personal views a little more forthrightly. I lack both the erudition and courage, however, to write quite so strongly as Gilbert (1600) in his great book on magnetism did of his contempories and predecessors, 'Who in all branches of learning are sure to hand on errors and occasionally did something false of their own'.

In SI units two field quantities $B$ and $H$ are defined. In describing magnetic materials $B$ is plotted against $H$, but as well as $B$, being a function of $H$, $H$ is also a component of $B$. There is no necessary connection between this traditional use of $B$ and $H$ in describing magnetic materials and the definitions of $B$ and $H$ in SI units. At least one other quantity is required to describe the state of a magnetic material, and it is even more difficult to see how this fits in with the SI definitions. Similarly geophysicists use the symbol $H$ for the horizontal intensity of the Earth's field, probably because $H$ is the first letter of horizontal, rather than because it should be measured in $Am^{-1}$.

I urge you not to read the rest of this Appendix, except to refer to Table A1.1, unless you are genuinely interested in the problems of units for their own sake, and are prepared to approach them with an open mind.

The CGS electromagnetic system of units started from the concept of a unit point magnetic pole, which exerts a force of one dyne on a similar pole 1 cm distant. The EMU of current was derived in various ways from the force which some specific shape of current carrying conductor exerts on such a pole. The details of this system of units are given in many classical textbooks and need not be repeated here. A point magnetic pole is entirely fictitious and its use in defining electric current was undesirable, and unnecessary. It is easy to rewrite the definition of the EMU of current without introducing magnetic poles. If currents $I_1$ and $I_2$ EMU flow in parallel conductors $r$ cm apart, the force per cm experienced by each conductor is

$$F = \frac{2I_1I_2}{r} \text{ dyne} \tag{A1.1}$$

One could have defined an MKS unit of current by this equation with the kilogram, metre and newton replacing the gram, centimetre and dyne. Such a definition would lead to a unit of current equal to $1000\sqrt{10}$ A, and has never been suggested.

Instead it was desired to incorporate the practical unit of electric current, the ampere in the system, making electric current a fourth fundamental quantity with similar status to mass, length and time. In addition an attempt was made to rationalize the system, by ensuring that the term $4\pi$ appears only in problems relating to spherical symmetry, and the term $2\pi$ in equations relating to cylindrical symmetry. Equation A1.1 related to cylindrical symmetry and was therefore rewritten

$$F = \frac{\mu_0 I_1 I_2}{2\pi r} \text{ newton} \tag{A1.2}$$

the units of $I_1$ and $I_2$ being ampere and of $r$ metre.

The value of $\mu_0$ is $4\pi \,.\, 10^{-7}$ and is determined by the value of the ampere. The attempt to achieve rationalization cannot be considered completely successful. The ampere originated as one tenth of the EMU of current and is therefore an unrationalized unit. Any system of units that includes the ampere must be in some ways unrationalized. The $4\pi$ included in the value of $\mu_0$ appears in circumstances that have no connection with spherical symmetry. The claim to have achieved rationalization by hiding this $4\pi$ in $\mu_0$ is not justified.

This supposedly rationalized MKSA system of units is now part of the officially approved SI system of units and we have to use it as far as possible. There are, however, certain points that are still unsettled.

The constant $\mu_0$ is not just a numerical factor, but has dimensions.

There is one school of thought that believes the dimensions of a physical quantity are a description of its physical nature. Now $\mu_0$ is the ratio of $B/H$ in the absence of matter or in free space, and in magnetic materials the ratio of $B/H$ is known as the permeability; hence it was suggested that $\mu_0$ was the permeability of free space and this name is used in many books. Those who hold this view tend to use one quantity for the absolute permeability of magnetic materials, rather than using the product of $\mu_0$ and the relative permeability.

The dimensions of $\mu_0$ can be derived in the following manner with T = tesla, A = ampere, m = metre, s = second, kg = kilogram, V = volt and H = henry.

$\mu_0$ is equal to $B/H$ and therefore has dimensions deduced from their definitions $\mathrm{TA^{-1}\,m = (Vs\,m^{-2})A^{-1}\,m = Vsm^{-1}\,A^{-1} = Hm^{-1}}$. It is possible to obtain an expression in terms of the four fundamental quantities kg, m, s, A but this is rarely done. Most books now use the fact that $\mathrm{Hm^{-1}}$ has the same dimensions as $\mathrm{Vsm^{-1}\,A^{-1}}$ and quote the units of $\mu_0$ as henry per metre. Although this usage is now almost universal, it seems to me to be a howler, indeed rather a good one, which if it had been spotted before people became accustomed to its use would have merited inclusion in a book such as 'A Random Walk in Physics' (Weber, 1973). The henry is a unit of self or mutual inductance, and it seems quite incongruous to me to associate a metre of free space with any number of henries. If one wishes to be silly, one can invent numerous absurdities of this kind, e.g. torque is measured in Nm or joule!

An alternative view of dimensions is put forward by Nicolson (1961) and Jeans (1933). They consider that the dimensions of a physical quantity are not in any sense a description of the quantity, but only of the method by which it is defined or measured. They point out for example that force is normally defined by Newton's second law of motion, which gives it the dimensions in terms of mass, length and time of $[\mathrm{MLT^{-2}}]$. If gravitational forces had been more easily observed in laboratory experiments, force might well have been defined by the inverse square law of attraction between two masses. It would then have had the dimensions $[\mathrm{M^2L^{-2}}]$.

It is not denied that the method of dimensions can be very useful in checking and even deducing the form of formulae, as well as calculating conversion factors between different systems of units. It should be noted that this last operation is not possible between CGS and SI units purely by the method of dimensions, partly because the SI electromagnetic system is a four unit system, and party because of rationalization. It is probably because of these complications that it was reported at the 1976 combined Intermag MMM Conference in Pittsburgh, that publications of tables of electromagnetic quantities in SI units seemed to contain an inordinately large number of errors.

From this point of view $B$ and $H$ in free space are the same quantity determined in two different ways. $B$ is measured by some method such as a coil and fluxmeter and $H$ is calculated from the current in a coil that prdouces it. Those who hold this view prefer to call $\mu_0$ the magnetic constant, and prefer to keep the relative permeability and $\mu_0$ separate.

I am now convinced that Jean's point of view is preferable. From this point of view $\mu_0$ is unity and dimensionless in the electromagnetic system of units, has the value $4\pi \times 10^{-7}$ and the dimensions $\mathrm{Vs(Am)^{-1}}$ in SI units and has the value and dimensions of the reciprocal of the velocity of light squared in electrostatic units. All of these statements are correct; they tell us nothing about the nature of $\mu_0$, but rather how $\mu_0$, $B$ and $H$ are defined and measured in the three systems.

The idea of attributing properties to free space appears to be a reversion to pre-relativistic ideas of an elastic solid ether. In SI units $\mu_0$ is not a physical quantity at all but a mathematical constant; it does not need to be measured, because its value is defined precisely by the value of $\pi$. Rather strangely the same is not true in SI units of the corresponding electrostatic constant $\epsilon_0$, which requires a knowledge of $c$, the velocity of light, for its determination. (In any system $\mu_0\epsilon_0 = 1/c^2$.)

Permeability in ferromagnetism is a quantity which can only be determined very roughly and it seems illogical to give it the same name as a mathematical constant.

In this book I have used relative recoil permeability rather than absolute permeability mainly for a practical reason. The permeability most commonly used in connection with permanent magnets is the recoil permeability determined by the slope of recoil lines. Typical relative recoil permeabilities are 3.0 for an Al–Ni–Co alloy and 1.05 for a ferrite. These are simple numbers that give a magnetician a mental picture of the nature of the material. Expressed as absolute permeabilities these numbers would become $3.77 \times 10^{-6}$ and $1.32 \times 10^{-6}$ respectively, and would be more difficult to comprehend.

The use of absolute permeability can also suggest false results. In SI units two alternative definitions of the moment of a magnet exist. The Kennelly system uses the magnetic dipole moment $\boldsymbol{j}$, which has the units weber metre. The magnetic dipole moment of a small volume $\Delta V$ of material with magnetic polarization $\boldsymbol{J}$ is given by $\boldsymbol{j} = \boldsymbol{J}\Delta V$. The Sommerfeld system uses the electromagnetic moment $\boldsymbol{m}$, which has the units ampere metre$^2$ and is related to the magnetization $\boldsymbol{M}$ by $\boldsymbol{m} = \boldsymbol{M}\Delta V$.

In the Kennelly system the torque on a magnet is given

$$\boldsymbol{T} = \boldsymbol{j} \times \boldsymbol{H} \tag{A1.3}$$

while in the Sommerfeld system

$$\boldsymbol{T} = \boldsymbol{m} \times \boldsymbol{B} \tag{A1.4}$$

or in free space

$$\boldsymbol{T} = \mu_0 \boldsymbol{m} \times \boldsymbol{H} \tag{A1.5}$$

Whitworth and Stopes-Roe (1971) argued that Equation A1.5 suggests that the torque on a magnet should be increased if it is immersed in a paramagnetic fluid with relative permeability greater than unity, because $\mu_0$ is replaced by a greater absolute permeability. They performed the experiment and found no change, interpreting their result as support for the Kennelly system. I believe their experimental results were correct, and cannot be ignored by those who formulate systems of units. However, I interpret their experiments slightly differently. They used long bar magnets which they regard as ideal dipoles. They believe that thin disc shaped magnets are models of a current loop and should obey the Sommerfeld Equation A1.4.

My opinion (McCaig, 1973) is that the ideal dipole is a sphere. According to a number of authors, of whom the first appears to have been Smythe (1939), the torque on a spherical magnet in a paramagnetic fluid is

$$\boldsymbol{T} = \boldsymbol{m} \times \boldsymbol{B} \frac{3}{2\mu_r + 1} = \boldsymbol{j} \times \boldsymbol{H} \frac{3\mu_r}{2\mu_r + 1} \tag{A1.6}$$

I do not know whether anyone has confirmed this equation experimentally, but if it is true it demolishes the argument of those who argue that relative permeability is unnecessary and only absolute permeability need be used, because it shows that relative permeability and $\mu_0$ do not always occur together as a product.

There is one school of thought that suggests that the moment of a magnet, rather than the law of force, should be considered to change when the magnet is immersed in a paramagnetic medium. In my view this procedure is legitimate but inconvenient and solves no problems. When the magnet is so immersed its moment would change not by a simple factor of $\mu_r$, but by a complicated function of $\mu_r$ and the shape of the magnet, similar to Equation A1.6.

The above arguments are rather abstract, but they do have a bearing on a very important practical choice of units that we have to make. There is an important relation that is becoming more and more necessary in permanent magnet technology as the coercivity of magnets increases and the difference between the flux density and intrinsic properties of a magnet become more significant. In the Kennelly system this equation is

$$\boldsymbol{B} = \mu_0 \boldsymbol{H} + \boldsymbol{J} \tag{A1.7}$$

In the Sommerfeld system it is

$$\boldsymbol{B} = \mu_0 (\boldsymbol{H} + \boldsymbol{M}) \tag{A1.8}$$

At the present time $\boldsymbol{J}$ is called the magnetic polarization and is measured in T and $\boldsymbol{M}$ is called the magnetization and is measured in $Am^{-1}$; both systems have international approval but there are intolerant lobbies on both sides seeking to have the other system banned.

In permanent magnetism there are very great practical reasons for preferring the magnetic polarization $\boldsymbol{J}$ of the Kennelly system to the magnetization $\boldsymbol{M}$ of the Sommerfeld formulation. Whitworth and Stopes-Roe believed they had shown that the Kennelly system was correct and the Sommerfeld system wrong. I do not believe they proved quite as much as this, but I do believe they had shown that there is no fundamental theoretical reason for imposing the Sommerfeld method. Personally I have no desire to prevent anyone using $\boldsymbol{M}$ in $Am^{-1}$ if they find it convenient, provided I can continue to use $\boldsymbol{J}$ in Tesla.

My first objection to the use of $\boldsymbol{M}$ in $Am^{-1}$ in connection with permanent magnets, is that it is grossly misleading. If one is replacing an electromagnet by a permanent magnet, it appears obvious that if the permanent magnet has a magnetization $\boldsymbol{M}$ $Am^{-1}$, and a length $\boldsymbol{L}$ m, it ought to replace a winding of $\boldsymbol{ML}$ ampere-turns. Anyone advocating this system is under an obligation to say how one should explain to a student that this expectation is completely wrong.

My second objection to using $\boldsymbol{M}$ is that it is often necessary to plot both $\boldsymbol{B}$ and whatever quantity one chooses for the intrinsic properties on the same scale. It is easy to plot $\boldsymbol{B}$ and $\boldsymbol{J}$ both in tesla on the same graph, and a small adjustment often allows the same equipment to be used. If one must plot $\boldsymbol{M}$, it will be necessary to use more paper, more complicated equipment or make more complicated calculations, and there is a greater risk of error.

In this book I have used the magnetic polarization $\boldsymbol{J}$. If it were a book dealing also with electrostatics or optics, it would be necessary to use the full name 'magnetic polarization' throughout. As this book is concerned primarily with magnetism, it seems sensible to drop the adjective magnetic and, most of the time, just call $\boldsymbol{J}$ the polarization.

In connection with soft magnetic materials a plot of $\boldsymbol{B}$ against $\boldsymbol{H}$ usually suffices, and $\boldsymbol{J}$ (or $\boldsymbol{M}$) is rarely needed. One great advantage of the SI system is that areas on such graphs are easily converted to energies in joule.

The requirements for modern permanent magnet materials is somewhat different. The product $\mu_0 H_{ci}$, where $H_{ci}$ is the intrinsic coercivity, i.e. the magnetizing force that reduces $\boldsymbol{J}$ to zero, is often greater than $B_r$. It is often important to the designer to know whether $\mu_0 H_{ci}$ is actually greater than $B_r$ or not. In order to calculate $\boldsymbol{B}$ from $\boldsymbol{J}$ or $\boldsymbol{J}$ knowing $\boldsymbol{B}$, the value of $\mu_0\boldsymbol{H}$ in tesla rather than $\boldsymbol{H}$ in $\mathrm{Am}^{-1}$ is required. Modern methods of measuring hysteresis loops of high coercivity materials do in fact depend upon a measurement of $\mu_0\boldsymbol{H}$ in T and not $\boldsymbol{H}$ in $\mathrm{Am}^{-1}$. When this situation is likely to arise it is rather a waste of time for the manufacturer to convert his measurements into $\mathrm{Am}^{-1}$, and the user to convert them back to T. It is surely more sensible to use the product $\mu_0\boldsymbol{H}$.

In most branches of science and mathematics it is considered quite usual to plot different quantities with different symbols but in the same units as the ordinate and abscissa of a graph. Graphical solutions to problems often require that the units are the same.

Magnet technologists for some obscure reason have tended to feel that the axes of a hysteresis loop ought to be labelled in different units. In CGS units they invented the name oersted for $\boldsymbol{H}$, although they sometimes used geometrical constructions that really depended on the gauss and oersted being the same. In SI units $\boldsymbol{B}$ and $\boldsymbol{H}$ are in different units, but the magnetizing force $\boldsymbol{H}$ is actually a component of $\boldsymbol{B}$, and many geometrical and other procedures require them to be converted into the same units.

I wish this book to be understandable to the many older technologists who have never used anything but the old CGS system, as well as to the rising generation of scientists and engineers who have been taught in the new SI system. At least in the UK the accent in teaching is on $\boldsymbol{B}$ in T, and $\boldsymbol{H}$ in $\mathrm{Am}^{-1}$ is introduced only at a fairly late stage.

Thus when there is a choice I tend to speak of $\boldsymbol{B}$ in T. I assume that no one should have any great difficulty in multiplying values in T by 10 000 to obtain values in gauss. On those occasions when it seems appropriate or necessary I quote $\boldsymbol{H}$ in $\mathrm{Am}^{-1}$. In such cases I try to give values in oersted as well. When I feel it appropriate, I quote values of $\mu_0\boldsymbol{H}$ in T, either instead of or as well as $\boldsymbol{H}$ in $\mathrm{Am}^{-1}$. In such circumstances I deem it unnecessary to give values in oersted, because I assume the reader can multiply by $10^4$.

$(BH)_{max}$ is rather troublesome. The general practice is to quote this important quantity in mega-gauss-oersted (MGOe) in CGS units and $\mathrm{Jm}^{-3}$ in SI. It seems necessary to give both units on most occasions. I am not quite happy about $\mathrm{Jm}^{-3}$, because although $(BH)_{max}$ is clearly related to energy its value in $\mathrm{Jm}^{-3}$ is actually double the available energy. It is even conceivable that its use could be challenged under the Trades Description Act in the UK. I have thought of giving the units of $(BH)_{max}$ as $\mathrm{TAm}^{-1}$, but have finally decided that it is best to fall in line with the general practice and use $\mathrm{Jm}^{-3}$, but I have tried not to describe it as the energy product.

Finally a word must be said about the concept of magnetic poles. It is unavoidable to speak of the north and south pole of a magnet, but the analogy between magnetic poles and electrostatic charges is misleading. There is a similarity between magnetic materials and dielectrics in the absence of free electric charges. At the boundary of a dielectric medium the normal component of the electric displacement $\boldsymbol{D}$ and the tangential component of the electric intensity $\boldsymbol{E}$ are unchanged, just as at the boundary of a magnetic material, the normal component of $\boldsymbol{B}$ and the tangential component of $\boldsymbol{H}$ are constant.

A free electric charge is, however, a source of the electric displacement $\boldsymbol{D}$ as well as of $\boldsymbol{E}$. If a sphere of radius $r$ is drawn round a small charge $Q$,

$$Q = 4\pi r^2 D$$

In a dielectric medium of permittivity $\epsilon_0 \epsilon_r$

$$D = \epsilon_0 \epsilon_r E$$

**Table A1.1** UNITS AND CONVERSION FACTORS

| *Quantity* | *SI Unit* | *CGS Unit* | *Conversion factor** |
|---|---|---|---|
| Length | metre, m | centimetre, cm | $10^{-2}$ |
| Mass | kilogram, kg | gram, g | $10^{-3}$ |
| Force | newton, N | dyne, dyn | $10^{-5}$ |
| Energy | joule, J | erg | $10^{-7}$ |
| Density | $\text{kgm}^{-3}$ | $\text{gcm}^{-3}$ | $10^{3}$ |
| Magnetic flux | weber, Wb | maxwell, Mx | $10^{-8}$ |
| Flux density, B | tesla, T | gauss, G | $10^{-4}$ |
| Magnetizing force (field strength), H | $\text{Am}^{-1}$ | oersted, Oe | $10^{3}/4\pi$ |
| Magnetic polarization | T | EMU intensity of magnetization | $4\pi \times 10^{-4}$ |
| Magnetic dipole moment | Wbm | EMU | $4\pi \times 10^{-10}$ |
| Magnetic pole strength | | EMU | $4\pi \times 10^{-8}$ |
| $(BH)_{\text{max}}$ | $\text{TAm}^{-1}$ or $\text{Jm}^{-3}$ | megagauss-oersted, MGOe | $10^{5}/4\pi$ |
| Magnetic constant | $\text{TmA}^{-1}$ ($\text{Hm}^{-1}$) | unity | $4\pi \times 10^{-7}$ |
| Magnetomotive force, MMF | A | Oe cm | $10/4\pi$ |
| Resistivity | Ωm | Ωcm (EMU rarely met) | $10^{-2}$ |
| Specific heat | $\text{Jkg}^{-1}\,\text{K}^{-1}$ | $\text{calg}^{-1}\,°\text{C}^{-1}$ | $4.18 \times 10^{3}$ |

* The number by which the quantity in CGS units must be multiplied to obtain the quantity in SI units: e.g. 1 cm × 0.01 = 0.01 m.

Thus if a charge $Q$ is placed in a dielectric medium

$$E = Q/(4\pi\epsilon_0\epsilon_r r^2)$$

Thus the electric intensity $\boldsymbol{E}$ is reduced by a factor equal to the relative permittivity $\epsilon_r$ of the dielectric.

It has often been assumed that the value of $\boldsymbol{H}$ due to a magnetic pole should be reduced by a factor equal to the relative permeability of a surrounding medium. There are no magnetic poles that are sources of $\boldsymbol{B}$ in the sense that free charges are sources of $\boldsymbol{D}$, so that analogous argument cannot be applied to magnetism.

Sometimes, however, it is found useful in electrostatics to postulate imaginary charges on the surface of a dielectric. These imaginary charges mark a discontinuity in the electric intensity $\boldsymbol{E}$, but are not sources of displacement $\boldsymbol{D}$. Their value depends on the discontinuity in the permittivity and the value of $\boldsymbol{E}$. It would be quite wrong to reduce $\boldsymbol{E}$ by a further factor equal to the relative permittivity which has already been used to calculate the value of the imaginary charge.

One can suppose that there are imaginary poles on the surface of a magnetic material. These imaginary poles indicate a discontinuity in $\boldsymbol{J}$ and $\boldsymbol{H}$ but not in $\boldsymbol{B}$ and do assist some magnetic calculations, particularly the calculation of self-demagnetizing fields. One may use these imaginary surface poles to calculate $\boldsymbol{H}$, but it is important to recognize that such calculations should not under any circumstances introduce the relative permeability of the medium, as to do so would give completely incorrect results.

Table A1.1 contains the names, symbols and conversion factors for a few of the more important SI and CGS units.

# Appendix 2

# Magnetic materials, composition and properties

At first sight permanent magnet materials can be grouped in a few well-defined classes, and magnet manufacturers in different countries make equivalent products, but a closer examination reveals a very complex system of grades with different combinations of properties. To list all the compositions used and combinations of properties offered would be impractical. The task is made more difficult because some manufacturers do not state the compositions they use and different policies are followed in quoting magnetic properties. The former Permanent Magnet Association attempted to determine mean properties from random production surveys. In Germany there is a move to state guaranteed minimum properties, while one suspects that in competitive situations, maximum properties have been quoted.

Magnetic properties can only be measured on special test pieces or on production magnets if they have simple shapes such as cylinders or cuboids. There is some reason to believe that more complicated production magnets, the properties of which cannot be measured, may also be more difficult to make with the standard properties.

In order to reduce the enormous amount of available data to manageable proportions code letters and numbers are used to describe the more common groups of materials. One such code may cover several different grades of material produced by one manufacturer, and an even greater number offered by manufacturers throughout the world.

As an example the code Al in Table A2.2 embraces all isotropic alloys of the Al–Ni–Co system with less than 13% Co. Some manufacturers make two well-defined materials such as Alni without Co and Alnico with 12% Co, but intermediate proportions of Co are used by other makers.

The range of weight percentages for each element and of magnetic properties are given for each class of material designated by a single code. Obviously all the maximum percentages for the ingredients are unlikely to be found in any one material. The maximum values of the magnetic parameters may also not occur together, as a high value of $B_r$ is often accompanied by a low $H_c$, and vice versa. A better idea of combinations of magnetic properties that may be expected in some of the more popular materials can be found from the demagnetization curves shown in Figs.A2.1 to 4.

In the tables A2.1 to A2.5 $B_r$ is given only in tesla, but $(BH)_{max}$ is given in both $Jm^{-3}$ and MGOe, while $H_c$ is given in $Am^{-1}$ and oersted. The intrinsic coercivity $H_{ci}$ differs from $H_c$ appreciably only in high coercivity materials such as ferrites and $RCo_5$. Here it is felt more useful to quote $\mu_0 H_{ci}$ in tesla in addition to $H_{ci}$ in $Am^{-1}$. The reader is reminded that to convert $B$ tesla to $B$ gauss or $\mu_0 H$ tesla to $H$ oersted it is only necessary to multiply by 10 000. In Figs.A2.1 and A2.2 the abscissa is $H$ ($Am^{-1}$) but in Figs.A2.3 and A2.4, which deal with high-coercivity materials, it is more convenient and useful to plot $B$ and $J$ against $\mu_0 H$ (T). Alternative units are marked on all the scales.

The values given in the Table A2.2 and Fig.A2.1 refer to normal cast rather than sintered magnets, for which the properties are rather lower.

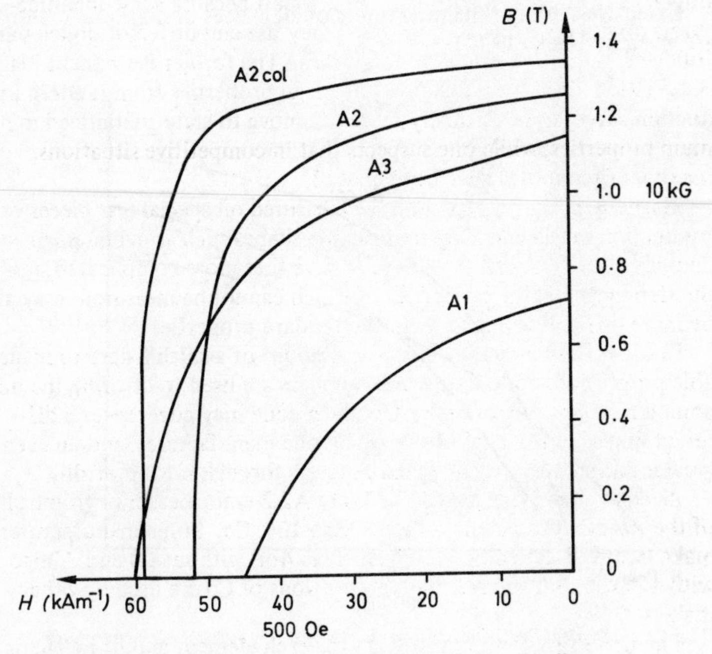


*Fig.A2.1. Typical demagnetization curves of some Al–Ni–Co alloys*

The alloy A2 can be cast in semicolumnar form with properties that are intermediate between those given for A2 and A2col. It seems too complicated to give these properties.

Table A2.6 lists other physical properties of permanent magnet materials. Bonded materials are not included in this table, because their physical properties depend very greatly on the nature and proportion of the bonding material.

A few values of the thermal conductivity of alloy magnets have been published. They are rather low, as would be expected for alloys with a high electrical resistivity. Typical values of the thermal conductivity are 10 JK $m^{-1}$ or 0.025 cal°C $cm^{-1}$. The conductivity of ferrite magnets is obviously much lower, but no precise figures are available. The specific heat of the alloys is about what would be expected from the average for the ingredients.

Some manufacturers quote mechanical properties such as tensile strength or transverse modulus of rupture. These values are liable to be misleading as the real strength of most magnets depends on the presence or absence of cracks. Thus according to some of these tables the high titanium A4 and A5 alloys are several times stronger than the standard A2, but it is common knowledge that the former give much more trouble from breakages.

Extensive tables of manufacturers of all classes of magnetic materials are given by Heck (1974).

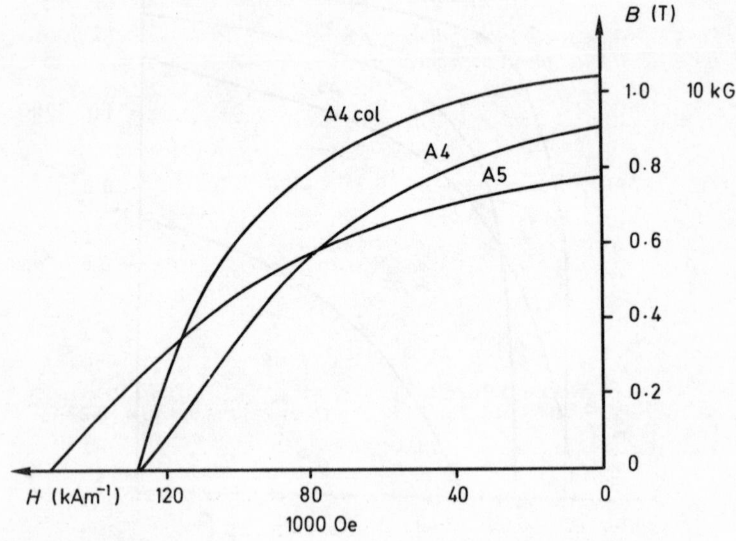


*Fig.A2.2. Typical demagnetization curves of a few typical high-coercivity Al–Ni–Co alloys*

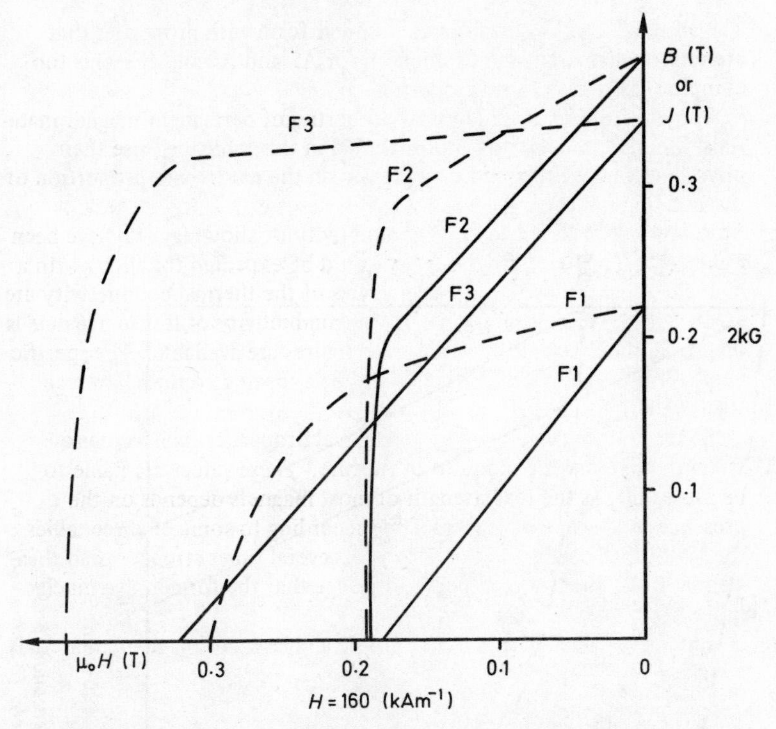


*Fig.A2.3. Demagnetization (full curves) and intrinsic demagnetization (broken curves) for typical ferrite permanent magnets; Abscissa $\mu_0 H$*

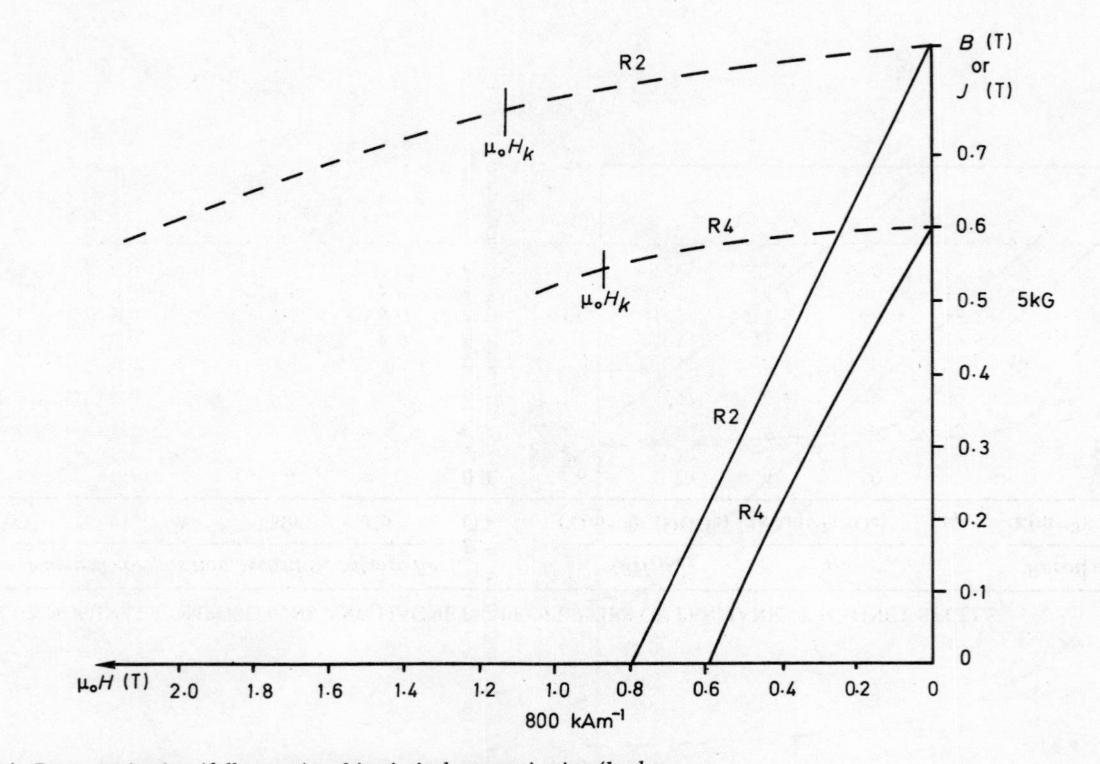


*Fig. A2.4. Demagnetization (full curves) and intrinsic demagnetization (broken curves) for typical sintered $R_2$ and bonded $R_4$ rare-earth magnets. (Abscissa $\mu_0 H$)*

**Table A2.1** NOMINAL COMPOSITIONS AND MAGNETIC PROPERTIES OF PERMANENT MAGNET STEELS

| | *Nominal composition, weight %, balance Fe* | | | | | $B_r$ | $(BH)_{max}$ | | $H_c$ | | | *Recoil product* | |
|---|---|---|---|---|---|---|---|---|---|---|---|---|---|
| *Material* | C | Cr | W | Mo | Co | (T) | ($kJm^{-3}$) | (MGOe) | ($kAm^{-1}$) | (Oe) | $\mu_r$ | ($kJm^{-3}$) | (MGOe) |
| 1% C | 1.0 | – | – | – | – | 0.9 | 1.6 | 0.20 | 4 | 50 | | | |
| 6% W | 0.7 | 0.9 | 6.0 | – | 0.5 | 1.05 | 2.4 | 0.30 | 5.2 | 65 | 39 | | |
| 6% Cr | 1.05 | 6.0 | – | – | – | 0.98 | 2.4 | 0.30 | 5.2 | 65 | 35 | | |
| 2% Co) 4% Cr ) | 0.9 | 4.0 | 0.45 | – | 2.0 | 0.98 | 2.5 | 0.32 | 6.4 | 80 | 30 | | |
| 3% Co | 1.05 | 9.0 | – | 1.5 | 3.0 | 0.72 | 2.8 | 0.35 | 10.4 | 130 | 18.5 | 1.0 | 0.124 |
| 6% Co | 1.05 | 9.0 | – | 1.5 | 6.0 | 0.75 | 3.5 | 0.44 | 11.6 | 145 | 17.5 | | |
| 9% Co | 1.05 | 9.0 | – | 1.5 | 9.0 | 0.78 | 4.0 | 0.50 | 12.8 | 160 | 16.5 | | |
| 15% Co | 1.05 | 9.0 | – | 1.5 | 15.0 | 0.82 | 5.0 | 0.62 | 14.4 | 180 | 15 | | |
| 35% Co | 0.85 | 6.0 | 4.0 | – | 35.0 | 0.90 | 7.6 | 0.95 | 20.0 | 250 | 12 | 2.6 | 0.332 |

**Table A2.2** NOMINAL COMPOSITIONS AND MAGNETIC PROPERTIES OF Al–Ni–Co PERMANENT MAGNET ALLOYS

| | *Nominal composition, weight %, balance Fe* | | | | | | $B_r$ | $(BH)_{max}$ | | $H_c$ | | | *Recoil product* | |
|---|---|---|---|---|---|---|---|---|---|---|---|---|---|---|
| *Grade* | Al | Ni | Co | Cu | Nb | Ti | (T) | (kJm$^{-3}$) | (MGOe) | (kAm$^{-1}$) | (Oe) | $\mu_r$ | (kJm$^{-3}$) | (MGOe) |
| Al | 10 | 16 | 0 | 0 | 0 | 0 | 0.55 | 10 | 1.25 | 38.5 | 470 | 4 | 4.3 | 0.53 |
| A1 | 13 | 28 | 13 | 6 | 0 | 1 | 0.75 | 13.5 | 1.7 | 58 | 720 | 7 | 5.4 | 0.67 |
| A2 | 8 | 12 | 23 | 3 | 0 | 0 | 1.2 | 41 | 4.5 | 46 | 580 | 2.5 | 12 | 1.5 |
| A2 | 9 | 16 | 26 | 4 | 1 | 0 | 1.3 | 44 | 5.5 | 52 | 650 | 4.5 | 13 | 1.6 |
| A2 col | 8 | 12 | 23 | 3 | 0 | 0 | 1.3 | 56 | 7.0 | 56 | 700 | 1.5 | 16 | 2.0 |
| A2 col | 9 | 16 | 26 | 4 | 1 | 0 | 1.35 | 60 | 7.5 | 59 | 740 | 4.0 | 16 | 2.0 |
| A3 | 8 | 13 | 23 | 3 | 0 | 0 | 1.05 | 31 | 3.9 | 58 | 720 | 4.0 | 12 | 1.5 |
| A3 | 9 | 16 | 26 | 4 | 3 | 1 | 1.15 | 36 | 4.5 | 62 | 780 | 5.5 | 12 | 1.5 |
| A3 iso | 8 | 13 | 23 | 3 | 0 | 0 | | >16 | >2.0 | | | | | |
| A3 iso | 9 | 16 | 26 | 4 | 3 | 1 | | | | | | | | |
| A4 | 6.5 | 13 | 30 | 2 | 0 | 3 | 0.75 | 32 | 4.0 | 96 | 1200 | 2.0 | 17 | 2.1 |
| A4 | 8 | 16 | 35 | 4 | 2 | 5 | 0.92 | 44 | 5.5 | 132 | 1650 | 3.0 | 17 | 2.1 |
| A4 col | 6.5 | 13 | 30 | 2 | 0 | 3 | 1.0 | 64 | 8.0 | 120 | 1500 | 1.5 | 21 | 2.6 |
| A4 col | 8 | 16 | 35 | 4 | 2 | 5 | 1.05 | 72 | 9.0 | 128 | 1600 | 2.0 | 21 | 2.6 |
| A5 | 7 | 14 | 38 | 2 | 0 | 6 | 0.72 | 40 | 5.0 | 152 | 1900 | 2.0 | 19 | 2.4 |
| A5 | 8 | 16 | 40 | 4 | 2 | 8 | 0.78 | 46 | 5.8 | 160 | 2000 | 2.0 | 19 | 2.4 |
| A6 | Al–Ni–Co powder with resin bond | | | | | | 0.28 | 3.1 | 0.25 | 37 | 460 | | | |
| A6 | | | | | | | 0.45 | 10.7 | 1.35 | 81 | 1010 | | | |
| A7 | Al–Ni–Co powder with resin bond anisotropic | | | | | | 0.47 | 16.8 | 2.4 | 120 | 1500 | | | |

The different values given for a given class of material are for different alloys made by different producers. Obviously all the high and all the low values cannot occur together. A1, $A3_{iso}$ and A6 are isotropic; the remainder are anisotropic

**Table A2.3** MAGNETIC PROPERTIES OF FERRITE PERMANENT MAGNETS

| Material | Description | $B_r$ (T) | $(BH)_{max}$ (kJm$^{-3}$) | $(BH)_{max}$ (MGOe) | $H_c$ (kAm$^{-1}$) | $H_c$ (Oe) | $H_{ci}$ (kAm$^{-1}$) | $\mu_0 H_{ci}$ (T) | $\mu_r$ | Recoil product (kJm$^{-3}$) | Recoil product (MGOe) |
|---|---|---|---|---|---|---|---|---|---|---|---|
| F1 | Sintered | 0.2 | 6.4 | 0.8 | 136 | 1700 | >240 | >0.3 | 1.2 | 6 | 0.8 |
| | isotropic | 0.23 | 8.4 | 1.05 | 152 | 1900 | | | | | |
| F2 | Sintered anisotropic, | 0.35 | 26.4 | 3.5 | 144 | 1800 | | | 1.05 | 13 | 1.6 |
| | high $B_r$ | 0.39 | 30.4 | 3.8 | 200 | 2500 | | | 1.05 | 12 | 1.6 |
| F3 | Sintered anisotropic, | 0.25 | 20 | 2.5 | 200 | 2500 | 240 | 0.3 | 1.05 | 14.4 | 1.8 |
| | high $H_c$ | 0.37 | 26.5 | 3.3 | 265 | 3200 | 320 | 0.4 | 1.05 | 19 | 2.35 |
| F4 | Bonded, flexible, | 0.1 | 2.4 | 0.3 | 76 | 900 | | | ~1.1 | | |
| | isotropic | 0.17 | 5.6 | 0.7 | 128 | 1600 | | | ~1.1 | | |
| F5 | Bonded, flexible, | 0.2 | 8.0 | 1.0 | 140 | 1750 | 200 | 0.25 | | | |
| | anisotropic | 0.25 | 12 | 1.5 | 168 | 2100 | 200 | 0.25 | | | |
| F6 | Bonded, rigid, | 0.13 | 2.8 | 0.35 | 72 | 900 | | | | | |
| | isotropic | 0.14 | 3.2 | 0.4 | 84 | 1050 | | | | | |
| F7 | Bonded, rigid, | 0.2 | 7.3 | 0.91 | 120 | 1500 | | | | | |
| | anisotropic | 0.26 | 10.4 | 1.3 | 147 | 1835 | | | | | |

All the ferrite permanent magnets listed above have a composition that can be represented as lying between $MO(Fe_2O_3)_{5.5}$ and $MO(Fe_2O_3)_6$ where M stands for Ba, Sr or Pb or any mixture of these elements. Some manufacturers do not state whether they use Ba or Sr, but magnets with the highest coercivity usually contain Sr. Magnets can be made wholly with Pb, but it is unlikely that Pb is used as more than a minor ingredient. Some makers list 7 or 8 grades of anisotropic sintered ferrites with different combinations of $B_r$, $(BH)_{max}$ and $H_c$ that do not fall precisely within the simplified classification shown above.

**Table A2.4** MAGNETIC PROPERTIES OF RARE–EARTH MAGNETS

| *Material* | *Composition* | $B_r$ (T) | $(BH)_{max}$ (kJm$^{-3}$) | $(BH)_{max}$ (MGOe) | $H_c$ (kAm$^{-1}$) | $H_c$ (Oe) | $H_{ci}$ (kAm$^{-1}$) | $\mu_0 H_{ci}$ (T) | $\mu_r$ |
|---|---|---|---|---|---|---|---|---|---|
| R1 | Sintered $RCo_5$ | 0.7 | 96 | 12 | 480 | 6000 | 560 | 0.7 | 1.05 |
| | R partly misch-metal | 0.75 | 112 | 14 | 560 | 7000 | 720 | 0.9 | 1.05 |
| R2 | Sintered $SmCo_5$ | 0.8 | 128 | 16 | 600 | 7500 | 1200 | 1.5 | 1.0 |
| | | 0.9 | 160 | 20 | 680 | 8500 | 3200 | 4.0 | 1.05 |
| R3 | New compositions with | 0.9 | 160 | 20 | New developments – precise properties not yet certain | | | | |
| | excess Co or Pr or Nd | 1.1 | 240 | 30 | | | | | |
| R4 | Polymer bonded, rigid | 0.55 | 56 | 7 | 400 | 5000 | 800 | 1.0 | 1.0 |
| | | 0.63 | 80 | 10 | 480 | 6000 | 800 | 1.0 | 1.0 |
| R5 | Polymer bonded, flexible | Variable properties up to those of R4 | | | | | | 1.0 | |

**Table A2.5** MAGNETIC PROPERTIES OF MISCELLANEOUS PERMANENT MAGNET MATERIALS

| *Material* | $B_r$ (T) | $(BH)_{max}$ (kJm$^{-3}$) | $(BH)_{max}$ (MGOe) | $H_c$ (kAm$^{-1}$) | $H_c$ (Oe) | $H_{ci}$ (kAm$^{-1}$) | $\mu_0 H_{ci}$ (T) | $\mu_r$ |
|---|---|---|---|---|---|---|---|---|
| CoPt (approximately stoichometric) | 0.6 | 74 | 9.2 | 368 | 4600 | 580 | 0.72 | 1.1 |
| | 0.64 | 76 | 9.5 | 400 | 5000 | 580 | 0.72 | 1.1 |
| Remalloy (Comol, Comalloy) 12% Co | 0.85 | 8 | 1.0 | 20 | 250 | | | |
| 15% Mo balance Fe | 1.0 | 10 | 1.25 | 28 | 350 | | | |
| Lodex, ESD Fe–Co particles in metal | 0.36 | 8.8 | 1.1 | 67 | 830 | | | 2.5 |
| binder. Isotropic | 0.48 | 10 | 1.25 | 78 | 970 | | | 3.8 |
| Lodex, ESD, Fe–Co particles in metal | 0.5 | 18 | 2.3 | 77 | 960 | | | 1.5 |
| binder. Anisotropic | 0.68 | 24 | 3.0 | 80 | 1000 | | | 3.0 |
| Vicalloy 1 10% V, 52% Co, balance Fe | 0.75 | 6.4 | 0.8 | 18.4 | 230 | | | |
| | 1.1 | 10.4 | 1.3 | 24.0 | 300 | | | |
| Vicalloy II, 13% V. Cold worked, | 1.1 | 13.6 | 1.7 | 22 | 280 | | | |
| anisotropic | 1.27 | 27 | 3.4 | 33 | 410 | | | |
| Cunife | 0.55 | 11.2 | 1.4 | 42 | 530 | | | |

**Table A2.6** PHYSICAL PROPERTIES OF PERMANENT MAGNETS

| *Material* | *Saturation $J_s$* (T) | *Curie temp.* (°C) | *Density* ($kgm^{-3}$) | ($gcm^{-3}$) | *Resistivity* ($\mu\Omega$ m) | ($\mu\Omega$ cm) | *Coeff. linear expansion* ($°C^{-1}$) |
|---|---|---|---|---|---|---|---|
| 35% Co | 1.5 | 890 | 8200 | 8.2 | 0.8 | 80 | $11 \times 10^{-6}$ |
| A1 | 0.9 | 700 | 6900 | 6.9 | 0.5 | 50 | $12 \times 10^{-6}$ |
| A1 | 1.2 | 800 | 7300 | 7.3 | 0.5 | 50 | $14 \times 10^{-6}$ |
| A2 | 1.4 | 850 | 7350 | 7.35 | 0.5 | 50 | $11 \times 10^{-6}$ |
| | | | | | | | $12 \times 10^{-6}$ |
| A4 | 1.1 | 850 | 7350 | 7.35 | 0.5 | 50 | $11 \times 10^{-6}$ |
| | | | | | | | $12 \times 10^{-6}$ |
| F1 | 0.42 | 450 | 4800 | 4.8 | $10^{10}$ | $10^{12}$ | $10 \times 10^{-6}$ |
| F2 | 0.43 | 450 | 5000 | 5.0 | $10^{10}$ | $10^{12}$ | $8 \times 10^{-6}$ * |
| | | | | | | | $13 \times 10^{-6}$ † |
| F3 | 0.42 | 450 | 4700 | 4.7 | $10^{10}$ | $10^{12}$ | $8 \times 10^{-6}$ * |
| | | | | | | | $13 \times 10^{-6}$ † |
| R2 | – | >700 | 8200 | 8.2 | 0.5 | 50 | $13 \times 10^{-6}$ * |
| | | >700 | 8300 | 8.3 | 0.5 | 50 | $5 \times 10^{-6}$ † |

* Perpendicular to axis
† Parallel to axis

# Appendix 3

# Manufacturers and trademarks

This appendix attempts to list magnet manufacturers throughout the world with their trade names and range of products, identified by means of the codes used in the previous tables. A list of this nature is almost inevitably incomplete and must become out-of-date. Apologies are offered to any manufacturer who is omitted or misrepresented. The table is divided into sections according to country in alphabetical order. In general, firms are arranged in alphabetical order within each country. There are two exceptions to this procedure. In America there is a Magnetic Materials Producers Association. With a few exceptions members of this Association use a common code for designating materials, and these firms are listed together. Similarly in Britain ex-members of the PMA are listed together, because they still use the same designations.

Little information is available about manufacturing facilities for permanent magnets in Russia, although there is abundant scientific literature which suggests that there is considerable activity and much the same materials are made as in the West. The standard A2 type material is sometimes called Alnico, while Ticonal seems to be used for Ti containing materials such as A4. Columnar materials are often called Magnico. Alternatively alloys are often specified by a code made up of the symbols for the chemical elements with numbers representing their nominal weight percentages. Professional translators usually give a literal translation of the Cyrillic letters for the chemical elements, so that the significance may not be immediately apparent to anyone not familiar with the Russian alphabet.

## AMERICAN MAGNET MANUFACTURERS, TRADE NAMES AND PRODUCTS

Aml   Magnetic Materials Producers Association, Evanston, Illinois

60202. Members of this association are listed first in alphabetical order. Unless otherwise stated they use the MMPA code, but often with their own Trade Name for ferrite magnets. The equivalence between the code used in this book and that of the MMPA is as follows.

| *This book* | *MMPA* |
|---|---|
| A1 | Alnico 1, 2, 3, 4 |
| A2 | Alnico 5 (DG = semi-columnar) |
| A2col | Alnico 5col (formerly Alnico V-7) |
| A3 | Alnico 6 |
| A4 | Alnico 8 |
| A4col | Alnico 9 |
| A5 | Alnico 8 HC |
| F1 | Ceramic 1 |
| F2 | Ceramic 3, 5 or 8 (ceramic 8 appears to combine the high $H_c$ of F3 with the high $B_r$ of F2) |
| F3 | Ceramic 2, 4, 6 or 8 |
| R1 | Rare-earth-cobalt 12 or 15 (the number indicates $(BH)_{max}$ in MGOe) |
| R2 | Rare-earth-cobalt 16 or 18 |

Certain steels, Remalloy, Vicalloy, ESD, and Cunife are also listed. (Bonded Alnico A6 and A7 not mentioned.)

AMa Allen-Bradley Co., Milwaukee, Wisconsin 53204. F1, 2, and 3

AMb Arnold Engineering Co., Marengo, Illinois 60152. Alnico A1, A2, A3, A4, and A5, ArKomax 800 = A2col, Arnox F2, F3 (n.b. Arnox 3 is bonded F6 not MMPA Cer.3)

AMc Colt Industries Crucible Magnetic Div. Elizabethtown, Kentucky 42701. Alnico all grades, Ferrimag F1, F2, and F3, Crucore R1, Vicalloy II

AMd Electron Energy Corp., Landisville, Pennsylvania 17538. Remco R2

AMe General Magnetic Co., Detroit, Michigan 48234. Genox F1, F2, and F3

AMf Hitachi Magnetics Corp., Edmore, Michigan 48829. Alnico most grades, Lodex ESD, Hicorex R2, YBM F1, F2, and F3 (probably as Hitachi Japan)

AMg Indiana General, Valparaiso, Indiana 46225. Alnico all grades, Indox F1, F2, and F3, Incor R2, Cunife

AMh Permanent Magnet Co., Indianopolis, Indiana 46226. Alnico

AMj Raytheon Co., Waltham, Massachusetts 02154. Raeco R2

AMk Spectra-Flux Inc., Watsonville, California 95076. Spectra R2

AMl Stackpole Carbon Co., Kane Pennsylvania 16735. Ceramagnet (own code) F1, F2, F3 and F6

AMm Thomas and Skinner Inc., Indianopolis, Indiana 46205. Alnico, most or all grades

**American manufacturers, non-members MMPA**

Am2 Ceramic Magnets, Fairfield NJ 07006. Import rare-earth magnets
Am3 General Motors. Ferriroll F4
Am4 B. F. Goodrich. Koroseal F4
Am5 Hamilton Technology Inc., Lancaster, Pennsylvania, Placovar PtCo, Trevar S, R2, Trevar, Ps, R2 (contains some Pr)
Am6 Leyman, Cincinnati (now 3M). Plastiform F5
Am7 Permag Corp. Jamaica, New York 11518 (and throughout USA). Official distributers and finishers for many manufacturers
Am8 D. M. Steward Manufacturing Co., Chattenooga, Tennessee. F or FS (own code), F1, F2 and F3
Am9 Symonds Steel Div., Lockport, New York 14094. A1, A2, A3, A4, A2col, most steels, Remalloy
Am10 Varian Associates, TWT Div. R2

BRITISH MAGNET MANUFACTURERS TRADE NAMES AND PRODUCTS

B1 Members of the former Permanent Magnet Association, who still use the old Trade Names and designations are listed first. The equivalence of the code used in this book and that of the former PMA is as follows

| *This book* | *former PMA* |
|---|---|
| A1 | Alni, Alnico |
| A2 | Alcomax II, Alcomax III |
| A2col | Columax |
| A3 | Alcomax IV (Hycomax I rarely made) |
| A4 | Hycomax II, Hycomax III |
| A4col | Columnar Hycomax III |
| A5 | Hycomax IV |
| F1 | Feroba I |
| F2 | Feroba II |
| F3 | Feroba III |
| F4 | Rubber-bonded Feroba |

B1a Balfour Darwins Ltd., Sheffield S9 1RL. Most Al–Ni–Co alloys, some steels F1, F2 and F3
B1b BOC Magnets, Rainham, Essex. Sintered Al–Ni–Co alloys, possibly Comalloy
B1c Firth Vickers Foundry Ltd., Sheffield S4 7QY. Cast A1, A2, A2col, A3, A4. Probably some steels
B1d Swift Levick and Sons Ltd., Sheffield S4 7YW. Most grades of

Al–Ni–Co, F1, 2, and 3. Probably some steels

B1e James Neill Ltd., Sheffield S11 8HB. Cast A1, A2, A3 and A4, F4. Full range of steels

B1f Ross and Catherall Ltd., Killamarsh S31 8BA. Cast A1, A2, A3 and A4. Probably some steels

B1g Sheffield Magnet Co. Ltd., Sheffield S1 3BZ. Cast A1, A2, and A3. Probably some steels

**Other British magnet manufacturers in alphabetical order**

B2 Alladin Industries Inc., Greenford, Middlesex. possibly F1

B3 Johnson Matthey Ltd., London, NW14. Platinax, PtCo

B4 Magnetic Polymers Ltd., Swindon. Hera R4, R5

B5 Mullard Ltd., London, WC1E 7HD. (Use German code see German manufacturers) Ticonal A2, A3, A4 Magnadur, F1, F2, F3, F4, F6, and F7

B6 Neepsend Castings Ltd., Sheffield S3 8ET. NL followed by own code, A1, A2, A2col, A3, A4, 35% Co steel

B7 Neosid Ltd., Welwyn Garden City. Neoperm (own code) F1, F2, F3 and F6

B8 Preformations Ltd., Swindon, Magloy (own code) A1, A2, A3 and A4. Possibly A2col. Supermagloy rare-earth-cobalt details not yet announced.

B9 H. Shaw Magnets Ltd., Sheffield. Most magnet steels

B10 Telcon Metals Ltd., Crawley, Sussex Vicalloy I and II

## FRENCH MAGNET MANUFACTURERS TRADE NAMES AND PRODUCTS

F1 Aimants Ugimag SA St Pierre d'Allevard. Nialco A1, Ticonal (followed by coercivity in Oe) A2, A3, A4, and A5. Superugimag A2col, Nialco A (agglomeres) A6 Spinal F1, Spinalor (own code), F2, F3, Ferriflex F4, F5, Coramag, R2

F2 Giffey Pretre, 75 Paris XIX. Nial, Nialco, Al, Ticonal G and E, A2, Ticonal M, A3, Ticonal X, A4, Ticonal GG, A2col (n.b. This classification was formerly used by Philips)

## GERMAN MAGNET MANUFACTURERS TRADE NAMES AND PRODUCTS

German manufacturers use different trade names for each class of material followed by a number, which indicates $(BH)_{max}$ in MGOe multiplied by 100. These numbers are laid down in DIN standards, but are likely to be changed in the near future in order to conform with SI units. These numbers are also used internationally by Philips (Mullard in Britain). They need care in their interpretation, because many materials with quite different coercivities have similar values of $(BH)_{max}$.

G1 Max Baermann Ltd. Bensberg. Tromalit A6, A7, F7, Tromaflex F4

G2 Robert Bosch, Stuttgart. A1, Roboxid F1, F2, and F3

G3 Thyssen Edelstahlwerke (TEW) Dortmund. Oerstit A1, A2, A2col, A3, A4, Oxit F1, F2, and F3, Oxilit F4, F5, and F6, PtCo

G4 KMH East Germany. Maniperm F1, F2, and F3

G5 Krupp Widia-Fabrik Essen. Koerzit A1, A2, A2col, A3, A4, Keorox F1, F2, F3, and F6, Koerflex Vicalloy I and II (also semi-hard alloys)

G6 Magnetfabrik, Bonn, Alnico A1, A2, A3, and A4. Possibly A2col. Placo PtCo, Prac A6, Ox F1, F2, and F3 Prox F7, Sprox F4

G7 Siemens, Munich. Ds, F2, F3

G8 Vacuum Schmelze, Hanau. Alnico A2, Vacomax R2

G9 Valvo Hamburg. *See* Philips, Holland.

## ITALIAN MAGNET MANUFACTURERS TRADE NAMES AND PRODUCTS

I1 C. M. P. Milan. Alni, Coalni A1, Coercimax A2, Ansemax A2col, Sinterox F1, F2 and F3

I2 I. O. S. Varesse. Oxymag F1, F3, Ferriplast F4

I3 Sampas. Alni, Alnico A1, Maxalco A2, Maxalco CC, A2col, Alnico VI or VII, A3, FXD1, F1, Plastomag F5

## JAPANESE MAGNET MANUFACTURERS TRADE NAMES AND PRODUCTS

Although standard codes have been suggested most Japanese manufacturers appear to prefer their own code. The details given here are far from complete.

J1 Hitachi Metals Ltd., Tokyo. YCM A1, A2, A3, A4, A2col, YBM F1, YRM F4, Hicorex R1, R2, R3

J2 Japan Special Steel Co. Ltd., Tokyo. NFW, A1, A2, A2col, A3, Ferromax F1, Ferrogum F4

J3 Mitsubishi Steel Manufacturing Co. Ltd., Tokyo. MK A1, A2, MRC, R2, MP, R5

J4 Namiki Precision Jewel Co. Ltd. Naminet R2

J5 Shinetsu Chemical Industry Co. Ltd. Rarenet R1, R2

J6 Sumitomo Special Metals Co. Osaka-fu. NKS, A1, A2, A2col, A3, A4, FXD, F1, CKS Cr–Fe–Co

J7 Suwa Seikosha Co. Ltd. (Seiko). Sam R4

J8 TDK Electronics Co. Ltd., Tokyo. RecR1

J9 Tohuku Metal Industries Ltd., Tokyo. Lanthanet R1, R4

## PERMANENT MAGNET MANUFACTURERS IN OTHER COUNTRIES

M1 Bohler, Vienna, Austria. Permanit Al–Ni–Co alloys

M2 General Electric Co., Toronto, Canada. Gecor R2

M3 Henricot Belgium. Nial, Nialco A1

M4 Philips, Eindhoeven, Holland. (The headquarters of Philips, who operate in many countries. German classification. Materials as for Mullard B5 Britain but Ferroxdure used instead of Magnadur. Formerly a letter code was used for Ticonal. *See* F2, Giffey Pretre, France)

M5 Hansa, Spain. Nialco A1, Supernialco A2, Superultra A2col Ferribarita F1, F2

M6 Brown Boveri and Co., Baden, Switzerland. Recoma R2, possibly R1, R3

M7 Von Roll, Switzerland. Alconit A1, Ticonal A2

M8 Surahammars, Bruks, Sweden. Sura A1, A2, A3, A4

# References

Page numbers of references to books are sometimes included for the convenience of readers, but are necessarily incomplete, and if difficulty is met in tracing such a reference the index of the book concerned should be consulted.


Albanese, G., Asti, G. and Criscouli, R. (1970). *IEEE Trans. Magn.*, **Mag-6**, 161

Alger, P. L. (1965). *The Nature of Induction Machines*, Gordon and Breach, New York

Anderson, J. C. (1968). *Magnetism and Magnetic Materials*, pp.192, 130–132, Chapman and Hall, London

Andrade, E. W. da C. (1958). *Endeavour*, **17**, 22

Bachmann, H. and Nagel, H. (1976). *2nd Int. Workshop on Rare-Earth-Cobalt Magnets*, University of Dayton, 347

Bailey, L. J. and Richter, E. (1976) *2nd Int. Workshop on Rare-Earth-Cobalt Magnets*, University of Dayton, 235

Baran, W. (1972). *Int. J. Magn.*, **3**, 103

Barbara, B., Becle, C., Lemaire, R. and Paccard, D. (1971). *IEEE Trans. Magn.*, **Mag.-7**, 654

Bates, L. F. (1961). *Modern Magnetism* (4th edn.), p.457, Cambridge University Press

Bean, C. P. and Jacobs, I. S. (1956). *J. appl. Phys.* **27**, 1448

Bean, C. P. and Livingston, J. D. (1959). *J. appl. Phys.* **30**, 120S

Becker, J. J. (1976). Paper 7E.4. MMM – *Intermag. Conf. Pittsburgh*, to be published

Becker, R. and Döring, W. (1939). *Ferromagnetismus*, pp.114–127, Springer, Berlin

Binns, K. J., Jabbar, M. A. and Barnard, W. R. (1975). *IEEE Trans. Magn.*, **Mag.-11**, 1538

Bohlmann, M. A. (1976). *2nd Int. Workshop on Rare-Earth-Cobalt Magnets*, University of Dayton, 31

Bozorth, R. M. (1951). *Ferromagnetism*, p.846, Van Nostrand, New York

Bradley, A. J. and Taylor, A. (1937). *Proc. R. Soc.* **A159**, 56

Bradley, A. J. and Taylor, A. (1938). *Magnetism*, Institute of Physics, London

Brailsford, F. (1960). *Magnetic Materials*, 3rd Edn., Methuen, London

den Broeder, F. J. A. and Zijlstra, H. (1974). *Proc. 3rd Eur. Conf. on Hard Magn. Mater.*, Amsterdam, 118

Brown jr., F. (1945). *Rev. mod. Phys.*, **17**, 15

Bulygina, T. I. and Seregev, V. V. (1969). *Physics Metals Metallogr.*, **27**, No.4, 132, (English translation)

Buschow, K. H. J., Naastépad, R. A. and Westendorp, F. F. (1969). *J. appl. Phys.*, **40**, 4029

Bye, G. C. and Howard, C. R. (1971). *J. appl. Chem. Biotechnol.*, **21**, 319

Carey, R. and Isaac, E. D. (1966). *Magnetic Domains and Techniques for their Observation,* English Universities Press, London.
Charles, R. J., Martin, D. L., Valentine, L. and Cech, R. E. (1972). *AIP Conf. Proc.,* No.5, 1072
Clegg, A. G. (1955). *Br. J. appl. Phys.,* **6**, 120
Clegg, A. G. and Keyworth, D. A. (1976). *2nd Conf. on Advances in Magn. Mater. and their Applic.,* IEE Conf. Pub. No.142, London, 143
Clegg, A. G. and McCaig, M. (1957). *Proc. Phys. Soc.* **B LXX,** 817
Clegg, A. G. and McCaig, M. (1958), *Br. J. appl. Phys.,* **9**, 194
Craik, D. J. and Harrison, A. J. (1974). *Proc. 3rd Eur. Conf. on Hard Magn. Mater.,* Amsterdam, 33
Craik, D. J. and Tebble, R. S. (1965). *Ferromagnetism and Ferromagnetic Domains,* North Holland, Amsterdam
Crangle, J. and Goodman, G. (1971). *Proc. R. Soc.* **A321**, 477
Dean, A. V. and Mason, J. J. (1969). *Cobalt,* No.43, 73
Dietrich, H. (1968). *Feinwerktechnik,* **72**, 313
Dietrich, H. (1970). *DEW Technische Berichte,* **10**, 219
Domazer, H. G. and Strnat, K. J. (1976). *2nd Int. Workshop on Rare-Earth-Cobalt Magnets,* University of Dayton, 348
Ebeling, D. R. G. (1948). US Patent 2 578 401
Fahlenbrach, H. and Baran, W. (1965). *Dauermagnete und ihre Anwendung in Betrieben,* Carl Hanser, Munich
Foner, S. (1959). *Rev. scient. Instrum.,* **30**, 548
Foner, S., McNiff jr., E. J., Martin, D. L. and Benz, M. G. (1972). *Appl. Phys. Lett.,* **30**, 447
Frei, E. H., Shtrikman, S. and Treves, D. (1957). *Phys. Rev.,* **106**, 446
Gadalla, A. M. and Hennicke, H. W. (1974). *Proc. 3rd Eur. Conf. on Hard Magn. Mater.*, Amsterdam, 62
Geary, P. G. (1964). *Magnetic and Electric Suspensions,* British Scientific Instrument Association
Gilbert, W. (1600). *De Magnete Magnetisque Corporibus et de Magno Magnete Tellure Physiologia Nova*
Gould, J. E. (1968). *Cobalt,* No.39, 75
Gould, J. E. (1971). *Cobalt Alloy Permanent Magnets,* Centre d'Information du Cobalt, Brussels
Gould, J. E. (1973). *Stability of Permanent Magnet Materials,* Bulletin No.2, PMA Sheffield
Gould, J. E. (1972). *Cobalt,* No.55, 79
Granovskii, Ye. B., Pashkov, P. P., Seregev, V. V. and Fridman, A. A. (1967). *Physics Metals Metallogr.,* **23**, No.3, 55 (English translation)
Hadfield, D. (1956). *The Magnetic Stability of some Permanent Magnet Materials,* Thesis, University of Sheffield
Hadfield, D. (1962). *Permanent Magnets and Magnetism,* pp.69, 132, 137, 314, 328, 334, Iliffe Books Ltd., London
Hansen, J. R. (1955). *Conf. Magn. Magnetic Mater. Am. Inst. elect. Engrs.,* 198
Harrison, J. (1962). Brit. Pat. 987 636 and 999 523
Harrison, J. (1966). *Zeits. für angewandte, Phys.,* **21**, 101
Heck, C. (Transl. Hill, S. S. ) (1974). *Magnetic Materials and their Applications,* Butterworth, London
Henkel, O. (1966). *Zeits. für angewandte Phys.,* **21**, 32, *Phys. stat. solidi* **15**, 211 (Paper 2.17)
Higuchi, A. (1966). *Zeits. für angewandte Phys.,* **21**, 80
Higuchi, A., Kamiya, M. and Suzuki, K. (1974). *Proc. 3rd Eur. Conf. on Hard Magn. Mater.,* Amsterdam, 197

Hollitscher, H. and Whiteley, E. (1976). *Proc. 2nd Int. Workshop on Rare-Earth-Cobalt Magnets,* University of Dayton, 174
Hoselitz, K. (1952). *Ferromagnetic Properties of Metals and Alloys,* pp.36, 40, 74, 274, Clarendon Press, Oxford
Hoselitz, K. and McCaig, M. (1951). *Proc. Phys. Soc.* **B64**, 549
Hoselitz, K. and McCaig, M. (1952). *Proc. Phys. Soc.* **B65**, 229
Howard, C. P. (1969). The Influence of Texture on the Reactants in the Synthesis of Barium Hexaferrite, Thesis, University of Sheffield
Hume-Rothery, W. (1962). *Atomic Theory for Students of Metallurgy, 4th Edn.*, Institute of Metals, London
Inoue-Japax Res. Inc. (1971). German Patent 2 165 052 (1972), French Patent 2 149 076 (Also British Patents (1974) 1 367 174, (1976) 1 435 684)
Ireland, J. R. (1968). *Ceramic Permanent Magnet Motors,* McGraw-Hill, New York
Jaffe, W. and Herr, J. (1976). *Proc. 2nd Int. Workshop on Rare-Earth-Cobalt Magnets,* University of Dayton, 100
Jakubovics, J. P., Jolly, T. W. and Lapworth, A. J. (1975). *Digest 4.3, Intermag Conference,* London.
Jeans, Sir J. (1933). *The Mathematical Theory of Electricity and Magnetism,* (5th edn.) Cambridge University Press
Jellinghaus, W. (1943). *Archiv. Eisenhuttenw.*, **16**, 247
Johnson, R. E. and Fellows, C. I. (1971). *Cobalt,* No.53, 191
Johnson, R. E. and Fellows, C. I. (1972). *Cobalt,* No.56, 141
Johnson, R. E. and Fellows, C. I. (1974). *Cobalt,* No.1, 21
Johnson, R. E., Rayson, H. W. and Wright, W. (1972). *Acta. Met.*, **20**, 387
Jones, F. G. and Tokunaga, M. (1976). *IEEE Trans. Magn.*, **Mag.-12**, 968
Kaneko, H., Homma, M., Fukunago, T. and Okado, M. (1975). *IEEE Trans. Magn.*, **Mag.-11**, 1440
Kaneko, H., Homma, M. and Nakamura, K. (1972). *AIP Conf. Proc.*, No.5, 1088
Kaneko, H., Homma, M., Okada, M., Nakamura, S. and Ikuta, N. (1976). *AIP Conf. Proc.*, No.29, 620
Kennedy, A. and Womach, G. J. (1964). *Electl. Rev., Lond.*, 9th October
Kittel, C. (1946). *Phys. Rev.*, **70**, 965
Klitzing, K. H. (1957). *Z. Inst.*, 4
Koch, J. (1974). *Proc. 3rd Eur. Conf. on Hard Magn. Mater.*, Amsterdam, 37
Kojima, S., Ohtani, T., Kato, N., Kojima, K., Sakamoto, Y., Kondo, I., Tsukaharu, M. and Kubo, T. (1975). *AIP Conf. Proc.*, No.24, 768
Krijtenburg, G. S. (1965). *Proc. 1st European Conf. on Hard Magnetic Mater.*, Paper 2-13
Ledeboer, W. A. and Schophuizen, P. J. (1974). *Proc. 3rd Eur. Conf. on Hard Magn. Mater.*, Amsterdam, 247
Livingston, J. D. (1973). *AIP Conf. Proc.*, No.10, 643
Luborsky, F. E., Paine, T. O. and Mendelsohn, L. I. (1959). *Powder Metall. Bull.*, No.4, 47
Luteijn, A. J. and de Vos, K. J. (1956). *Philips Res. Rep.*, **11**, 489
McCaig, M. (1952). *Nature,* **169**, 889
McCaig, M. (1956). *J. scient. Instrum.*, **33**, 311
McCaig, M. (1957). *Proc. Phys. Soc.* **B70**, 823
McCaig, M. (1961). *Electrical Rev.*, **169**, 425
McCaig, M. (1966a). *Zeits für angewandte. Phys.*, **21**, 66
McCaig, M. (1966b). *Cobalt,* No.31, 83
McCaig, M. (1967). *Attraction and Repulsion,* Oliver and Boyd, Edinburgh (now obtainable only from the author)
McCaig, M. (1968). *IEEE Trans. Magn.*, **Mag.-4**, 221

McCaig, M. (1970). *IEEE Trans. Magn.*, **Mag.-6**, 198
McCaig, M. (1973). *Nature*, **242**, 112
McCaig, M. (1975). *IEEE Trans. Magn.*, **Mag.-11**, 1443 (Chap.6), 1458 (Chap.7)
McCaig, M. (1976). *IEEE Trans. Magn.* **Mag.12**, 986 (Chap.6), *Proc. 2nd Int. Workshop on Rare-Earth Cobalt Magnets,* University of Dayton, 129 (Chap.7)
McCaig, M. and Wright, W. (1960). *Br. J. appl. Phys.* **11**, 279
McCurrie, R. A. and Gaunt, P. (1964). *Proc. Int. Conf. Magn.*, Institute of Physics and Physical Soc., London, 780
McCurrie, R. A. and Mitchell, R. K. (1975). *IEEE Trans. Magn.*, **Mag.-11,** 1408
Magnetic Materials Producers Association (1975). *Testing and Measurement of Permanent Magnets,* Evanston, Illinois
Martin, D. H. (1967). *Magnetism in Solids,* Iliffe, London
Martin, D. L., Laforce, R. P., Rockwood, A. C., McFarland, C. H. and Valentine, L. (1974). *Rare-Earth Research Conf.* (1973) or G. E. Report No.73 CRD 068
Martin, D. L., Sneggill, J. G. Hatfield, W. and Bolon, R. (1975). *IEEE Trans. Magn.* **Mag.-11**, 1420
Meiklejohn, W. H. and Bean, C. P. (1956). *Phys. Rev.*, **102**, 1413
Meiklejohn, W. H. and Bean, C. P. (1957). *Phys. Rev.* **105**, 904.
Mildrum, H. F., Hartings, M. F. and Strnat, K. J. (1974). *AIP Conf. Proc.*, No.18, 1163
Mildrum, H. F., Hartings, M. F., Wong, K. D. and Strnat, K. J. (1974). *IEEE Trans. Magn.* **Mag-10**, 723
Mildrum, H. F. and Wong, K. M. D. (1976). *Proc. 2nd Int. Workshop on Rare-Earth Cobalt Magnets,* University of Dayton, 35
Mishima, T. (1931). Brit. Pat. 378 478 and 392 658
de Mott, E. G. (1970). *IEEE Trans. Magn.*, **Mag.-6**, 269
Nagel, H. (1974). *Proc. 3rd Eur. Conf. on Hard Magn. Mater.*, Amsterdam, 153
Nagel, H. and Klein, H. P. (1975). *AIP Conf. Proc.*, No.24, 695
Nagel, H., Klein, H. P. and Menth, A. (1975). *IEEE Trans. Magn.*, **Mag.-11**, 1426
Narita, K. (1976). *2nd Int. Workshop on Rare-Earth-Cobalt Magnets,* University of Dayton, 55
Néel, L. (1946). *Annales de l'Université de Grenoble,* **22**, 321
Nesbitt, E. A. and Wernick, J. H. (1973). *Rare-Earth Permanent Magnets,* Academic Press, New York
Nesbitt, E. A. and Williams, A. J. (1955). *J. appl. Phys.*, **26**, 1217
Nicholson, R. B. and Tufton, P. J. (1966). *Zeits. für Angewandte Physik,* **21**, 59
Nicolson, M. M. (1961). *Fundamentals and Techniques of Mathematics for Scientists,* Longmans, London
Noodleman, S. (1976). *2nd Int. Workshop on Rare-Earth Cobalt Magnets,* University of Dayton, 214
Oliver, D. A. (1938). *Magnetism,* Institute of Physics, London, 69
Oliver, D. A. and Goldschmidt, H. J. (1946). *An X-Ray Investigation of Anisotropic Permanent Magnets,* Electrical Research Association, Rep., N/T41
Oliver, D. A. and Sheddon, H. J. (1938). *Nature,* **142**, 209
Osborn, J. H. (1945). *Phys. Rev.* **37**, 351
Paine, T. O. and Luborsky, F. E. (1960). *J. appl. Phys.*, **31** Suppl.785
Palmer, D. J. and Shaw, S. W. K. (1969). *Cobalt,* No.43, 53
Parker, R. J. and Studders, R. J. (1962). *Permanent Magnets and their Applications,* pp.158, 171, 225, 245, 345, 347, Wiley, New York
Perkins, R. S., Gaiffi, S. and Menth, A. (1975). *IEEE Trans. Magn.*, **Mag.-11**, 1431
Perry, A. J. and Menth, A. (1975). *IEEE Trans. Magn.*, **Mag.-11**, 1423
Polgreen, G. R. (1966). *New Applications of Modern Magnets,* Macdonald, London
Ray, A. E. and Strnat, K. J. (1975). *IEEE Trans. Magn.*, **Mag.-11**, 1429

Rayner, G. H. and Drake, A. E. (1970). *SI Units in Electricity and Magnetism,* HMSO, London
Roters, H. C. (1941). *Electromagnetic Devices,* Chapman and Hall, London
Roth, W. L. and Luborsky, F. E. (1964). *J. appl. Phys.,* **35**, 966
Schemman, H. (1973). *Philips Tech. Rev.* **33**, 235
Scholes, R. (1968). *J. Phys. E.,* **1**, 1016
Scholes, R. (1970). *IEEE Trans. Magn.,* **Mag.-6**, 289
Schüler, K. and Brinkmann, K. (1970). *Dauermagnete,* pp.119, 159, 170, 182, 236, 400, 463, 425 482, 556, Springer, Berlin
Schweizer, J., Strnat, K. J. and Tsui, J. B. Y. (1971). *IEEE Trans. Magn.,* **Mag-7**, 429
Sixtus, K. J., Kronenberg, K. J. and Tenzer, R. K. (1956). *J. appl. Phys.,* **27**, 1051
Smythe, W. R. (1939). *Static and Dynamic Electricity,* McGraw-Hill, New York
van der Steeg, H. G. and de Vos, K. J. (1956). *J. appl. Phys.* **26**, 1217
Steingroever, E. (1974). *AIP Conf. Proc.,* No.18, 725, *Proc. 3rd Eur. Conf. on Hard Magn. Mater.,* Amsterdam, 295
Steinort, E. (1974). *Proc. 3rd Eur. Conf. on Hard Magn. Mater.,* Amsterdam, 66
Steinort, E., Cronk, E. R., Garvin, S. J. and Tiderman, H. (1962). *J. appl. Phys. Suppl.* **33**, 1441
Stewart, K. H. (1954). *Ferromagnetic Domains,* Cambridge University Press
Stoner, E. C. (1945). *Phil. Mag.* **Ser.7, 36**, 803.
Stoner, E. C. (1950). *Rep. Prog. Phys.,* **13**, 83
Stoner, E. C. and Wohlfarth, E. P. (1947). *Nature,* **160**, 650
Stoner, E. C. and Wohlfarth, E. P. (1948). *Phil. Trans. R. Soc.,* **A240**, 599
Strnat, K. J. (1971). *AIP Conf. Proc.,* No.5, 1047
Strnat, K. J. and Hoffer, G. (1966). *USAF Materials Lab. Rep.,* AFML TR-65-446
Strnat, K. J., Hoffer, G., Olsen, J., Ostertag, W. and Becker, J. J. (1967). *J. appl. Phys.,* **38**, 1674
Swift, W. M., Reynolds, W. T., Schrecengost, R. M. and Ratnam, D. V. (1976). *AIP Conf. Proc.,* No.29, 612
Tenzer, R. K. (1957). *Conf. Magn. Magnetic Mater., AIEE,* 203
Tsui, J. B. Y., Strnat, K. J. and Schweizer, J. (1972). *Appl. Phys. Lett.,* **21**, 446
Turelli, M. and Leopardi, J. (1971). *Zeits. für Angewandte, Phys.,* **32**, 247
Velge, W. A. J. J. and Buschow, K. H. J. (1967). *Proc. Conf. Magn. Mater. and their Applic.,* IEE, London, 45
Voigt, A. A. (1976). *2nd Int. Workshop on Rare-Earth-Cobalt Magnets,* University of Dayton, 259
de Vos, K. J. (1966). *The Relationship between Microstructure and Magnetic Properties of Alnico Alloys,* Thesis, Eindhoeven
Weber, R. L. (1973). *A Random Walk in Physics,* Institute of Physics, London
Weihrauch, P. F., Paladino, A. E., Das, D. K., Reid, W. R., Wettstein, E. C. and Gale, A. A. (1973). *AIP Conf. Proc.,* No.10, 638
Wells, R. G. and Ratnam, D. V. (1974). *IEEE Trans. Magn.,* **Mag.-10**, 720
Wells, R. G. and Ratnam, D. V. (1976). *IEEE Trans. Magn.,* **Mag.-12**, 971
Went, J. J., Rathenau, G. W., Gorter, E. W. and van Oosterhout, G. W. (1952). *Philips Tech. Rev.,* **13**, 194
Whiteley, E. (1976). *2nd Int. Workshop on Rare-Earth-Cobalt Magnets,* University of Dayton, 167
Whitworth, R. W. and Stopes-Roe, H. V. (1971). *Nature,* **234**, 32
Wohlfarth, E. P. (1958). *J. appl. Phys.,* **29**, 595
Wohlfarth, E. P. (1959). *Hard Magnetic Materials, Phil. Mag. Suppl.* **8**, 87
Wright, W., Johnson, R. E. and Burkinshaw, P. L. (1974). *Proc. 3rd Eur. Conf. on Hard Magn. Mater.,* Amsterdam, 197
Wright, W. and Roberts, R. J. (1976). Private communication
Zijlstra, H. (1972). *Chemtech.,* May, 280

# Index